Application of Sampling and Detection Methods in Agricultural Plant Biotechnology

Application of Sampling and Detection Methods in Agricultural Plant Biotechnology

Edited by

Ray Shillito
BASF Corporation, Morrisville, NC, United States

Guomin Shan
Corteva Agriscience, Indianapolis, IN, United States

ELSEVIER

WP
WOODHEAD
PUBLISHING
An imprint of Elsevier

Woodhead Publishing is an imprint of Elsevier
The Officers' Mess Business Centre, Royston Road, Duxford, CB22 4QH, UK
225 Wyman Street, Waltham, MA 02451, USA
Langford Lane, Kidlington, OX5 1GB, UK

ISBN: 978-0-323-99293-0

For information on all Woodhead Publ. & Cereals & Grains Assoc. Bookstore publications visit our website at https://www.elsevier.com/books-and-journals

Publisher: Nikki Levy
Acquisitions Editor: Nancy Maragioglio
Editorial Project Manager: Moises Carlo P Catain
Production Project Manager: Paul Prasad Chandramohan
Cover Designer: Mark Rogers

Typeset by TNQ Technologies

Contents

List of contributors

Frederic Bois, Eurofins Scientific Group, Brussels, Belgium

Nancy Brooker, Recursion Pharmaceuticals, York, Alabama, United States; Abbott Toxicology Laboratories, Gretna, LA, United States

Jingwen Chen, SeqMatic, Fremont, CA, United States

Tom Currier, Bayer CropScience, Morrisville, NC, United States

Larry Freese, United States Department of Agriculture Grain Inspection, Packers and Stockyards Administration, Washington, DC, United States

Farhad Ghavami, Eurofins BioDiagnostics Inc., River Falls, WI, United States

Rong Guo, BASF Corporation, Morrisville, NC, United States

Rod A. Herman, Corteva Agriscience, Indianapolis, IN, United States

Sonia Herrero, Syngenta Crop Protection, LLC, Durham, NC, United States

Ryan Hill, Eli Lilly and Company, Lilly Corporate Center, Indianapolis IN, United States

Donna Houchins, Romer Labs, Inc., Union, MO, United States

Yanshan Ji, Syngenta Crop Protection, LLC., Durham, NC, United States

Wujun Jin, Biotechnology Research Institute, Chinese Academy of Agricultural Sciences, Beijing, Beijing, China

Kazumi Kitta, National Agriculture and Food Research Organization, Tsukuba, Ibaraki, Japan

Dean Layton, EnviroLogix Inc., Portland, ME, United States

Liang Li, Institute of Quality Standard and Testing Technology for Agro-Products, Chinese Academy of Agricultural Sciences, Beijing, Beijing, China

Lucy Liu, Bayer Crop Science, Chesterfield, MO, United States

Junichi Mano, Institute of Food Research, National Agriculture and Food Research Organization, Tsukuba, Ibaraki, Japan

Helen Mero, Bayer Crop Science, Chesterfield, MO, United States

Doug Miller, Illinois Crop Improvement Association, Champaign, IL, United States

Karen Moore, Syngenta, Greensboro, NC, United States

Tim Perez, Conagra Brands, Omaha, NE, United States

Reetika Rawat, Joyn Bio, Boston, MA, United States

Margit Ross, Corteva Agriscience, Johnston, IA, United States

Michael Schneider, Corteva Agriscience, Indianapolis, IN, United States

Guomin Shan, Corteva Agriscience, Indianapolis, IN, United States

Ray Shillito, BASF Corporation, Morrisville, NC, United States

Amy Smith, Formerly Corteva Agriscience, Indianapolis, IN, United States

Yusong Wan, Biotechnology Research Institute, Chinese Academy of Agricultural Sciences, Beijing, China

Yongcheng Wang, Bayer Crop Science, Chesterfield, MO, United States

Sherry Whitt, BASF Corporation, Morrisville, NC, United States

Kerrm Y. Yau, Corteva Agriscience, Johnston, IA, United States

Cathy Xiaoyan Zhong, ABC Consulting LLC, Hockessin, DE, United States

Ning Zhou, Corteva Agriscience, Johnston, IA, United States

Foreword

The first genetically engineered or modified (GM) crops were commercialized in 1996. Since then, the global adoption of GM crops has been one of the most exciting revolutions in agricultural history. They have entered almost every country around the world either through cultivation or by food and/or feed imports. Not surprisingly, countries have regulated the technology to suit their own needs at different stages along the food supply chains.

New international regulations in food and feed led to the need for robust and validated analytical methods to generate characterization data. These data are used in the trade of commodity crops for the detection and quantification of the presence of GM crops.

An initial request to AACC (now Cereals and Grains Association) from the key US regulatory agencies covering food and feed provided the incentive to develop a new training workshop. The workshop was designed to introduce the agency scientists to the characteristics of GM crops and the specific protocols for GM sampling and analytical methods.

The workshops introduced scientists with chemical and biological laboratory skills to the full range of challenges in sampling and the use of GM detection methods. Participants used examples of raw grains and finished foods in the practical application of both protein and DNA detection methods. The experiments invariably demonstrated many of the challenges and pitfalls in the use and validation of methods (with high sensitivity) in the local laboratory environment.

Following the introductory workshops, many international requests for similar (customized to individual country needs) training programs led to AACC teaming up with the International Life Sciences Institute to provide a series of capacity-building workshops that were delivered at the invitation of governments between 2001 and 2017 in more than 20 different countries in the Americas and East, South-East, and South Asia.

Many of the interested countries are parties to the Cartagena Protocol on Biosafety and had identified a need for their analysts, regulators, and decision makers to understand the nature of the specific analytical methods for GM detection. It was also clear that there was a need for a consolidated source of information on sampling and analysis that was appropriate for the detection of the low-level presence of GM materials.

This book brings together the information delivered in these workshops, updated to reflect current scientific knowledge. It includes information on internationally accepted and consistent standards for sampling and detection of GM traits in a wide range of products. This information is indispensable to those who need to implement standardized methods and practices essential to operating in the current trade and regulatory environment.

Chapter 1

Introduction: genetically modified crops and their detection

Rod A. Herman and Guomin Shan
Corteva Agriscience, Indianapolis, IN, United States

Transgenic crop varieties are a relatively recent development in the long history of crop improvement, having only been commercially available for just over two decades (Fernandez-Cornejo et al., 2014). These crops are typically referred to as genetically modified (GM) crops. Other terms that have been used are genetically engineered, bioengineered, and biotechnology-derived crops. These terms are often used interchangeably. Authors in this book may therefore use one or more of these terms.

Crop improvement is the effort to improve the attributes (phenotype) of the crop such that it performs in a superior manner under cultivation or has an improved nutritional profile. Such improvements might include higher yield, tolerance to pests or diseases, or improved drought tolerance (Ricroch and Hénard-Damave, 2015). As DNA is the template for the phenotype of plants (and animals), all crop improvement involves modifying the genetic material to obtain the improved plant. Nontransgenic methods for genetic modification have included crossing different crop varieties, selecting natural genetic mutations, inducing mutation with chemicals or radiation, and selecting desirable mutants. Other approaches include breeding crop plants with wild relatives (sometimes of different but related species) and creating new crops by crossing different species of existing crops. All these approaches are followed by selection and breeding of desirable traits originating from these operations. These processes all cause dramatic genetic changes, both intended and unintended (Herman and Price, 2013). Some of the most dramatic modifications of crop plants through non-GM methods are those that occurred during crop domestication. Many crop plants are so morphologically distinct from the wild relatives from which they were domesticated, that the origin of those crops was not even apparent until studied at the genetic level (Parrott, 2010). The modern technique of marker-assisted breeding has allowed the selection of

Application of Sampling and Detection Methods in Agricultural Plant Biotechnology
https://doi.org/10.1016/B978-0-323-99293-0.00013-1

desirable traits and reduction of undesirable traits to proceed more quickly through selection of specific genetic markers associated with the desirable traits, and culling of plants that have markers for undesirable traits. Marker-assisted breeding allows the breeder to select plants with certain traits that are closely associated with a specific genetic element that can be detected independent of the phenotype for the trait. This technique is especially useful for polygenic traits where many genes control a phenotype, such as drought tolerance (Mir et al., 2012).

However, GM crops are commonly considered to be only those created using transgenic techniques. Transgenesis is typically the insertion of one or more selected genes from one species into another species, although genes can also be moved within a single species using transgenic methodology. The movement of transgenes can be carried out by various techniques, one of the most common being *Agrobacterium*-mediated transformation. This method uses the natural ability of *Agrobacterium tumefaciens* to insert its DNA into plants as part of its life cycle. Recently, it was found that such transfer actually happened without human intervention approximately 8000 years ago in sweet potato, potentially leading to its selection by ancient people as a food crop (and making it the first known transgenic crop) (Kyndt et al., 2015). In addition, recent research has shown that a significant percentage of dicot species carry *Agrobacterium*-derived sequences (Matveeva and Otton, 2019).

One common application of modern transgenesis has been the movement of genes that code for highly specific insecticidal proteins from the *Bacillus thuringiensis* (Bt) bacteria (used as an insecticide in organic farming) into crop plants so that these plants produce the insecticidal proteins themselves and do not need an insecticide spray to protect them from the pests targeted by the Bt proteins (Sanahuja et al., 2011). The most comprehensive application of this technology so far has been the commercial release of the corn product SmartStax^{tm} which is a breeding stack of multiple independent transgenic events to control both caterpillar and soil-grub pests using a combination of different Bt proteins. This product was developed with the intent of delaying the onset of insect-resistant pest populations (Head et al., 2014; Rule et al., 2014). Crops containing Bt proteins have been very successful and are credited with reducing insecticide use by millions of pounds per year. Another common application of transgenesis is to impart tolerance to herbicides. The most widely planted herbicide-tolerant crop varieties are tolerant to the broad-spectrum herbicide, glyphosate. Glyphosate is considered more effective and environmentally friendly than the herbicides it replaced, thus providing direct environmental benefits. Additional indirect environmental benefits occurred due to its use enabling greater adoption of conservation tillage leading to soil conservation benefits (Fernandez-Cornejo et al., 2014).

New GM traits in new crops are in development. These include resistance to abiotic stress (e.g., drought and heat), improved nutrition, and production of therapeutic molecules (Ricroch and Hénard-Damave, 2015). These will be enabled by more precise gene insertion techniques that are able to target insertion sites within the crop plant (Fichtner et al., 2014). Such techniques will improve the efficiency of inserting new genes and editing the existing genomes of crop plants. Application of gene-editing technologies will add further tools to the toolbox for plant improvement (Jaganathan et al., 2018; Chen et al., 2019).

Despite these widely understood benefits, the technology to move specific isolated genes from one species to another is relatively new, and as a result, strong regulatory safeguards were put in place to evaluate the environmental safety and human health effects of GM crop varieties. However, after over 25 years of commercial adoption, the scientific consensus is that deregulated GM varieties are not per se less safe than those developed through traditional breeding. Even so, regulatory oversight of GM crops has not yet aligned with this scientific understanding, and many members of the public are highly suspicious of the technology. As a result, governments have implemented various regulations for evaluating GM crops, which have created additional hurdles for approval for cultivation and importation (Waters et al., 2021). Furthermore, along with growing globalization and the increasing need for international trade, GM products have entered almost every country either through cultivation or food and feed import. Asynchronous approvals across the globe have thus driven a need to identify specific crops, agricultural commodities, and food items that might contain GM crops and may require labels or are not approved in certain geographies. In many countries, current legislative regimes require that the novel genes and proteins in GM plants and their products be monitored and tracked in every phase of development, production, commercialization, and the supply chain. Appropriate sampling processes and accurate and reliable qualitative or quantitative detection methods are required from early discovery through product development, farmer cropping, food and feed processing, grain import and export, environmental monitoring, and risk assessment. Antibody-based protein detection (immunoassay) and PCR-based nucleic acid detection technologies have played a pivotal role in this field and both are methods of choice for the detection of specific GM products (Alarcon et al., 2019). Immunoassay and PCR are not new technologies. They have been widely used in basic research, pharmaceutical, clinical diagnostic, and agricultural fields, and several books have been published regarding these technologies and their application in basic research. However, the application of these technologies in GM detection in the international community is in some cases hampered by a lack of technical training and consistency in practice and data interpretation. This poses a potential risk of disrupting the movement of crops and food due to imperfect testing procedures. The purpose of this book is to provide practical and technical guidance on science-based sampling and detection methods and their application to agricultural botechnology products.

In this book, we focus on three key areas of sampling and detection of GM products: (1) sampling, (2) detection technologies, and (3) laboratory design and testing strategies.

Chapters 2 and 3 introduce the background of testing, and how it is applied throughout the product cycle. Chapter 2 describes specific applications of GM detection in seed, in particular for seed purity and low-level presence testing. The focus of chapter 3 is on GM testing in the grain supply chain, which is an important enabler for international trade and border control. The basic principles and detailed procedures of PCR and immunoassay method development are discussed in Chapters 4 and 6, which provide readers with practical guidance on assay development including reagent generation and screening, assay format and design, assay optimization, and troubleshooting. To ensure the developed assay is robust and reliable for its intended use, thorough validation is required. Chapters 5 and 7 describe method validation criteria and process steps for both DNA and protein detection methods in plant matrices. Reference materials are critical for the validation of both qualitative and quantitative detection methods. A thorough overview of sources and uses of reference materials is described in Chapter 8, which also provides practical guidance on considerations of reference material selection.

The second focus of this book is the application of these validated methods to field sites and harvested crops. Sampling is a key step. For seed and grain testing, a subsample from the lot is collected and analyzed. Sampling is a significant source of uncertainty with regards to its representativeness of the greater lot from which it originated, and thus this uncertainty needs to be considered in an analytical measurement. Therefore, appropriate sampling is important to ensure that the sample is a good representation of the lot, which contributes to accurate testing. The principles of sampling and details of sampling plans and procedures are discussed in Chapter 9. Chapter 10 describes key elements and considerations for testing plant materials in field and laboratory settings, including sampling, detection, data interpretation, and sources of contamination.

Good laboratory design and management practices are critical to enabling a testing laboratory to use available technologies/methods and deliver reliable and high-quality results. The third focus of this book is therefore to provide guidance and practical steps for developing a GM testing laboratory. General guidance and practice for laboratory design and management are discussed in Chapter 11, including lab design, lab workflow and process, equipment management, reagent and control, and personnel management. Harmonization and adherence to standards are important to facilitate quality control and proper use of methods. Chapter 12 provides a review of the international harmonization of the laboratory application of GM detection methods. Chapter 13 describes the analytical strategies for designing efficient procedures to carry out GM detection and for interpreting laboratory testing results.

The increasing number of applications of genome-editing technology in agriculture has generated interest in the detection of genome-edited products. Although these are generally not considered GM products, we consider this technology to be an important development in agricultural biotechnology. Chapter 14 gives an overview of genome-editing technology, detection method approaches, and challenges in the detection of genome-edited products.

Recent trends in agricultural biotechnology with more complex and sophisticated stacked products, asynchronous regulatory approval worldwide, and new emerging gene-editing breeding technologies have led to new challenges for detection methods. Chapter 15 provides a perspective on emerging analytical technologies, which will enable testing laboratories to meet ever-changing needs.

These chapters cover important aspects of GM-crop detection and thus serve as a current comprehensive reference for those involved in this field. Technicians, scientists, laboratory directors, and others responsible for designing and carrying out GM detection activities should benefit from this book's broad and in-depth coverage of this topic.

References

Alarcon, C.M., Shan, G., Layton, D.T., Bell, T.A., Whipkey, S., Shillito, R.D., 2019. Application of DNA- and protein-based detection methods in agricultural biotechnology. Journal of Agricultural and Food Chemistry 67 (4), 1019–1028.

Chen, K., Wang, Y., Zhang, R., Zhang, H., Gao, C., 2019. CRISPR/Cas genome editing and precision plant breeding in agriculture. Annual Reviews in Plant Biology 70, 667–697. https://doi.org/10.1146/annurev-arplant-050718-100049.

Fernandez-Cornejo, Wechsler, L., Mitchell, 2014. USDA Economic Research Report 162 viewed on January 13th 2020. www.ers.usda.gov/publications/pub-details/?pubid=45182.

Fichtner, F., Urrea Castellanos, R., Ülker, B., 2014. Precision genetic modifications: a new era in molecular biology and crop improvement. Planta 239 (4), 921–939.

Head, G., Carroll, M., Clark, T., Galvan, T., Huckaba, R.M., Price, P., Samuel, L., . Storer, N.P., 2014. Efficacy of SmartStax® insect-protected corn hybrids against corn rootworm: the value of pyramiding the Cry3Bb1 and Cry34/35Ab1 proteins. Crop Protection 57 (0), 38–47.

Herman, R.A., Price, W.D., 2013. Unintended compositional changes in genetically modified (GM) crops: 20 years of research. Journal of Agricultural and Food Chemistry 61 (48), 11695–11701.

Jaganathan, D., Ramasamy, K., Sellamuthu, G., Venkataraman, G., 2018. CRISPR for crop improvement: An update review. Frontiers in Plant Science 9, 985. https://doi.org/10.3389/fpls.2018.00985, 985.

Kyndt, T., Quispe, D., Zhai, H., Jarret, R., Ghislain, M., Liu, Q., Gheysen, G., Kreuze, J.F., 2015. The genome of cultivated sweet potato contains Agrobacterium T-DNAs with expressed genes: an example of a naturally transgenic food crop. Proceedings of the National Academy of Sciences 112 (18), 5844–5849.

Matveeva, T.V., Otton, L., 2019. Widespread occurrence of natural genetic transformation of plants by *Agrobacterium*. Plant Molecular Biology 101, 415–437. https://doi.org/10.1007/s11103-019-00913-y.

Mir, R.R., Zaman-Allah, M., Sreenivasulu, N., Trethowan, R., Varshney, R.K., 2012. Integrated genomics, physiology and breeding approaches for improving drought tolerance in crops. Theoretical and Applied Genetics 125 (4), 625–645.

Parrott, W., 2010. Genetically modified myths and realities. New Biotechnology 27, 545–551.

Ricroch, A.E., Hénard-Damave, M.C., 2015. Next biotech plants: new traits, crops, developers and technologies for addressing global challenges. Critical Reviews in Biotechnology 1.

Rule, D.M., Nolting, S.P., Prasifka, P.L., Storer, N.P., Hopkins, B.W., Scherder, E.F., Siebert, M.W., Hendrix, W.H., 2014. Efficacy of pyramided Bt proteins Cry1F, Cry1A.105, and Cry2Ab2 expressed in SmartStax corn hybrids against Lepidopteran insect pests in the Northern United States. Journal of Economic Entomology 107 (1), 403–409.

Sanahuja, G., Banakar, R., Twyman, R.M., Capell, T., Paul, C., 2011. Bacillus thuringiensis: a century of research, development and commercial applications. Plant Biotechnology Journal 9 (3), 283–300.

Waters, S., Ramos, A., Hendrickson Culler, A., Hunst, P., Lawrence, Z., Gast, R., Mahadeo, D., Jordan, S., Huber, S., Shan, G., Chakravarthy, S., Goodwin, L., 2021. Recommendations for science-based safety assessment of genetically modified (GM) plants for food and feed uses. Journal of Regulatory Science 9 (1), 16–21.

Further reading

Deepa, J., Ramasamy, K., Sellamuthu, G., Jayabalan, S., Venkataraman, G., July 17, 2018. CRISPR for crop improvement: an update review. Frontiers of Plant Science. https://doi.org/10.3389/fpls.2018.00985.

Chapter 2

Seed purity testing and low-level presence

Cathy Xiaoyan Zhong
ABC Consulting LLC, Hockessin, DE, United States

2.1 Background

Grain and seed trade is a global business that involves complex regulatory rules and requirements that can vary significantly among countries. Seed purity is an extremely important subject as both seed and grain move around the world. Seed purity (in the context of this chapter) is a measure of seed lot quality. Specifically, when the purity of a seed lot is mentioned, it refers to the amount of a seed lot that contains the expected genetic background and contains only the expected traits or transgenic elements. This discussion of seed purity does not provide any perspective on the safety of the seed or its genetically modified (GM) components. Purity has two major components, genetic background and the presence or absence of traits or transgenic events. The first measure is the seed lot's genotypic background; a seed lot should be homogenous with all of the seed coming from the same genetic source. In some cases, different backgrounds can be identified through phenotypic means, but more often, confirming genotypic purity involves analysis by molecular markers. The second measure of seed purity is molecular/trait purity, which is the primary focus of this chapter. In the framework of this discussion, molecular purity is defined as the extent to which the desired trait is present and undesired traits are absent. Testing for molecular purity also includes testing for low-level presence (LLP). LLP will be described in more detail later in the chapter. Testing the purity of a seed lot is a complex process, and the information contained herein is intended to provide some guidance on common techniques, technologies, and processes.

One hundred percent seed purity in an agriculture production system is an unattainable goal due to the nature of the biological systems and the mechanistic environment of farming. The biology of plants in a cultivated environment makes pollen movement a potential source of contamination for many crops. Pollen from unwanted sources can be moved by wind, insects, people,

Application of Sampling and Detection Methods in Agricultural Plant Biotechnology
https://doi.org/10.1016/B978-0-323-99293-0.00001-5

and animals, and result in accidental pollination. Mechanical handling of seed and grain throughout the production process from packing, planting, harvesting, sorting, and repacking creates opportunities for mistakes to occur, or for seed to end up in unintended places. Because of the challenging system, and the potential rate at which contamination can occur, it is important to monitor seed lots throughout their life cycle.

Acceptable purity thresholds must be set with consideration of the end usage of the seed as well as cost. Producing pure seed requires testing, isolation, dedicated equipment, and handling procedures. Seed intended for regulatory studies supporting global registration submissions for transgenic plant products has a high standard for purity. This stringent criterion is not surprising considering that data generated in regulatory studies are used to assess the safety, composition, transgenic protein expression, and agronomics of the biotech product. In addition to the assessment aspect of the regulatory environment, regulatory studies, by design due to global data requirements, contain unapproved transgenic traits or products whose use is strictly monitored. Seed produced for commercial use has a relatively lower purity threshold. This is due to the large volume of seed required, the need to limit production costs, and the fact that produced seed contains approved or deregulated traits.

Definitions of common terms:

- **Seed lot**
  - **A uniquely identified unit of seed.**
- **LLP**
  - **Low-level presence—Unintended presence of a transgenic trait that has been assessed for food, feed, and environmental safety and approved for cultivation and commercial sale in one or more export countries, but not for import into other countries.**
- **Pooled/bulk sample**
  - **A group of seeds from one seed lot ground and tested together.**
- **Limit of detection**
  - **The lowest level at which the transgenic trait or target has a high probability of being detected. This level can vary based on detection method.**
- **Limit of quantification**
  - **The lowest level of the target that can be reliably quantified accurately. This level can vary based on detection method.**

This chapter will cover several different factors to consider when assessing seed purity and LLP detection. This is not a comprehensive source for regulations and testing information as those requirements can vary based on geographic region.

2.2 Sampling

The only way to guarantee that 100% of a seed lot is pure is to analyze every seed using a method with a zero error rate. It is not feasible to evaluate every individual in a lot as most methods for molecular characterization of seed are destructive, and the cost of processes involved in sampling and testing can be prohibitive. Sampling is the act of selecting a portion of a seed lot that serves as a representative of the whole. Sampling assumes the acceptance of a certain amount of risk because not every seed is being tested. When sampling for genotypic purity and LLP, it is important to balance the purity needs with the associated testing costs. Increasing the sampling intensity allows the analyst to effectively increase the likelihood of detecting a contaminant thereby increasing the statistical likelihood that the lot is pure. The benefit to purity comes at a cost in terms of labor, reagents, equipment, etc. It is important to find a sampling scheme that meets the needs of the end-use of the seed lot without increasing the cost unnecessarily. This section discusses a few parameters to consider when sampling for purity testing. More information about sampling schemes can be found in Chapter 7.

Taking a representative sample for purity testing incorporates a few considerations. The first concept is a representative sampling. This presumes that any individual in the lot has an equal probability of being sampled (Remund et al., 2001 and Chapter 9). To aid in representative sampling, multiple samples should be taken from representative locations within a container, and all portions of the seed lot should be available for sampling. A second consideration is the uniformity of the sample. Heterogeneous samples should not be tested as they are not representative of the whole seed lot (ISTA, 2020). Sampling intensity is a further concern that should be addressed based on the final purpose of the seed lot and the detection method to be used. As mentioned earlier, it is important to sample to a degree where the seed lot is well represented and there is high confidence in detecting the target, but it is also important to consider costs.

Testing for the presence of a transgenic target versus absence presents different considerations. When testing for the presence of GM target in seed, the goal is to confirm, with a high level of confidence, that the seed lot has all the intended traits. Conversely, the purpose of testing for the absence of a GM target is to confirm that unapproved or undesired transgenic targets are absent from the seed lot. Many countries have defined restrictions on the quantity of LLP that can be present in a seed lot. These rules and the end purpose of the seed should be considered when establishing sampling schemes. More information about LLP testing is presented later in this chapter.

How samples are collected and processed is also an important factor. The decision to sample and test individual seeds versus a pool is one based on cost and data needs (Freese et al., 2015). Testing individual seed samples gives an accurate picture of the presence or absence of the target in each individual.

Individual seed testing increases costs, as each sample is independent and more testing is necessary to meet statistical requirements but is often the only way to measure purity. The testing of independent samples can happen in a couple of ways. Single seeds can be ground, and nucleic acid extracted, or the seeds can be germinated. The resulting seedlings can be treated with herbicide or sampled for extraction of DNA or protein. When testing individual seeds there is no quantification involved—each seed is a qualitative test and the methods are operating well above any limit of detection. Sampling plants after germination to assay them has a greater space requirement than testing seeds directly and increases turnaround times but decreases components like starches and oils that are more concentrated in the seed and can be inhibitory during testing.

Testing a pool of ground seed can give a picture of the presence or absence of a target in a group. This is a cost-effective way to sample a larger and more representative portion of the seed lot. It does not however allow for the purity of every seed to be assessed. Individual seed testing is the appropriate approach when confirming the presence of more than 1 GM target or when trying to detect null seeds (ISTA, 2020). When testing a non-GM seed lot, or confirming the absence of a target, it is critical to consider the limit of detection and/or quantification of the assay when determining the size of the pool to test. Tools are available to determine the best strategy (Kirk et al). It is also essential to follow careful sample preparation practices to assure that the sampling, grinding, and processing of the seed or plant tissue does not introduce contaminants that lead to false-positive scores.

Sampling is the first step in characterization, and its importance cannot be overlooked. It is essential for testing laboratories to adapt their sampling schemes to match the end purpose for the seed. A proper sampling strategy sets the stage for the rest of the characterization process.

2.3 Detection methods and techniques

A general examination of purity involves the direct detection of target nucleic acids or proteins and in some instances involves phenotypic bioassays (Alarcon et al., 2018; CXG 74-2010, 2010). A number of different techniques can be used to determine purity. Enzyme-Linked Immunosorbent Assay (ELISA) and other protein-based methods are used for the detection and sometimes quantification of the transgenic protein in a sample. When it is necessary to quantify the amount of a transgene or transgenic event that is present, DNA-based methods, primarily PCR assays, are most commonly used. Phenotypic bioassays are generally used to detect the plant phenotype produced by the transgenic trait such as the presence or absence of herbicide tolerance. All of these detection methods have a role in purity testing as they will be described further in this section.

2.4 Protein-based detection methods

Most protein detection methods for purity testing are based on the binding of an antibody to a target protein or antigen (Chapter 6). The most common method is ELISA. An ELISA uses an antibody to detect a protein or antigen through a coupled enzymatic reaction. Following extraction of protein from the sample, the method initiates by binding the target antigen to a surface. An antibody with a bound enzyme is then allowed to bind the antigen. When the substrate for the enzyme is added, the reaction produces a measurable effect. This generally occurs in the form of a color or absorbance change in the substrate. For seed purity testing the available ELISA kits are generally a qualitative test and are generally not recommended as a quantitative assay (ISTA, 2020).

Lateral flow strips are a convenient method for protein detection as the entire assay system is contained and enclosed in one device (Grothaus et al., 2006; Van den Bulcke et al., 2007; Alarcon et al., 2018). Similar to traditional ELISA tests, the lateral flow devices contain antibodies that will bind the target protein. The lateral flow strips generally come in the form of capillary paper with antibodies bound to it. The strip is dipped in an extract from the sample and each strip is scored based on the presence of the control band and the presence or absence of the band representing the target.

The sensitivity of these protein detection methods can be affected by many factors (see Chapter 6). The antibodies used, assay setup and format, sample matrices, and the quantity and the overall quality of the protein in a sample will all effect the assay. Many proteins degrade rapidly, and care should be taken to minimize those effects. Extracting proteins can be difficult depending on the properties of the specific protein, sample type, and buffer compatibility. Both ELISA and lateral flow methods can present challenges with false positive and false negative scores, so it is important for the laboratory to validate the extraction and detection components of each detection method (ISTA, 2020). Another consideration for protein-based methods is that they are not event-specific detection methods. Many transgenic events contain the same transgenes and proteins as discussed in Chapter 6.

2.5 DNA-based detection methods

The most common detection methods for purity testing use Polymerase Chain Reaction (PCR) (Holst-Jensen et al., 2003; Lipp et al., 2005). DNA detection lends itself very well to purity testing due to its stability, abundance, and the fact that it is relatively easy to extract. The principles of DNA detection methods are discussed in Chapter 4, and only an overview is included here. The basic premise of PCR is the use of fragments of DNA complimentary to the target that act as primer guides for Taq Polymerase. The polymerase makes copies of the target sequence during a series of heating and cooling steps. The result is a logarithmic amplification of the target end-product. The DNA testing can also present challenges. Contamination can be a significant problem and a well-defined workflow is essential to ensure the laboratory remains free of contamination. The use of dedicated laboratory rooms for different processing steps is preferred and is detailed in Chapter 11. There are several different PCR-based techniques that can be used for DNA detection. Some of the common ones are described here in more detail, and more detail is given as well in Chapters 4, 5, 14, and 15.

Which PCR chemistry is to be used in the detection method is one of the first decisions that must be made. This choice affects how the data are collected and analyzed. The three most common techniques are Gel electrophoresis, TaqMan chemistry, and intercalating dyes. Gel-based methods take the end product, following thermal cycling, and load it into an agarose gel that contains a stain to allow the DNA to be visualized. An electric current is applied drawing the negatively charged DNA molecules through the gel. As they travel through the gel, the smaller DNA fragments move faster allowing the product to separate by size. It is possible to differentiate between the intended product and off target amplification based on their size when a DNA ladder with known fragment sizes is run along with the samples. Gel-based methods can be used to make decisions of presence/absence of the specific DNA targets, but this approach has lower throughput, less sensitivity, and is more labor intensive than other methods.

TaqMan methods use fluorescent probes complementary to the end-product in addition to the normal oligonucleotide primers used in other PCR reactions. When the target is amplified, the exonuclease activity of the Taq polymerase hydrolyzes the fluorescent probe releasing the fluorophore. With each end product that is produced, more fluorophore is released by the hydrolysis of the probe. The hydrolysis probe-based technology increases the specificity of the reaction, reducing false-positive results. The increased specificity comes at a cost of time and money that go into manufacturing the fluorescent probe. This method is the industry standard for the detection of specific DNA targets and lends itself to multiplex assays that will be discussed in more detail later.

Intercalating dyes are used to stain and quantify double-stranded DNA. In principle, as the target of interest increases during PCR, it creates more double-stranded DNA for the dye to bind. Intercalating dyes such as SYBR[1] Green or EvaGreen[2] are less expensive and require less optimization and design input compared with fluorescent probe-based techniques. Both hydrolysis probe and intercalating dye technology systems work well to measure the amplification of the target either during the run after each heating and cooling cycle (real-time PCR) or following the completed run (endpoint PCR). The decision of which chemistry should be used is contingent on the judgment and needs of the laboratory.

1. SYBR is a registered trademark of Life Technologies Corporation.
2. EvaGreen is a registered trademark of Biotium, Inc.

There are several ways to analyze the data generated during a PCR run. The amplification of the target is measured based on the accumulation of fluorescence from either the probe or the intercalating dye. In an endpoint run, the fluorescence is measured following cycling and compared with the controls. A positive or negative call is based on the accumulation of raw fluorescence. In a real-time run, the fluorescence is measured following each heating and cooling cycle. This allows the amplification to be plotted as a sigmoidal curve that shows the logarithmic amplification in the early PCR cycles and a plateau in the latter stages. The amplification of a target in a PCR reaction is recorded as a quantification cycle or Cq (Bustin et al., 2009). The Cq is the cycle at which the amplification crosses a threshold (this is also known as a Ct or Cp value). When using PCR as a detection method, it is important to have an internal control or endogenous reference that shows that the quality of the DNA extracted from the substance being analyzed is suitable. A common way of scoring real-time methods is to use a comparative Cq method in which the Cq of the target is compared with that of the reference (delta Cq; Bustin et al., 2009). This allows the analyst to normalize for the amount of template DNA included in the reaction.

Qualitative analysis is the testing for the presence or absence of a trait. This can be accomplished with endpoint PCR as described earlier or through real-time PCR by setting thresholds for the delta Cq. For qualitative analyses, it is important to consider the lower limit of detection for an assay. The limit of detection becomes a critical factor in detection methods being used on pooled seed samples. The more seeds in the pool, the more challenging it becomes to detect a single seed containing the target.

Quantitative analysis involves not only the detection of the target but also quantifying the amount that is present in the sample. In an individual plant or seed sample, a quantitative assay might be used to assess the number of genomic copies of a target in that sample. It is also commonly used to quantify the amount of a target to determine the number of off-type seeds within a pooled sample, by estimating the amount of a transgenic trait within a seed lot. Due to its sensitivity, throughput, and reproducibility, real-time PCR is the most common tool for quantitative analysis. Quantification experiments require controls of known values in the form of reference calibrators that are used to construct a standard curve. Quantification of transgenic seed in a lot can also be achieved using a subsampling approach (Freese et al., 2015; Remund et al., 2001).

Digital PCR is increasingly used for quantitative detection. In principle, digital PCR uses the same fluorescence-based methods described previously (hydrolysis probes, or intercalating dyes), but breaks the PCR reaction into many small reactions (in the range of hundreds to thousands). This allows for the template to be partitioned across the reactions so that some reactions contain the target of interest and some do not. The ratio of positive and negative reactions combined with a Poisson correction allows for the accurate quantification of the target in the reaction (Huggett et al., 2013). Digital PCR is most commonly used to detect and quantify low-level targets such as off-types in pool of seed, measure subtle gene expression differences, and conduct high precision experiments like measuring copy number variability. Due to how the target is quantified, a standard curve or control calibrator is not required for digital PCR. Currently, the partitioning of digital PCR reactions is done either with a chip or with a water/oil emulsion method. The best instrument and method depend on the application and the budgetary constraints of the end user.

2.6 Managing costs in a detection laboratory

Real-time PCR can be an expensive detection method. One way to reduce the cost of testing and increase throughput per instrument is to use multiplex reactions. Multiplexing allows for the amplification of multiple targets in a single reaction (Elnifro et al., 2000). This is accomplished using hydrolysis probes that fluoresce at different wavelengths. Each target in a reaction is paired with a probe that fluoresces at a unique wavelength. The simplest form of multiplexing is a duplex reaction with the target of interest and an endogenous reference. A basic duplex reaction reduces the cost of each data point by almost half. Many newer real-time PCR instruments allow for multiplexing of four to six different fluorophores. An important requirement for multiplex PCR is making sure that the assays perform the same in a multiplex as they do when run alone (Bustin et al., 2009; Wittwer et al., 2001; Eum et al., 2019). Because of this requirement, developing a multiplex PCR assay is more labor intensive. The inputs required in the development and validation of multiplex analyses make them most economical with assays that are run with high frequency. A purity testing lab is an ideal environment for running multiplex PCR due to the frequency at which targets are tested. As long as the limit of detection for each target within the multiplex assay meets the standards of the laboratory, this can be an important tool to save substantially on reagent costs, maximize instrument capacity, and reduce processing time.

There are several factors that limit throughput and productivity in seed purity testing labs using PCR methods. Some of the common ones are cost, sample capacity, turn-around time, and quality. Previously, we described multiplex PCR as a way to reduce reagent costs and increase instrument throughput and capacity. Another way to increase throughput is by using array and chip-based platforms. In these platforms, arrays or chips are used to reduce reaction volumes, often to nanoliter levels, and increase the number of total reactions per instrument run. On platforms like the OpenArray from Life Technologies, the arrays can be preloaded with the assays of interest and then the nucleic acid sample is partitioned into the wells. This allows for samples to be efficiently tested against numerous targets. The drawbacks of this platform are that the custom preloaded arrays take time to manufacture and can be expensive. Once the arrays are produced, they cannot be modified, thus reducing the flexibility of the system. Other platforms like *EvaGreen* allow the analyst to prepare the PCR master mix and samples and combine the components in nanoliter-sized reactions using fluidics. There are also several companies that offer custom premixed oligonucleotides to make PCR setup easy. This can come in the form of bulk premixed oligonucleotides or having the oligonucleotides dried down in the reaction plate. These premixed assays can save time in the preparation of the master mix, but they also reduce the analyst's ability to optimize the assay. There is a wide range of real-time PCR equipment and methods. All of these approaches have strengths and weaknesses that must be evaluated to see if they are appropriate for a laboratory's processes and objectives.

2.7 Bioassays

Bioassays are visual methods of assessing the presence or absence of a phenotypic trait (ISTA, 2020). The most common form of bioassays is the assessment of herbicide resistance. This generally occurs on seedlings or plants and can occur in the form of a spray assessment or a leaf painting assay. Plants without the trait die, show symptoms of the herbicide treatment, or fail to develop normally. This form of assay can help determine the presence of a genetically modified (GMO) trait, but it is not possible to determine the identity of the transgenic event producing the phenotype if multiple events have the same trait.

2.8 Low-level presence and purity

Due to the complexity of global regulatory requirements, it is common for cultivation and import approvals to occur asynchronously creating an environment where transgenic products are approved for cultivation in some countries, but not for import in others (Davison, 2010). In some cases, a GMO product might also undergo asynchronous registration and only achieve approval in a limited number of countries. LLP is the occurrence of small quantities of transgenic grain or seed in a large shipment where the transgenic event has been assessed for food, feed, and environmental safety and is approved for cultivation in the export country, but is not approved in the import country (Davison, 2010; Krueger and Le Buanec, 2008; OECD, 2013). The tolerance for LLP and the response to detection can vary significantly between geographies and only a few countries have legal thresholds for LLP. LLP can occur during various phases of the seed production process such as cultivation, harvesting, cleaning, sorting, packing, and transporting. Limiting the LLP in the seed and commodity stream is important for both exporting and importing countries.

Detecting LLP is a balance between the ability to detect a contaminant and the resources invested. As described by the ENGL ad hoc working group on "unauthorized GMOs" (ENGL, 2011), GMOs can be classified into four main categories. Knowledge level 1 constitutes fully characterized GMOs where the event, construct, and genes are well understood. Often, level 1 events are approved, have been approved in the past, or are somewhere in the approval process. Level 2 events use the same constructs as in level 1 but are independent insertions and are therefore located in different places in the genome. These are commonly known as sister events. Level 3 events are GMOs that have constructs with novel groupings of genetic elements, but at least one element in common with Level 1. Level 4 events are those that contain entirely novel genetic elements and genes. This chapter addresses LLP and therefore primarily deals with knowledge level 1.

DNA detection methods are the most established technique for detecting LLP. There are several different ways to detect the presence of transgenic events with PCR. The most general way is to use assays that detect genetic elements such as promoter or terminator elements that are common to many of the plasmids used to produce transgenic plants. Some common examples are the detection of the cauliflower mosaic virus 35S promoter (P—35S) or the *Agrobacterium tumefaciens* nopaline synthase terminator (T-nos) (Waiblinger et al., 2008, 2010). These assays are sufficient for identifying the presence of transgenic material, but they give very little information about what transgene or event is present. For that reason, they cannot differentiate between approved and unapproved events.

The next level of detection is to look at gene-specific assays. These assays detect the presence of a specific gene within a transgenic construct. Common examples of gene-specific assays are those that detect the presence of genes involved in herbicide resistance like *pat*, *bar*, or *cp4-epsps* (Grohmann et al., 2009). These assays give significantly more information about what is being detected than the construct-specific assays, but they still do not provide enough information to distinguish between different events.

The most specific form of detection is the event-specific assay. These assays generally span the junction between the plasmid and the genomic DNA that borders the unique insertion site. The specificity is gained from the genomic DNA that borders the transgene insertion. They detect only the intended event and no other events containing the same introduced DNA cassette. This assay type is the gold standard for the detection of a specific transgenic event (ENGL ad hoc working group on "unauthorized GMOs, 2011").

As mentioned previously, the detection of transgenic contamination is a balance between the ability to detect contaminants and the resources needed. For this reason, a matrix based approach is often used (ENGL ad hoc working group on "unauthorized GMOs, 2011"). When screening for the absence of LLP, the first step is often to test using common element-specific assays such as P-35S and T-nos, as these assays occur in a wide range of transgenic events and are thus the most cost-effective method for screening. If something is detected, then further tests can be run to confirm the result and/or to determine the LLP source event (Scholtens et al., 2013). However, the use of 35S and nos is becoming less common, and these element-specific assays might not cover the full range of possible GMO events for a crop. In these cases, it is necessary to use event-specific or gene-specific assays to supplement the screening. The testing becomes more complicated when attempting to detect LLP in seed containing authorized GMO events. In these cases, it is no longer sufficient to detect the genetic elements as they are often components of many approved events. With a matrix-based approach to testing (Morisset et al., 2014), the results of all the tests can be compared to determine what GMOs are present and/or what follow-up tests need to be run to identify the GMOs present.

2.9 Data analysis and reporting

Upon completion of testing, the results must be interpreted (Chapter 13) and reported to the customer. It is recommended to report all data with similar templates. This report must contain sample identification, the method used and limit of detection (ISO, 24276). The number of extracts per sample and the number of replicates per extract may be reported (ISTA, 2020). Detection of a signal lower than the limit of detection should be reported as "the target was not detected above the limit of detection" (ISO, 24276) and a positive result as "the target was detected above the limit of detection." Quantitative results should, in addition, include the percentage of the target detected as number of seeds, mass, or number of DNA copies, and the limit of quantification and the confidence interval (as an estimate of the uncertainty of the result). Results should not be presented for detection. Detection of a signal lower than the limit of quantification should be presented as "The target was detected below the limit of quantification (ISTA, 2020; ISO, 24276)." The program SeedCalc (Remund et al., 2001) can be a useful tool when calculating the percentage of seeds containing a target using multiple seed pools or when setting up sampling schemes and determining confidence intervals.

2.10 Conclusion

Agriculture is a global industry with seed and grain constantly being transported across borders. The evolution of biotechnology will continue, and GMO traits are becoming more common and more diverse. Seed purity and testing techniques will continue to be paramount for the industry. As countries continue to approve traits asynchronously, it is important that laboratories keep up with changing techniques and technologies. Here we presented several different techniques and methods for detecting GMO targets as well as ways to manage the limitations and challenges of each detection approach and the inherent costs of each. New and long-established laboratories need to regularly evaluate technology options and find the techniques most suited for their needs.

References

Alarcon, C.M., Shan, G., Layton, D.T., Bell, T.A., Whipkey, S., Shillito, R.D., 2019. Journal of Agricultural and Food Chemistry 67 (4), 1019–1028. https://doi.org/10.1021/acs.jafc.8b05157.

Bustin, S.A., Benes, V., Garson, J.A., Hellemans, J., Huggett, J., Kubista, M., Mueller, R., Nolan, T., Pfaffl, M.W., Shipley, G.L., Vandesompele, J., Wittwer, C.T., 2009. The MIQE Guidelines: minimum information for publication of quantitative real-time PCR experiments. Clinical Chemistry 55, 611–622.

CXG 74-2010, 2010. Guidelines on Performance Criteria and Validation of Methods for Detection, Identification and Quantification of Specific DNA Sequences and Specific Proteins in Foods. Codex alimentarius FAO Rome, Italy accessed October 7th 2020. http://www.fao.org/fao-who-codexalimentarius/codex-texts/guidelines/en.

Davison, J., 2010. GM plants: science, politics and EC regulations. Plant Science 178, 94–98.

Elnifro, E.M., Ashshi, A.M., Cooper, R.J., Klapper, P.E., 2000. Multiplex PCR: optimization and application in diagnostic virology. Clinical Microbiology Reviews 13, 559–570.

ENGL, 2011. Overview on the Detection, Interpretation and Reporting on the Presence of Unauthorised Genetically Modified Materials. Joint Research Centre – Institute for Health and Consumer Protection, JRC Scientific and Technical Report No. EUR 25008 EN - 2011.

Eum, S.-J., Kim, I.R., Lim, H.S., Ro Lee, J., Choi, W., 2019. Event-specific multiplex PCR method for four genetically modified cotton varieties, and its application. Applied Biological Chemistry 62, 52. https://doi.org/10.1186/s13765-019-0459-8.

Freese, L., Chen, J.W., Shillito, R.D., 2015. Sampling of grain and seed to estimate the adventitious presence of biotechnology-derived seeds in a lot. Cereal Foods World 60 (1), 9–15. https://doi.org/10.1094/CFW-60-1-0009.

Grohmann, L., Brunen-Nieweler, C., Nemeth, A., Waiblinger, H.-U., 2009. Collaborative trial validation studies of real-time PCR-based GMO screening methods for detection of the bar gene and the ctp2-cp4epsps construct. Journal of Agricultural and Food Chemistry 57, 8913–8920.

Grothaus, G.D., Bandla, M., Currier, T., Giroux, R., Jenkins, G.R., Lipp, M., Shan, G., Stave, J.W., Pantella, V., 2006. Immunoassay as an analytical tool in agricultural biotechnology. Journal of AOAC International 89 (4), 913–928.

Holst-Jensen, A., Ronning, S.B., Lovseth, A., Berdal, K.G., 2003. PCR technology for screening and quantification of genetically modified organisms (GMOs). Analytical and Bioanalytical Chemistry 375, 985–993.

Huggett, J.F., Foy, C.A., Benes, V., Emslie, K., Garson, J.A., Haynes, R., Hellemans, J., Kubista, M., Mueller, R.D., Nolan, T., Pfaffl, M.W., Shipley, G.L., Vandesompele, J., Wittwer, C.T., Bustin, S.A., 2013. The digital MIQE Guidelines: minimum information for publication of quantitative digital PCR experiments. Clinical Chemistry 59, 892–902.

ISO 24276: 2020 Foodstuffs — Methods of Analysis for the Detection of Genetically Modified Organisms and Derived Products — General Requirements and Definitions. International Organization for Standardization, Geneva, Switzerland.

ISTA, 2020. International Rules for Seed Testing 2020. The International Seed Testing Association, Bassersdorf, Switzerland.

Kirk, R., Simpson, R., Laffont, J L., Wright, D., Gregoire, S. SeedCalc. Available online at: http://www.seedtest.org/en/statistical-tools-for-seed-testing-_content—1–1143–279.html.

Krueger, R., Le Buanec, B., 2008. Action needed to harmonize regulation of low-level presence of biotech traits. Nature Biotechnology 26, 161–162.

Lipp, M., Shillito, R., Giroux, R., Spiegelhalter, F., Charlton, S., Pinero, D., Song, P., 2005. Polymerase Chain reaction technology as analytical tool in agricultural biotechnology. Journal of the Association of Official Analytical Chemists 88 (1), 136–155.

Morisset, D., Novak, P.K., Zupanič, D., et al., 2014. GMOseek: a user friendly tool for optimized GMO testing. BMC Bioinformatics 15, 258. https://doi.org/10.1186/1471-2105-15-258.

OECD, 2013. Low Level Presence of Transgenic Plants in Seed and Grain Commodities: Environmental Risk/Safety Assessment, and Availability and Use of Information. Organisation for Economic Cooperation and Development, ENV/JM/MONO(2013)19.

Remund, K.M., Dixon, D.A., Wright, D.L., Holden, L.R., 2001. Statistical considerations in seed purity testing for transgenic traits. Seed Science Research 11, 101–119.

Scholtens, I., Laurensse, E., Molenaar, B., Zaaijer, S., Gaballo, H., Boleij, P., Bak, A., Kok, E., 2013. Practical experiences with an extended screening strategy for genetically modified organisms (GMOs) in real-life samples. Journal of Agricultural and Food Chemistry 61, 9097–9109.

Van den Bulcke, M., De Schrijver, A., De Bernardi, D., Devos, Y., MbongoMbella, G., Casi, A.L., Moens, W., Sneyers, M., 2007. Detection of genetically modified plant products by protein strip testing: an evaluation of real-life samples. European Food Research and Technology 225, 49–57.

Waiblinger, H.-U., Ernst, B., Anderson, A., Pietsch, K., 2008. Validation and collaborative study of a P35S and T-nos duplex real-time PCR screening method to detect genetically modified organisms in food products. European Food Research and Technology 226, 1221–1228.

Waiblinger, H.-U., Grohmann, L., Mankertz, J., Engelbert, D., Pietsch, K., 2010. A practical approach to screen for authorised and unauthorised genetically modified plants. Analytical and Bioanalytical Chemistry 396, 2065–2072.

Wittwer, C.T., Herrmann, M.G., Gundry, C.N., Elenitoba-Johnson, K.S.J., 2001. Real-time multiplex PCR assays. Methods 25, 430–442.

Chapter 3

Grain supply chain stewardship and testing—import and export

Amy Smith[1], Frederic Bois[2] and Dean Layton[3]
[1]*Formerly Corteva Agriscience, Indianapolis, IN, United States;* [2]*Eurofins Scientific Group, Brussels, Belgium;* [3]*EnviroLogix Inc., Portland, ME, United States*

3.1 Introduction

Plant breeding has dramatically increased grain harvests, productivity, and sustainable farming practices. Productivity has been driven by the commercialization of technologies such as mutagenic breeding beginning in the 1920s, hybrids in the 1930s, marker-assisted selection in the 1980s and genetic engineering beginning in the 1990s. For the future, plant breeding innovations, such as gene editing, show great promise. Just in the last 50 years, average US yields for corn, the most valuable crop in the United States, have more than doubled. Other field crops share a similar success story. Innovation-driven productivity has given rise to whole industries including feed, exports, biofuels, transportation, and logistics. Innovation has enabled the United States, Brazil, and other countries to become a bedrock of global food security.

The first genetically modified plants were developed in the 1980s and, the first launch of commercial products occurred a decade later. Since then, Genetically modified organisms (GMO), also called biotech crops, have consistently added value and benefits to farmers and the food supply chain. It is projected that in 2050 there will be nine billion people to feed, and with that, increased use of biotech to meet the doubled demand for food. The technology is well adapted to many different plants that are used at different steps of our food chain (direct use in feed/food industries, animal feed, etc.).

During the last decades, the acreages cultivated with biotech crops have been intensively increased. In 2018 a total of 191.7 million hectares were cultivated with biotech crops worldwide in 26 countries planted by 17 million farmers (ISAAA, 2018; Pocket K16: Biotech Crop highlights, 2018). Such an increase mainly focused on specific countries. Main nations growing

Application of Sampling and Detection Methods in Agricultural Plant Biotechnology
https://doi.org/10.1016/B978-0-323-99293-0.00003-9

genetically modified (GM) plants are the United States (75 m ha), Brazil (51.3 m ha), Argentina (23.9 m ha), India (11.6 m ha), Canada (14.7 m ha), and Paraguay (3.8 m ha), although many other countries are adopting the technology.

Among these 191.7 million hectares, a large diversity of biotech crops is cultivated (e.g., squash, apples, sugar beet, brinjal, alfalfa, papaya, etc.), but the main crops in terms of acreage are soybean, corn, cotton, and rapeseed (Table 3.1; (ISAA Pocket K16, 2019)).

Increased field production of biotech crops suggests wide acceptance of these crops by farmers. At the final consumer, there continues to be varied acceptance and knowledge regarding the safety and use of biotech crops.

The adoption and use of biotech crops are often impacted by complex and sometimes disconnected regulations across countries and regions. As the agricultural industry complies with the complex regulatory environment for commercializing, producing, and marketing crops and products, detailed supply chains, product launch and stewardship plans, and testing methods are important to support grower and consumer access to innovation and continued worldwide exchanges of goods.

The lack of alignment on regulations across countries and regions requires complex management of biotech products by the industry. In addition, the situation is becoming more and more complex by asynchronous regulations between crops, GM traits, and product purposes (ISAAA, 2019. Accomplishment report) To meet market needs, the industry has established processes to handle a variety of products and meet evolving regulatory requirements.

Stewardship and quality management plans, based on appropriate considerations for each product and intended market, are important for helping enable worldwide agribusiness. Based on the consideration of appropriate factors for each product and intended market, appropriate management structures, communications, or other practices can then be defined,

TABLE 3.1 Market percentage of the major biotechnology crops.

	Total acreage worldwide (millions of hectares)	Biotech plant cultivated acreage worldwide (millions of hectares)	Proportion of biotech plant
Soybean	123.5	96.3	78%
Maize	197.2	59.2	30%
Cotton	32.9	25.0	76%
Canola	34.7	10.1	29%

Extracted from annual report published by International Service for Acquisition of Agri-biotech Applications (ISAAA).

communicated, and implemented for addressing the regulatory and stewardship requirements by each member of the value chain for the product life cycle. For biotech products and byproducts, such as meal or processed oil, appropriate stewardship plans may be developed to manage the product across its lifecycle.

It may take 10–13 years or more to develop a biotech crop, and during this time different countries may continually refine their regulatory requirements and processes. Such crops and their final products are tested for food and feed safety as part of the development and before commercialization. With such a long development cycle, there is a challenge to ensure that the products will comply with the final market regulation, including delays due to new regulatory study requirements.

Testing for the presence of traits occurs at many points in the life cycle, from product development through commercialization, and during grain or product movement and utilization. Different analytical strategies exist, using different tools and targeting different components such as protein and DNA molecules. It may be difficult to compare and align the results from these methods. DNA-based methods may be used as screening tools or for specific event identification, adding complexity to interpreting the results from the different tests. The lack of defined, authoritative tests adds complexity to regulatory requirements and product management.

A statistical approach to ensure the robustness and the reliability of the laboratory results is one of the key elements in the quality management system and stewardship program that the stakeholders have to implement.

In addition to understanding authoritative testing, points of testing, and interpreting results, agribusiness stakeholders must also understand sampling requirements to ensure representative samples and points of testing, which will be discussed in Chapter 9.

In this chapter, we will focus on different aspects of the testing for import and export, the control at the borders, analytical strategies, and stewardship.

3.2 Testing for import and export

Because there is no single worldwide standard regarding the acceptance of GM traits, international grain traders must manage varying requirements regarding confirmation of GM presence in their shipments.

3.2.1 Biotechnology traits in the movement of grain globally

Although traits are evaluated for food and feed safety before commercialization, asynchronous regulatory approvals may impact how grain handlers may source, commingle, and move grain and products. For countries desiring to import conventional (nonbiotech) products, it may be increasingly difficult to import products that meet testing requirements if the country has established a zero tolerance for low-level presence (LLP: the presence of a trait not approved in the country of import but is approved in the country in which it was cultivated, and often, approved in multiple cultivation countries).

When developing new biotech products, companies that are members of Excellence Through Stewardship conduct export market assessments to identify key import activities before commercial launch of any new biotech plant product (crop by event) in intended markets (ETS, 2018). Due to changes in trade climate, countries may dramatically increase or decrease imports of a crop or byproduct year over year. Market assessments are therefore regularly updated during the development and consider additional factors including whether or not a market has a functional regulatory system. Each company determines what, if any, regulatory approvals must be obtained before commercialization, and use for food or feed, including import, of biotech products. For example, some countries have requirements only for cultivation, whereas other countries also require food and feed regulatory approvals. Some countries allow import if the products are approved for cultivation in the country of production, with no additional regulatory requirements. Asynchronous regulatory requirements and timelines for approval of biotech products, as well as different methods used to analyze for the presence of traits, add significant complexity to the supply chain, but should not be permitted to inhibit innovation.

3.2.2 Available methods and suitability for each testing application

There are several protein and DNA detection methods for detecting and quantifying biotech traits in grain, each with its own benefits and limitations. Both types of detection assays have well-established chemistries and offer qualitative, semiquantitative, and quantitative detection platforms that are widely used in the domestic and international testing laboratories. The reliability of test methods is based on many factors, such as the validation of the methods, training of personnel conducting the tests, quality of samples, and the testing environment. Although there is a reasonably good qualitative agreement between the results of protein and DNA analyses, quantitative results from such analyses may differ; both are routinely used in the analysis of biotechnology-derived traits in plants.

3.2.2.1 Protein-based detection methods

Immunoassays, or protein-based detection methods, include technologies such as Enzyme-Linked Immunosorbent Assays and lateral flow devices (LFDs), as described in Chapter 6. Immunoassays provide a flexible format for diverse applications, offering simple, specific, and sensitive protein detection methods to address a wide range of needs.

The most common immunoassay method used in the grain supply chain is LFDs, also known as dipsticks. LFDs are currently robust and well-adapted to use in nonlaboratory settings by nonlaboratory personnel.

While having the advantage of simplicity, protein-based methods cannot discern between different biotechnology-derived traits that express the same protein. In addition, detection and measurement can be hampered by low levels of expression of transgenic proteins, as in the case of the Cry1Ab protein expression in the Bt176 trait. Furthermore, GMO proteins that mimic proteins native to the grain are indistinguishable.

3.2.2.2 DNA-based methods

While DNA-based technologies, such as polymerase chain reaction (PCR), are widely used as the "gold standard" in testing, the technology does have limitations, including substantial instrumentation and consumables costs, the necessity for highly trained personnel, and laboratory conditions for conducting analysis (Chapters 4 and 5).

PCR is the technique most commonly used to verify the presence or absence of a specific DNA sequence in a particular biotechnology-derived plant product. PCR methods are used qualitatively, to determine the presence or absence of a biotechnology-derived trait in a sample, and can also be used quantitatively, to quantify the amount of biotechnology-derived DNA present in a sample. DNA-based methods can be used to detect multiple distinct targets, for example, the transgene itself, its promoter, and/or border region sequences.

Traditional PCR requires multiple heating and cooling steps, which can be extremely time-consuming. Another limitation of PCR testing is the extreme sensitivity of the assay, making testing highly susceptible to sample contamination. For this reason, most assays are suitable for laboratory settings only. And lastly, due to this sensitivity, PCR testing can potentially provide false positives and data interpretation challenges.

3.2.3 Grain supply chain critical testing points and associated methods

The grain supply chain encompasses origination, movement and storage, export loading, and importation Figs. 3.1 and 3.2 illustrate the type of testing conducted at some of the testing points in the grain supply chain.

- Origination—Growers must consider their desired market and the grain channel in their selection of products to produce. For example, a grower producing for a non-GMO market will select different hybrids or varieties than those grown by growers producing biotech crops. Grain testing typically begins when the farmer/producer takes the grain to an elevator nearby. The elevator, typically set up to segregate non-GMO materials,

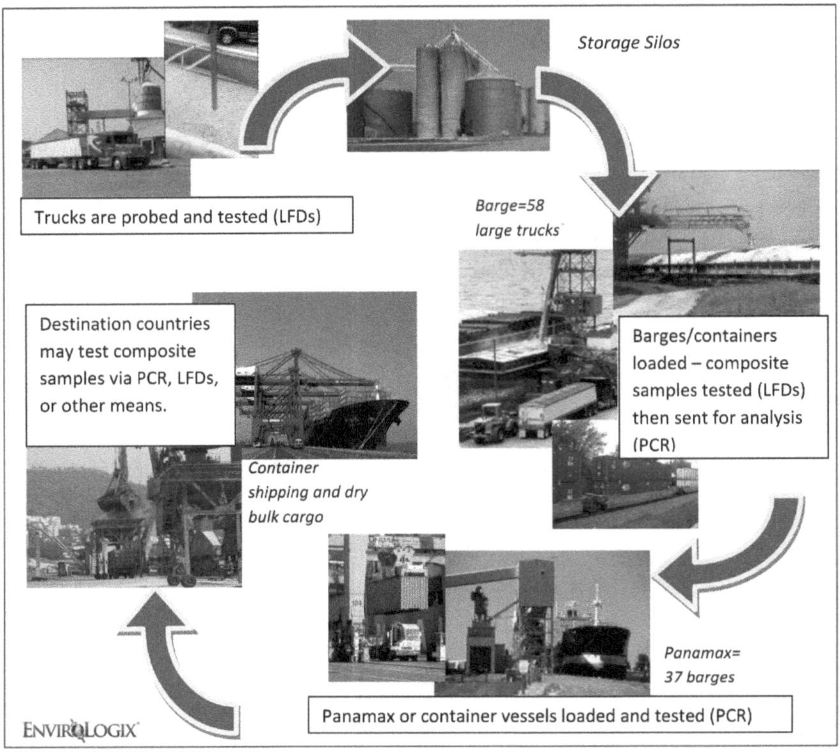

FIGURE 3.1 Grain supply chain, showing areas where testing may occur.

may probe each truck and test the incoming commodity for biotech traits. This must be done quickly, as during harvest these trucks must rapidly unload the grain and return to the field for more loads. Because of the need for speed and a quick screen to confirm what, if any, GMOs are present in each load and at what level, LFDs are the test of choice. They can be performed in rugged conditions, by nonlaboratory personnel, with results in minutes. LFDs may also be deployed as combs containing multiple test strips to detect a number of different traits. Recently, optical strip readers and software have been introduced that estimate the level of the presence in the lot. This allows the elevator to apply a semiquantitative (numerical) result to the delivery, producing a record that is particularly useful in making decisions regarding load disposition.

- Barge or rail loading—When the elevator has a critical mass of grain and an order for export, it will begin to aggregate it for export. Depending on the trade route and vessel specifications, elevators will generally load either railcars or barges. At this time, a composite sample representing the entire load of the grain is typically tested again via LFD and, in addition, a

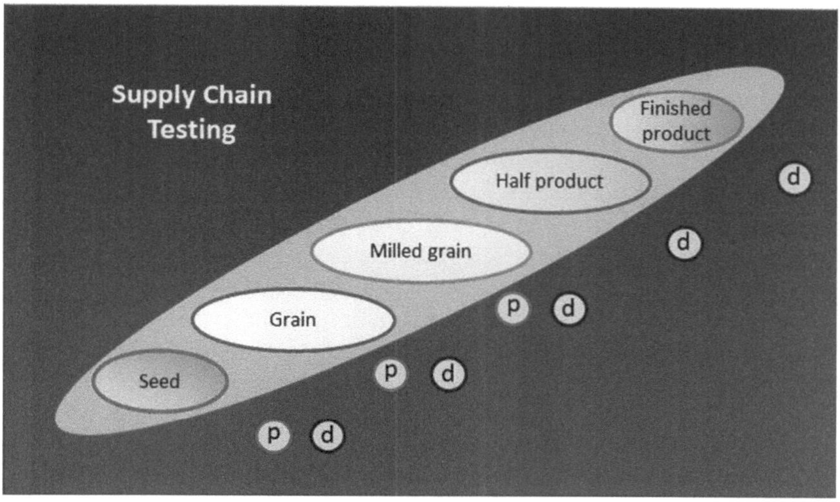

FIGURE 3.2 The type of tests applied changes as products move through the supply chain (d, DNA-based test; p, protein-based test).

sample of the grain may be sent to a laboratory for DNA analysis, generally via PCR. The protein test with its immediate results reduces the risk of moving the grain to an inappropriate destination. The PCR results, although taking longer to obtain, more closely align to testing conducted by many international trading partners, and help confirm that the final load will meet the agreed-upon specifications of grain origination contracts.

- Vessel loading—The barges and rail/container cars reach a port or point of departure and are loaded onto container ships or bulk bin vessels for export. A composite of all loads comprising the shipment may be tested via PCR per agreement with both parties.
- Incoming/import—Vessels arrive at their destination and a representative sample is tested according to the purchaser's incoming protocols, generally aligned with each country's import specifications. When conducted, these tests are typically molecular based, usually PCR.

3.2.4 Managing challenges in global grain movement

As mentioned earlier, companies commercializing traits typically conduct export market assessments before product launch. Asynchronous regulatory requirements and approvals, and the varying standards regarding grain testing continue to add complexity to the grain channel. The grain channel typically tests at multiple points throughout the channel to help ensure the product meets the regulatory and/or contractual requirements of the importing country.

3.2.5 Emerging challenges and future testing technologies

As technology evolves, two emerging challenges involve traits that incorporate no, or multiple copies, of common markers relied upon for PCR testing, and traits that incorporate no unique proteins into the grain or final product.

Traits that incorporate multiple copies of an event into a crop or stacked traits often make it difficult to correlate test results between different testing platforms. In the most common PCR tests, a promoter such as 35S and a terminator such as NOS are benchmarks used in quantification. According to published documentation regarding many new stacked traits, one transformation event may contain four or five times the content of these markers, making a load appear to have a higher percentage of biotechnology-derived traits than it actually does on a weight-to-weight basis. To complicate matters further, some of the newer traits in development may not include either of these two common markers. For this reason, it is critical that grain originators have a complete understanding of the crops planted in their area and sourced for export and that grain purchasers and trading partners clearly communicate the expectations of their non-GMO standards and thresholds.

Commercialized (and pipeline) traits that do not incorporate a unique protein, such as GA21 corn, whose protein signature is indistinguishable from the native corn protein, present another challenge for grain originators using protein methods for detection and quantification of biotech traits. Such products cannot be detected through standard protein immunoassay screens. To address this need, simplified DNA detection methods based on isothermal amplification techniques (Chapters 4 and 14) that enable use in nonlaboratory settings such as grain elevators are being developed.

In some biotech products, the same protein is developed and expressed through different transformation events. Regulatory and testing requirements may drive the need to differentiate between the two events expressing the same protein in biotechnology-derived grains if one transformation event is approved for import, while another is still in development or has pending approvals. The inability of protein detection methods to distinguish between the two events makes it even more important that DNA detection methods be adapted for speedy screening in nonlaboratory settings.

3.2.5.1 Low-level presence of genetically engineered material

A regulatory issue closely related to trade and testing is the definition of "mixing" of biotech and nonbiotech crops and whether there should be a threshold deminimis amount of genetically engineered (GE) material permissible in non-GE material. This is especially important in the global trade in grain, where numerous sources of grain are comingled into a commodity stream that may be treated as commercially interchangeable. "Adventitious presence" (AP), or low-level presence (LLP), refers to any incidental appearance of very small amounts of foreign material in a commodity, food, or feedstuff. This can occur at any time during production, harvesting, storage, or marketing.

No internationally recognized standards have existed for what amounts, if any, of GE material should be permitted in a non-GE crop, especially if that crop or a food derived from it will be labeled as non-GE. In the absence of international standards (and given the increasing global sourcing of food), individual countries are establishing their own, often varying, AP and LLP thresholds. The lack of consistent, scientifically sound standards is confusing consumers and disrupting trade, the biotech industry has asserted.

Regulatory approaches that create a zero-tolerance policy for LLP may not be practicable under current food and/or feed handling systems and will become increasingly trade-disruptive as new rDNA plant products are approved and commercialized.

One plurilateral organization working on LLP policy is the Global Low Level Presence Initiative. Its 2017 "Practical Approaches" document (https://www.fas.usda.gov/sites/default/files/2017-09/gli_principlescriteriaapproaches_with_date.pdf) outlines several principles, including:

- Safety: Before structuring an approach to mitigate the trade impact of LLP incidents, countries should consider how safety can be addressed. The food and feed safety of the rDNA plant product should be established, for example, by consideration of already completed safety assessments done either domestically or by other countries consistent with the Codex Alimentarius "Guideline for the Conduct of Food Safety Assessment of Foods Derived from Recombinant-DNA Plants" or by conducting a science-based assessment taking into account, as applicable from the same guideline, "Annex 3: Food Safety Assessment in Situations of Low-Level Presence of Recombinant-DNA Plant Material in Food." These safety assessments can inform risk management decisions by regulatory authorities and determination of appropriate measures for bringing an LLP occurrence into compliance.
- Compliance: Before an LLP occurrence, countries should put mechanisms in place to ensure that legal or compliance requirements to protect public health can be satisfied without unnecessarily disrupting trade, and,
- Industry Stewardship: Industry has an active and ongoing stewardship responsibility to ensure that only lawful products are traded in commerce. Practical approaches to addressing LLP developed by countries should fully take into account industry stewardship efforts and, where appropriate, build upon these efforts.

3.3 Regulation and official controls

Regulations for biotech crops are quite complex and vary considerably from country to country. Approval of a trait or event for cultivation is only one part of requirements. Additional regulatory requirements must often be met for food or feed consumption, or other use of the grain and byproducts. These regulations often include requirements, such as thresholds, for labeling and import, adding complexity to the entire grain channel.

With 195 countries globally, not all countries importing commodities have regulatory systems for the review and potential approval of biotech agricultural or food and feed imports. Of those countries that do have regulatory systems, their practices vary widely. Practices range from predictable, science-based, and timely reviews to unpredictable, politically driven, and lengthy reviews that may never culminate in an import approval. A global synchronous, science-based and predictable regulatory system would be ideal to harmonize trade as growers adopt innovative new technologies to increase productivity. Since there is not yet regulatory harmony, demand is created for testing services.

There are multiple challenges with testing and thresholds of detection. First, the thresholds are not tied to the safety of the grain, or food or feed byproducts, it is simply an indication of a level that is present. Second, as testing methods evolve, smaller quantities can be detected, leading to difficulty in establishing a meaningful threshold. It is therefore important to understand what the threshold is designed to control.

The first regulation for labeling GMO products in the EU was published in 1998 (Commission Regulation (EC) No 1139/98.). In addition to this regulation, some countries adopted additional thresholds for import or labeling GMO or GMO-containing products.

Borges et al. (2018) described different labeling thresholds and policies globally, illustrating the lack of harmonization. Some of the most restrictive countries are China and the EU. In China, the threshold is zero and labeling is mandatory for process products, whereas in the EU, the labeling threshold has been established at 0.9% per vegetal species, provided additional criteria have been met. Other countries do not require labeling provided the foods derived from biotechnology are not different from other foods, in alignment with Codex Alimentarius Commission (CAC, 2011a). Other countries do not require labeling, based on the risk assessments that foods derived from modern biotechnology are not necessarily different As part of stewardship programs, industries establish best practices, but it is important that testing is performed on representative samples. As testing becomes more robust, limits of detection have decreased, meaning it is easier to find trace levels of materials. Asynchronous testing and thresholds add unnecessary complexity to the channel on products that are shown to be safe for food and feed.

International harmonization of regulatory requirements, including areas such as testing and thresholds, and classifying material (such as genetically modified products), would greatly assist the grain channel and international movement of grain and byproducts.

Even though genetically modified materials have been tested and shown to be equivalent to conventional for food and feed safety, there are varying regulations in place for labeling genetically modified products. For example, in

Australia and Japan labeling is required only if genetically modified material is in the final product. In other countries, labeling is required if genetically modified material has been used in the process, even if there is no trace of the DNA or protein in the final material, such as refined oil. Labeling requirements are often tied to a country-specific, product-specific testing threshold. The complexity and asynchronicity of such regulations must be considered in export-import market transactions.

Industry and channel members design and implement processes for meeting the thresholds and the testing to ensure within thresholds. It first starts with defined stewardship and production processes to minimize the potential of commingling in production, processing, and transport. It also includes defining testing processes and the relevant processing stage for testing the different components of the final product. In addition, the testing strategy (DNA vs. protein, molecular targets, etc.), sampling processes, and sample sizes must be defined.

Some regions and countries, such as the EU, require test methods to be provided by the applicant and then validated by a commissioned lab. For example, the Joint Research Center (JRC) acts as the EU commission laboratory, with a defined validation process and minimum performance criteria defined Commission Regulation (EC) No 641/2004 (2004).

In the EU, these methods are DNA based and will be used by the official laboratories (one or more per country) belonging to European Network of GMO Laboratories (ENGL). Competent authorities define a testing plan for the year, stating the number of products that will be tested during the year, as well as the sampling process. Products may be sampled at different stages, including in the supermarket or in warehouses, and they are sent to the official lab that will carry out the test.

These laboratories are ISO/IEC 17025 certified and comply with the ISO standards for GMO detection (ISO24276, 21569, 21570, 21571). Most of the time, a screening matrix for limiting the number of molecular tests to be performed is first used and then, according to the obtained results, additional tests using event-specific methods, are conducted to determine which GM events are present in the product and at what quantity.

If an unauthorized event or level above the threshold is found, a specific system called "rapid-alert" is started. All ENGL members are informed, and the information is also published by the EU commission on their website. The member countries may then activate some dedicated measure for controlling the dissemination of the product containing the unauthorized event and may also perform additional investigations, and potentially initiate legal actions. In 2019, there were fewer than 20 alerts regarding genetically modified food or feed (RASFF, 2019), compared to 217 for allergens and 584 for mycotoxins.

The consequences of such discovery can be costly and negatively impact the industry. Unfortunately, there are many factors, including the testing strategy used by the official laboratories, that could contribute to a wrong conclusion. If a product contains some traces of an authorized event that contains several copies of the P35S target, then the official lab will find these DNA copies and continue the investigation to identify the specific event(s). Even if the quantity of GM material is below the labeling threshold, the GM event(s) detected must be authorized. If the quantity of the GM event in this product is very low, the identification-specific methods may not provide results due to the level being below the limit of detection of the method. For example, the screening element P35S may test positive because this element can be repeated in the GM construct, whereas the identification-specific target, which is required to be only one copy, maybe below the detection limit. Official laboratories have implemented specific procedures for avoiding misinterpretation of the results but dealing with a very low level of GM in a transformed product can be very difficult.

3.4 Analytical strategy

It is important to select an appropriate detection method for the purpose. In grain and food chain testing, both DNA-based and protein-based methods can be performed as qualitative or quantitative determination. While both protein and DNA-based tests can be used in seed and grain, and some partly processed products, once processing occurs, then a DNA test must be used due to denaturation of the proteins (Fig. 3.2). To determine the presence or absence of a GM trait, a qualitative method is sufficient. Qualitative determinations are usually used to detect unapproved traits where there is a zero tolerance. A quantitative testing method is needed where the purpose is to meet a regulatory or contractual nonzero threshold. As described in Section 2, a protein-based method is simple, quick, and cost-effective; however, it is usually not as sensitive as a DNA-based method. The biggest advantage of DNA-based methods is their ability to differentiate between events even if the same gene is inserted.

As described earlier, elevators need rapid methods to effectively segregate non-GM from biotech grain, and identify what traits, if any are present. The protein-based LFS testing is commonly used in the elevator due to its speed and simplicity. Testing of grain upon import is generally performed using DNA-based PCR methods to verify it meets thresholds and regulatory requirements. In grain supply chain testing, LLP, in which the trait is approved in the country of cultivation, has been well noted in internationally traded food crops. Many countries have established concentration thresholds/levels at which products must be labeled as containing GM materials, with thresholds ranging from 0.9% to 5.0% w/w.

The principle of technology for both protein-based and DNA-based assays are thoroughly discussed in Chapters 6 and 4. The strategy for seed and grain testing is discussed in more depth in Chapter 13.

3.4.1 GMO detection using real-time PCR method

Producers, importers, or testing authorities often test a sample against a library of many traits or events. The specificity of currently available DNA based methods can be divided into four levels (Fig. 3.3), with testing specificity increasing at each level: (1) screening methods that can detect a wide range of GMO without identifying the specific trait or event, (2) screening methods that can detect a certain type of genetic modifications (gene-specific detection), (3) construct-specific methods that sometimes can be used to identify the specific trait, and (4) transformation event-specific methods that can be used to identify the event with special exceptions mentioned below. Event-specific is the highest GMO-specific detection method and has the highest specificity, enabling identifying a unique event.

3.4.2 GMO screening

The promoter and terminator elements used to transform most of the currently approved genetically modified plants are the Cauliflower Mosaic Virus promoter (P-35S) and the *Agrobacterium tumefaciens* nopaline synthase terminator (T-Nos). Although other promoters and terminators have also been used, almost all GM plants contain at least one copy of the P-35S, T-35S, and/or the T-Nos as a part of the gene construct integrated into its genome. Consequently,

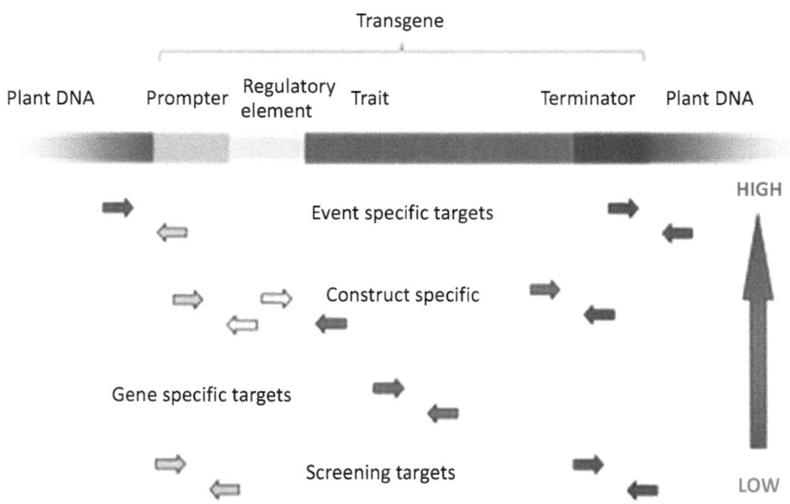

FIGURE 3.3 Levels of GMO-specific detection. *Adapted from Holst-Jensen, A., Rønning, S.B., Løvseth, A., Berdal, K.G. 2003. PCR technology for screening and quantification of genetically modified organisms (GMOs). Analytical and Bioanalytical Chemistry. 375(8), 985–993. https://doi.org/10.1007/s00216-003-1767-67.*

methods detecting one of these elements are popular for screening purposes. Because these methods detect elements that are from naturally occurring virus and bacteria, which are often present in fresh vegetables or the environment, there is a significant risk of false-positive results.

3.4.3 Gene-specific detection

The various genes inserted in a GMO may characterize a group of GMOs, although they may not identify the specific GMO. Detection of the synthetic specific gene coding for the *Bacillus thuringiensis* endotoxin CryIA(b) demonstrates the presence of a genetically modified maize, but the gene has been used in more than one GMO. Therefore, gene-specific detection methods may not be suitable for the identification of the specific event.

3.4.4 Construct-specific detection

The transgene has been integrated with different specific regulatory elements in the various GMOs containing the gene. Currently, it is therefore possible to identify the GMO targeting the junctions where the gene and regulatory elements are fused. For technical reasons, certain DNA sequences are shared by several GMOs. Construct-specific detection can test a sample for several GMOs in one step, but it is unable to identify precisely which of the specific events are present. In the future, even these junctions may be found in more than one GMO.

3.4.5 Event-specific detection

The junction between the transgene and the organism's original DNA is unique for each transformation event. Event-specific detection methods detect these regions for the presence of a DNA sequence unique to a certain GMO. This approach is ideal for precisely identifying the event present. To perform this level of detection, it is necessary to identify the precise inserted genome site of exogenous genes and isolate the flanking sequence.

3.4.6 Source of reference material

Reference materials are critical for testing in grain channel and food chain; details are provided in Chapter 8. Certified reference materials for GM events are available to testing and control laboratories, mainly from two sources, AOAC and JRC, that work in collaboration with the developers of GM crops: AOAC (https://www.aocs.org/crm) and JRC (https://crm.jrc.ec.europa.eu/).

3.5 Stewardship

According to a study conducted by CropLife International (https://croplife.org/wp-content/uploads/2014/04/Fact-Sheet-Getting-a-Biotech-Crop-to-Market.pdf), it takes an average of 13.1 years from initiation of a discovery project to the commercial launch of a trait. Because many products grown today contain stacked traits, and the grain channel is complex, additional years are required in the development and regulatory approval process. The extended time required before commercialization means that there can be changes in regulatory requirements and markets. For example, a country may implement or change regulatory requirements for import approval for a grain or its byproducts. Other changes may occur in the grain channel; economic factors may lead to increased import into an emerging market. CropLife International (CLI) and Biotechnology Industry Organization have adopted product launch stewardship policies to provide guidance to the industry on product launch to assist developers in the commercial launch.

To establish an effective stewardship program, it is important to understand the product life cycle, including its production, the products and byproducts, and the ultimate use and disposition of the products. Stewardship programs, in conjunction with quality management systems, are important to managing plant biotechnology products throughout the product life cycle. Stewardship practices also enable coexistence, or the presence of products such as conventional, organic, and biotech, within the same production and grain channels. Stewardship begins with discovery or development, having appropriate practices in place to maintain product integrity, starting in the lab and greenhouse setting. Practices such as protocols, material traceability, segregation, documentation and training, and testing at critical points in the product life cycle and grain channel are example components of an effective stewardship program.

Stewardship programs must be developed based on both relevant global and local needs. While striving for global harmonization, stewardship requirements may be locally applied based on local regulatory requirements and industry practices. Organizations such as CLI and its network of regional associations have established guiding philosophies widely adopted by plant biotech agribusinesses to support the establishment of stewardship programs.

The Excellence through Stewardship program (www.ExcellenceThroughSteawrdship.org), an industry-driven initiative, has established guidelines to assist the plant science industry in developing appropriate stewardship programs. As a member of the Excellence through Stewardship program, members commit to establishing and maintaining stewardship programs and quality management systems and participate in third-party audits to verify systems are in place.

Stewardship begins in discovery, at the lab level and continues throughout the product life cycle. Stewardship programs are designed to evaluate and assess various considerations for a specific product and its intended market during the development and use of plant biotechnology products. During development, researchers and industry must maintain product integrity and prevent the unintentional introduction of the trait into the market or grain channel. Due to asynchronous regulatory approvals, a product may be deregulated in the country in which it was developed well before other market approvals are received.

During development, confined (often referred to as regulated) field trials are required to evaluate the efficacy, performance, and safety of the trait. These studies are often complex and conducted at many field trial locations, with the acreage and volumes of materials increasing as development progresses. Field trials must often be conducted not only in the country in which the product will be commercialized, but also in countries that will be importing grain or byproducts. Confined field trials have been conducted for decades, and the combination of regulatory requirements, stewardship programs, and quality management systems has reduced the potential risk of material accidently entering the channel. CLI has provided extensive global workshops, training, and guidance for managing confined field trials (CropLife Intl 2014). Excellence Through Stewardship program also provides guidance in its "Guide for Maintaining Plant Product Integrity of Biotechnology-Derived Plant Products," including determining critical control points for consideration in stewardship programs (ETS, 2016). Regulatory agencies often define minimum requirements or protocols that must be followed for confined field trials and may require regulatory permits before initiating field trials. When conducting field trials, some of the items that need to be addressed include maintaining the integrity and preventing unintended release or comingling. An effective stewardship program must address how integrity and segregation will be maintained, including the disposition of material to prevent unintended release. Unintended release of a trait could be through outcrossing, or pollen movement to the same, or sexually compatible species, not considered part of the confined field trial. Unintended release could also happen through physical means, such as material comingling or improperly planting/harvesting the field trial. Protocols such as material transport, equipment cleaning, and traceability are typically common to all crops. Other requirements, such as isolation, subsequent use of the field trial site, and volunteer monitoring must be established with consideration of the crop and environmental factors to prevent unintended release of the trait. Field trial sites must be managed postharvest to destroy volunteer plants that emerge, creating the potential for pollen dissemination. The duration of volunteer monitoring is typically tied to physiological properties of the crop as well as environmental conditions, and is often defined as part of regulatory requirements.

In addition to establishing procedures to prevent unintended release, quality management systems must be established to verify the product integrity and purity at critical control points. Different types of testing, such as DNA-based or protein-based, and the specificity of the testing, will typically evolve as the product is developed. During introgression and breeding, it is important to verify product integrity and purity before advancing materials for seed increase. To prevent unintended release into the grain channel, each organization must establish stewardship processes and quality management systems based on the types of materials being developed and handled.

Seed multiplication for commercial crop production is typically a multi-year process. The complexity and uncertainty of regulatory approval timing may result in volumes of traited material that require the development of a

RASFF, 2019. European Commission RASFF Portal. https://webgate.ec.europa.eu/rasff-window/portal/.

Commission Regulation (EC) No 1139/98 of 26 May 1998. Concerning the compulsory indication of the labeling of certain foodstuffs prodcued from genetically modified organisms of particulars other than those provided for in Directive 79/112/EEC. Official Journal of the European Communities L 159/4. https://eur-lex.europa.eu/legal-content/EN/TXT/?uri=CELEX%3A31998R1139.

Commission Regulation (EC) No 641/2004 of 6 April 2004 on detailed rules for the implementation of Regulation (EC) No 1829/2003 of the European Parliament and of the Council as regards the application for the authorization of new genetically modified food and feed, the notification of existing products and adventitious or technically unavoidable presence of genetically modified material which has benefited from a favorable risk evaluation (Text with EEA relevance). Official Journal L 102 , 07/04/2004 P. 0014 - 0025. https://eur-lex.europa.eu/legal-content/EN/TXT/?uri=CELEX%3A32004R0641.

ETS, 2016. Guide for Maintaining Plant Product Integrity of Biotechnology-Derived Plant Products. © 2016 Excellence through Stewardship. All Rights Reserved.

ETS, 2017. Guide for Product Discontinuation of Biotechnology-Derived Plant Products. (C) 2017 Excellence through Stewardship. All Rights Reserved.

ETS, 2018. Guide for Product Launch of Biotechnology-Derived Plant Products. (C) 2018. Excellence Through Stewardship. All Rights Reserved.

ISAA Pocket K16: Biotech Crop Highlights in 2018-2019 ISAAA. Ithaca, NY. This material is published by ISAAA. www.isaaa.orghttps://www.isaaa.org/resources/publications/pocketk/document/Doc-Pocket%20K16-2019.pdf. 2019.

Further reading

2004/787/EC: commission Recommendation of 4 October 2004 on technical guidance for sampling and detection of genetically modified organisms and material produced from genetically modified organisms as or in products in the context of Regulation (EC) No 1830/2003Text with EEA relevance. Official Journal L 348 (24.11), 2004, 18–26. https://eur-lex.europa.eu/legal-content/EN/ALL/?uri=CELEX%3A32004H0787.

Commission Regulation (EC) No 641/2004 of 6 April 2004 on detailed rules for the implementation of Regulation (EC) No 1829/2003 of the European Parliament and of the Council as regards the application for the authorization of new genetically modified food and feed, the notification of existing products and adventitious or technically unavoidable presence of genetically modified material which has benefited from a favorable risk evaluation (Text with EEA relevance). Official Journal L 102 , 07/04/2004 P. 0014 - 0025. https://eur-lex.europa.eu/legal-content/EN/TXT/?uri=CELEX%3A32004R0641.

Commission Regulation (EU) No 619/2011 of 24 June 2011 laying down the methods of sampling and analysis for the official control of feed as regards presence of genetically modified material for which an authorization procedure is pending or the authorization of which has expired Text with EEA relevance. Official Journal L 166 (25.26), 2011, 9–*15*. https://eur-lex.europa.eu/legal-content/EN/TXT/?qid=1601458872475&uri=CELEX:32011R0619.

Commission Regulation (EC) No 1981/2006 of 22 December 2006 on detailed rules for the implementation of Article 32 of Regulation (EC) No 1829/2003 of the European Parliament and of the Council as regards the Community reference laboratory for genetically modified organisms (Text with EEA relevance). Official Journal L 368 (23.12), 2006, 99–109. https://

eur-lex.europa.eu/legal-content/EN/TXT/?qid=1601459028272&uri=CELEX:32006R1981. https://croplife.org/wp-content/uploads/2014/04/Fact-Sheet-Getting-a-Biotech-Crop-to-Market.pdf.

Holst-Jensen, A., De Loose, M., Van den Eede, G., 2006. Coherence between legal requirements and approaches for detection of genetically modified organisms (GMOs) and their derived products. Journal of Agricultural and Food Chemistry 54, 2799–2809.

ISAA, 2019. Pocket K16: Biotech Crop Highlights in 2018. ISAAA, Ithaca, NY. This material is published by ISAAA. www.isaaa.org. https://www.isaaa.org/resources/publications/pocketk/document/Doc-Pocket%20K16-2019.pdf.

Jana, Ž., Milavec, M., Morisset, D., Plan, D., Van den Eede, G., Gruden, K., 2012. How to Reliably Test for GMOs. Springer US.

PG economics, 2020. GM Crops: Global Socio-Economic and Environmental Impacts 1996-2018. https://pgeconomics.co.uk/pdf/globalimpactfinalreportJuly2020.pdf. (Accessed 27 July 2020).

Regulation (EC) No 1829/2003 of the European Parliament and of the Council of 22 September 2003 on Genetically Modified Food and Feed. OJ L 268, 18.10, 2003, pp. 1–23. https://eur-lex.europa.eu/legal-content/EN/ALL/?uri=CELEX%3A32003R1829.

Regulation (EC) No 1830/2003 of the European Parliament and of the Council of 22 September 2003 concerning the traceability and labelling of genetically modified organisms and the traceability of food and feed products produced from genetically modified organisms and amending Directive 2001/18/EC. Official Journal L 268 (18.10), 2003, 24–28. https://eur-lex.europa.eu/legal-content/EN/ALL/?uri=CELEX:32003R1830.

Regulation (EC) No 882/2004 of the European Parliament and of the Council of 29 April 2004 on official controls performed to ensure the verification of compliance with feed and food law, animal health and animal welfare rules. Official Journal L 165 (30.34), 2004, 1–141. https://eur-lex.europa.eu/legal-content/EN/TXT/?qid=1601458955604&uri=CELEX:32004R0882.

Chapter 4

Principles of nucleic acid-based detection methods

Yanshan Ji[1], Yusong Wan[2] and Jingwen Chen[3]
[1]*Syngenta Crop Protection, LLC., Durham, NC, United States;* [2]*Biotechnology Research Institute, Chinese Academy of Agricultural Sciences, Beijing, China;* [3]*SeqMatic, Fremont, CA, United States*

4.1 Introduction

For biotechnology applications in agriculture, changes introduced at the DNA level to produce genetically modified (GM) crops result in distinct genetic differences between the GM crop and its wild type or conventional crop. During the development of a GM crop, several detection methods—both protein-based and nucleic acid-based—are usually required for GM event characterization and selection, as well as for product quality testing (Holst-Jensen et al., 2012). GM crops typically contain at least one inserted transgene cassette (aka T-DNA), and commonly, an event-specific assay spanning the genome-insert junction is developed to detect and track the GM event throughout the product life cycle and to enforce relevant regulations. Depending on the applications, detection methods can target different DNA sequences. They can be used for qualitative analysis (i.e., presence or absence) or quantitative detection needed for labeling genetically modified organism (GMO) contents and for determining low-level presence in commercial products containing GMOs.

Genes introduced in GM crops to confer new traits are derived from different organisms that already exist in nature; however, they are considered exogenous genes to the target crops (vs. endogenous genes). GM products are the result of biotechnological manipulations and are regulated in various jurisdictions. Therefore, to determine that they do not pose any significant risk to consumers, live stocks, or to the environment, GM products are assessed for their safety. In addition, and depending on the region, monitoring systems must be in place. New GM crop registration and regulatory approval require product developers to provide an acceptable detection method for testing and tracking the new GM event.

Application of Sampling and Detection Methods in Agricultural Plant Biotechnology
https://doi.org/10.1016/B978-0-323-99293-0.00007-6

DNA is typically the most common target for detecting GM contents with molecular methods (Fraiture et al., 2015). DNA-based methods have broader applicability than phenotypic analysis or protein-based methods. DNA molecules and fragments are relatively stable and can endure much harsh environmental conditions, particularly more reliable than RNA or proteins. DNA-based methods are generally a better choice than other alternative methods for detection of GM contents, especially for GMOs in highly processed food and feed matrices.

Polymerase chain reaction (PCR) amplification is the most widely applied method to detect specific genes or junctions of the transgene and genomic DNA (Safaei et al., 2019). Assays targeting the junction of an inserted transgene and the adjacent plant genomic DNA are usually developed for a specific GM event. This assay is referred to as an event-specific assay and can be used for the identification and quantification of the GM event during the product development and commercialization processes. Other methods are also used for different purposes in different laboratories and countries, for example, Southern blot analysis for transgene characterization and microarray analysis for GMO detection. For detection and monitoring of GM-derived products in a complex matrix, for example, highly processed food or animal feed, a very sensitive and specifically validated method is required.

Significant progress has been made in developing new methods and technologies for the detection of GMOs in biotechnology products, particularly at the DNA level (Marmiroli, 2008) and using PCR amplification (Gachet et al., 1999). The current quantitative PCR (qPCR) system is the gold standard and has been widely accepted by enforcement laboratories for tracing and tracking authorized GMOs. Other approaches have been also developed for GMO detection targeting single or multiple components, with or without PCR amplification, as alternatives (Fraiture et al., 2015). However, current GMO detection systems are limited in their ability to detect or quantify GM adventitious presence, especially if found below the limit of detection, or gene-edited products with very small nucleotide changes as described in Chapter 14.

New detection methods based on next-generation sequencing (NGS) technology open the possibility of enabling the GMO detection systems to monitor virtually any type of biotechnology product at the DNA level, if necessary. Therefore, NGS-based detection methods could be adopted by more enforcements and research laboratories in the coming years and may become an enabling tool for product development and commercialization in agricultural biotechnology. NGS technology has been proposed by several laboratories for detection and identification of crops and products (Debode et al., 2019; Haynes et al., 2019). This may indicate that the technology is becoming more practical, although its wide and official adoption as a GMO detection method is pending (see more details in Chapter 15).

4.2 DNA preparation and requirements

The typical first step in most DNA-based analyses is the preparation of DNA from a biological sample. A fit-for-purpose DNA extraction protocol usually depends on the requirements of relevant applications and the appropriate detection method selected for the application. The requirements for DNA purity can be significantly different according to a particular detection method, and the quality of DNA does not necessarily have to be the same or the highest available for some applications. A fit-for-purpose method is an overall strategy for preparing DNA of which with "good enough" quantity and quality therefore suitable for specific DNA-based detection methods.

4.2.1 DNA extraction methods

The basic principle of DNA extraction is largely about releasing the DNA present in the biological sample into an aqueous solution for subsequent analysis. However, further treatments such as purification and/or dilution are often required to make the DNA quality suitable for specific applications. Many methods have been employed for DNA extraction from plant materials—ranging from low throughput and high-quality DNA preparation to high throughput and cost-effective operation for a variety of applications. DNA extraction from plant materials is generally more challenging than from animal cells. Most plants carry high levels of metabolites and other macromolecules, which can be difficult to remove without using hazardous chemicals such as phenol and/or chloroform. The major plant-originated metabolites and macromolecules include phenolic compounds, oils/lipids, proteins, and viscous polysaccharides. If these contaminants remain in DNA, they can interfere with subsequent analysis and even cause DNA damage if not properly treated (Sahu et al., 2012). There are many DNA extraction methods and commercial kits available to choose from. Here only four types of DNA extraction methods for plant materials will be discussed primarily based on popularity, principle, and operational scales.

The Cetyltrimethylammonium bromide (CTAB) method has been widely adopted by researchers and enforcement laboratories for DNA extraction from leaves, grains, and processed food and feed samples (Doyle et al., 1987). CTAB is an ionic detergent used to disrupt cell membranes and to release DNA. Organic solvents such as a chloroform-isoamyl alcohol mixture are then used to separate contaminants into the organic phase and nucleic acids into the aqueous phase. RNA is then digested with RNase. Leftover chemicals and remaining contaminants in the aqueous phase can be removed through DNA precipitation in isopropanol and with alcohol wash. The resulting DNA is usually in high concentration and quality and thus suitable for most subsequent molecular analyses. High DNA quality and concentration are required for GM event characterization and for the quantitative determination of GM content.

Quantitative analysis is necessary to monitor GM products and implement labeling mandates in certain regions. The CTAB method is usually robust for DNA extraction from complex matrices and the DNA quality is reliable (Turkec et al., 2015). However, the method is very laborious and with limited sample throughput. Therefore, it is not suitable for automation or high throughput operation for processing a large number of samples.

The sodium dodecyl sulfate (SDS)-based DNA extraction methods are also commonly used in various molecular analyses. A typical lysis buffer consists of 1% SDS in 100 mM Tris-HCl, pH 8.0, 25 mM EDTA, and 0.5 M NaCl (Dellaporta et al., 1983). Chemical concentrations may vary in between methods modified for different materials. Proteins and polysaccharides in the cell lysate are mainly removed as megacomplex with insoluble SDS in the presence of high concentration of salts. The SDS-based DNA extraction in combination with removal of impurities by organic solvents can also generate relatively high-quality DNA. The SDS-based method has also been validated by the EU Reference Laboratory for GM Food and Feed (EURL-GMFF), although it is less cited than the CTAB method (Jacchia et al., 2012). Simplified SDS-based methods without toxic organic solvents have been used in various molecular analyses where DNA purity is not so critical.

Magnetic beads-based methods are also simple and reliable methods for DNA and RNA extractions. The principle is based on the high affinity of the negatively charged DNA and RNA molecules to the positively charged silica coated on the magnetic particles. Under high salt and optimal pH, the DNA or RNA molecules are tightly bound on the silica surface of the particles. An alcohol wash removes most contaminates and subsequently, DNA or RNA molecules can be eluted off the silica surface under low ionic strength (Esser et al., 2006). The eluted DNA or RNA is of relatively good quality and suitable for many applications in molecular biology. The major advantage of this method is that there is no need for centrifugation or vacuum manifolds, which can be a bottleneck in some automated processes. The major drawback may be the limited DNA concentration for some subsequent analyses that require high DNA input such as in low-level GMO detection.

Rapid DNA extraction is particularly desirable for high throughput molecular analysis, such as high throughput screening and selection by a specific assay or maker analysis where throughput and running costs are major considerations. Simple and fast DNA extraction methods have been developed for processing a large number of plant samples, for example, by end-point PCR-based detection. The most popular choice of high throughput DNA extraction is alkaline solution-based methods for quick DNA preparation through efficient DNA releasing from various sample matrices. These methods usually involve dilution and/or neutralization to lower the pH of crude DNA extracts before subsequent PCR analysis (Porcar et al., 2007). These quick methods can be further tailored toward specific applications at different scales, for example, small number samples for rapid point-of-care detection or high throughput DNA preparation such as greater than 10,000 samples per person in a single day. The crude DNA extracts are usually "good enough" so suitable for qualitative detections such as genotyping and end-point PCR analyses. To make the crude DNA suitable for qPCR analyses, major plant-originated inhibitors including proteins, acidic polysaccharides, and phenolic compounds need to be treated, and these can be done with chemical combinations. On the other hand, chemicals to be used for such a quick method should be carefully selected to limit potential inhibitory effects in subsequent molecular analysis (Ji, 2017). The concentrations of the crude DNA extracts are usually low so that the DNA extracts may not be appropriate for low-level GMO detection that typically requires high DNA input, especially for bulk samples.

4.2.2 Requirements of DNA quality

Traditionally, DNA quality mostly means purity and integrity. However, the acceptance criteria of DNA quality for different molecular analyses can be significantly diverse. Following are general performance characteristics of a DNA extraction method that should be considered:

- DNA yield and concentration.
- Homogeneity of extracted DNA.
- Fragmentation state of recovered DNA.
- Purity and absence of inhibitory effect to subsequent application.

A prerequisite for accurate quantification of DNA is a sufficient amount of input DNA that is of appropriate quality. Accurate estimation of DNA concentrations is an important component of many molecular analyses (Demeke et al., 2010). Not all these performance characteristics are equally important to a specific detection method. But it is recognized that any of these situations, including insufficient DNA recovery, highly fragmented DNA, lots of impurities, or suboptimal representation of original DNA, may hinder the normal DNA detection process. The characteristics of DNA from highly processed food and feed samples made of various ingredients in complex matrices are particularly critical to subsequent analysis.

The quality of DNA extracted using a validated method from a given sample matrix should be reliable and suitable for the intended detection method and purpose, in terms of repeatability and reproducibility. Hence the DNA quality should be assessed using an analytical method that is influenced by the same quality parameters as for the analysis to be performed on the extracted DNA, wherever possible. A typical example of GMO quantification by qPCR analysis is that an endogenous taxon-specific TaqMan assay is used alongside of event-specific TaqMan assay for the same DNA sample. The taxon-specific assay can serve as a quality check purpose including DNA quality regarding concentration and absence of inhibitory effects to similar qPCR analysis of a GMO target.

The requirements of DNA quality largely depend on detection methods. They can be significantly different but must be suitable for the intended purpose of the method. Appropriate DNA quality for a detection method and relevant purpose does not necessarily mean that the DNA must be in high concentration or highly pure. For quantitative analysis of GMOs, the acceptance criteria of DNA have been well defined by EURL-GMFF and widely adopted (Mazzara et al., 2015). The criteria for qualitative analysis of GMOs can be relatively loosened. In addition to DNA quality, another main consideration and common practice in production are how to prepare suitable DNA efficiently. In large-scale applications, operational considerations become more critical, for example, running cost, throughput, and turnaround time. An efficient extraction method is usually preferred in high throughput operation, which may result in relatively crude DNA but still meet the minimum performance requirements in relevant analysis.

4.3 Nucleic acid hybridization

The base-pair matches (i.e., A paired with T or (U) and G paired with C) between nucleic acids is the fundamental property that enables various nucleic acid analysis techniques such as hybridization and PCR. Molecular hybridization of nucleic acids is a phenomenon in which single-stranded nucleic acid molecules (DNA or RNA) anneal to complementary base sequences forming double-stranded nucleic acid molecules (Felsenfeld et al., 1967). Hydrogen bonds formed in double-stranded nucleic acids can be disrupted by extreme physicochemical conditions such as high pH or high surrounding temperatures, such that the two strands of nucleic acids can be separated into single strands. The technique of nucleic acid hybridization is established and developed based on the denaturation and renaturation of nucleic acids. Many molecular analyses including amplification and sequencing also involve nucleic acid hybridization such as primer annealing to the target DNA molecule as starting point for the subsequent reactions.

4.3.1 Southern blot analysis

The Southern blot hybridization technique had been widely applied in molecular biology for decades since its invention (Southern, 1975). Among other applications, it has been used for screening of recombinant plasmids and analyses of gene expression and gene mutation. In agricultural biotechnology specifically, it has been used to characterize T-DNA in GM plants in combination with restriction enzyme mapping. Southern blot analysis was the most reliable method for determining the copy number of T-DNA insertions in a GM genome.

The basic principle behind the Southern blot analysis is the hybridization of nucleic acids (i.e., probes) with DNA fragments (i.e., targets in samples) that are separated by size and then immobilized on a membrane. Southern blot is a technique employed for the detection of specific DNA sequence(s) in DNA samples that are complementary to given RNA or DNA sequence(s) as probe(s). Southern blot involves the following steps:

- Separation of digested DNA fragments by electrophoresis in an agarose gel.
- Transfer of DNA separated by sizes in the gel and fixation of the DNA to a nitrocellulose or nylon membrane.
- Hybridization with a labeled probe that is complementary to the target DNA sequence and visualization by a radioactive or nonradioactive tracer.

A Southern blot can detect specific DNA fragments against a background of many other restricted DNA fragments, and it is used as a scientifically sound molecular technique in GMO characterization. Southern blot is also a prerequisite for other molecular techniques such as restriction fragment length polymorphism (RFLP) analysis (Jeffreys et al., 1985). However, the technique is very laborious and time-consuming. Its low sensitivity and low throughput among other considerations limit its applications in GMO detection and it is not suitable for bulked samples.

4.3.2 Probes

Nucleic acid hybridization as a technique typically uses a labeled single-stranded nucleic acid fragment as a probe, which is DNA or RNA of a known sequence, to bind to a complementary strand of target nucleic acids in a heterogeneous population of nucleic acids. The probe can be labeled with different detectable tracers, that can be either isotopic or nonisotopic, to determine the presence of a specific nucleic acid sequence in a sample via hybridization. The probe can be genomic DNA, cDNA (complementary DNA), RNA, or oligonucleotides with sequence specificity to the intended target, thus avoiding off-target and backgrounds issues. The size of the probe is often in 0.1−1 kb but can be shorter or longer. DNA probes are usually labeled in the process of DNA synthesis in vitro or by direct amplification by incorporating labeled substrate(s) into the DNA sequence. RNA probe is usually synthesized by RNA polymerase in the presence of ATP, GTP, CTP, and labeled UTP.

DNA and RNA probes generally provide greater specificity than shorter oligonucleotide probes because of their longer and specific sequences. Oligonucleotide probes may be used to detect small differences between targets, as the probe can be as short as 20 bases. However, the hybridization conditions using oligonucleotide probes will need to be carefully optimized. Several factors may contribute to the design of an appropriate probe for nucleic acid hybridization analysis:

- A longer sequence with a higher Tm (melting temperature) value is typically preferred, in order to eliminate background issues.
- Removal of any nonspecific portion of probe sequences to avoid potential nonspecific hybridization to off-target sequences.
- In GM event characterization, single or partial gene sequences alone or in combination with genomic sequence flanking the insert can be used as probes. For probe sequences used in the detection of GMOs in complex matrices, it is necessary to check any potential cross-hybridization to unintended targets by sequence alignment analysis, and to exclude repeat sequences and regulatory elements derived from other species.
- Multiple tracers may be incorporated in the probe sequence(s) to increase the sensitivity of detection, if necessary.

4.3.3 Choice of target nucleic acids

Target nucleic acids for hybridization analysis in conventional GMO characterization may include the entire T-DNA (transferred DNA), and/or the backbone of the construct used in plant transformation, but not any endogenous sequences or native regulatory elements present in the GM plant. Southern blot analysis, in combination with different restriction enzyme strategies, can be used to determine the copy number of the T-DNA in GM crops. For such an application, multiple elements or regions of the T-DNA can be used as targets. The experiments must be carefully designed as such results will not be ambiguous. Similarly, the backbone region of the construct used in transformation can be used as a probe to determine the presence or absence of even partial backbone sequences in GM events. For GM stacks created through conventional breeding techniques, most gene-of-interest or selectable markers can be used as targets to determine whether the breeding process has affected the integrity or copy number of the T-DNAs in corresponding GM events.

4.4 Amplification methods

Nucleic acids, especially DNA, are the targets of qualitative and quantitative diagnostics for GMOs in seeds, food, and feed. The amplification of the nucleic acids is an essential step for further analyses of the target sequence. Nucleic acids amplification methods can be divided into two classes: target amplification and signal amplification methods (Schweitzer and Kingsmore, 2001). Signal amplification, in which specific detection methods such as branched DNA, hybrid capture, and invader assay are used to increase the signal in proportion to the amount of the target, will not be discussed in this section. Target amplification methods, which will be discussed in this section, are used to increase the amount of target nucleic acid, thus providing an increased number of template molecules. The existing target amplification methods, which include PCR and its various forms, ligase amplification reaction, and several isothermal amplifications, are either based on primer extension, strand displacement, or use of alternative enzymes to DNA polymerases such as DNA ligase or RNA polymerase.

4.4.1 Polymerase chain reaction

PCR is a simple process of amplifying specific nucleic acid fragments in vitro, which was developed (Saiki et al., 1985; Mullis and Faloona, 1987) based on a concept developed by Mullis in 1983 (Mullis, 1990). The use of a thermostable enzyme, Taq DNA polymerase, which allows semiautomation and simplification of the process, has made PCR the most frequently used method for amplifying nucleic acids, especially DNA. The PCR technique is based on the reiteration of a three-step process: denaturing double-stranded DNA (dsDNA) into single-stranded DNA (ssDNA), annealing primers (specific synthetic oligonucleotides) to the ssDNA, and enzymatically extending the primers complementary to the ssDNA templates. After the primers are annealed to the denatured DNA, the ssDNA segment becomes the template for the extension reaction. Nucleotides, present in the solution in excess, are enzymatically incorporated to synthesize the complementary DNA (cDNA) sequences. During the second and subsequent cycles, the original DNA segment and the newly generated cDNA strands become templates. Each cycle of PCR, therefore, doubles the amount of specific DNA present (Wolcott, 1992). With two primers, each corresponding to one end of the DNA segment to be amplified by PCR, the DNA segment can be amplified exponentially within a reasonable time (usually less than 3 h to obtain 10^9 copies) (Holst-Jensen et al., 2003).

4.4.1.1 Targets for GMO detection

Currently, PCR is the most widely accepted GMO detection technique both by the laboratories and the regulatory authorities for its simplicity, easier methodology, extensively validated standard operating procedure and methods, and availability of reagents and equipment (Marmiroli et al., 2008). PCR-based GMO detection methods can be classified into four categories (Fig. 4.1) corresponding to their level of specificity (Holst-Jensen et al. (2003). Each category corresponds to the composition of the DNA fragment that is amplified in the PCR. The first category includes assays that are referred to as species-specific or taxon-specific assays, in which a housekeeping gene specific to the GM crop is amplified. The second category includes screening methods that amplify (i) widely used sequences especially transgenic regulatory elements such as P–35S (CaMV 35S promoter), T-35S (CaMV 35S terminator), and T-Nos (terminator of the nopaline synthase gene), or (ii) a specific transgene such as the Cry1A(b) gene or the EPSPS gene, aka

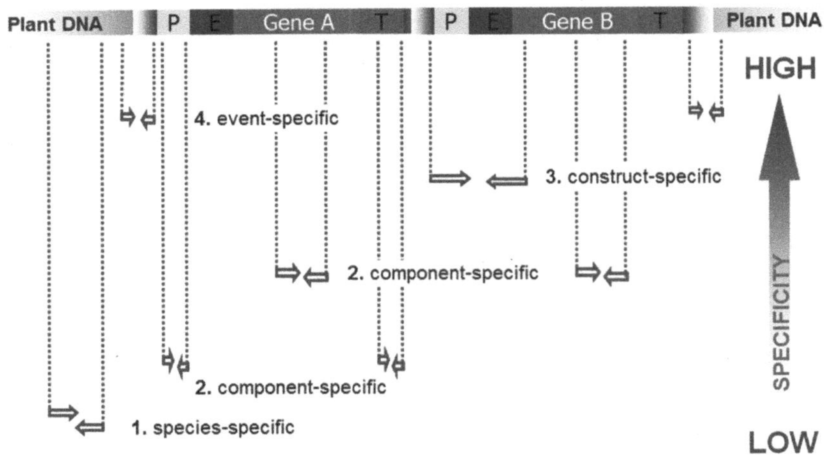

FIGURE 4.1 A schematic representation of a typical gene construct and four types of PCR-based assays showing increasing specificity (from top to bottom). P, E, G, and T stand for promoter element, enhancer element, gene of interest, and terminator, respectively. Each gene expression cassette is composed of P through T and has been inserted into plant genomic DNA (Holst-Jensen, 2003).

gene-specific methods. These assays are listed as component-specific assays in Fig. 4.1. In addition, when experiments are carefully designed, screening can be performed for the presence of multiple GMOs in a complex matrix. Popular promoter and terminator elements such as p35S and tNOS can be used as targets to indicate the presence or absence of multiple GMOs in a cost-efficient way. The third category is construct-specific assays that amplify, for example, the junction between transgenic regulatory elements such as the promoter sequence used to drive the expression of a transgene and the transgene itself. The fourth category is event-specific assays, which amplify the transgene integration site. Due to the molecular nature of transgenic DNA and the specific footprints created at the junctions between the transgene and the crop genome, any insertion site of the transgene on the plant genomic DNA can be used as target(s) for the development of an event-specific PCR assay. Hence the event-specific sequence is the most desirable target for specific GMO detection.

Typically, plant taxon-specific assays can be used for developing species-specific assay(s) that are used in parallel as control to avoid false-negative GMO results and in quantitative PCR (qPCR) to estimate the percentage of GM content in a sample. Similarly, sequences derived from donor organisms can be used to rule out false-positive GMO testing results, if necessary. These native sequences have been selected as targets for endogenous controls in GMO detections.

Based on the various purposes and the material matrix used in GMO detection, the end user can choose a single target for confirmation of any GMO event, or choose many targets for more complete results on the presence or absence of known GMOs. Some detection systems are able to perform singleplex reactions, duplex reactions, and even multiplex reactions in a high throughput manner. Combination of different targets in GMO screening has become routine, especially for product quality control and enforcement purposes.

4.4.1.2 Qualitative and quantitative PCRs

Qualitative methods detect the presence of the target sequence post PCR, which is also known as end-point PCR. Amplicons are then analyzed by gel electrophoresis. The fragment of amplified DNA corresponding to the inserted DNA in the GMO can be visualized through color detection using a dye that binds to double-stranded DNA and fluoresces under ultraviolet light. Some dedicated end-point PCR methods such as competitive PCR may be also used as quantitative methods (Gilliland et al., 1990).

Real-time PCR technology is most widely used in quantitative GMO methods. In real-time PCR, the amount of product synthesized during the PCR cycles is measured in real time by the detection of a fluorescent signal. By recording the amount of fluorescence emission at each cycle, it is possible to monitor the reaction during its exponential phase and subsequently to correlate the fluorescent signal with the initial amount of target template (Higuchi et al., 1993). There are two main methods of DNA analysis in real-time quantitative PCR: nonspecific detection of amplified products using double-strand DNA binding dyes, and specific detection that only detect specific PCR products via employing fluorophore-linked oligonucleotides as probe (Navarro et al., 2015). The most commonly used dye for nonspecific detection is dsDNA-binding fluorescent dye such as SYBR Green I, and the most commonly used probe is TaqMan hydrolysis probe. Some other fluorescent probes used in GMO detection are also discussed in the next section.

SYBR Green dye binds to the minor groove of dsDNA. The major limitation of this system is that the dye will bind to every dsDNA, including primer dimers or other nonspecific PCR products. To circumvent this problem, a melting curve analysis can be performed at the end of the PCR. This melting curve analysis measures the DNA dissociation as the temperature ramping, and it is based on each dsDNA having a specific melting temperature as a direct property of its nucleotide content. Based on this melting temperature, it is possible to distinguish the nonspecific fragments from the specific PCR products. This analysis allows not only to detect the specific products but also to get an idea on the presence of closely related targets. In addition, the SYBR Green-based methods are cost-effective as there is no need for the use of a fluorescent-labeled probe.

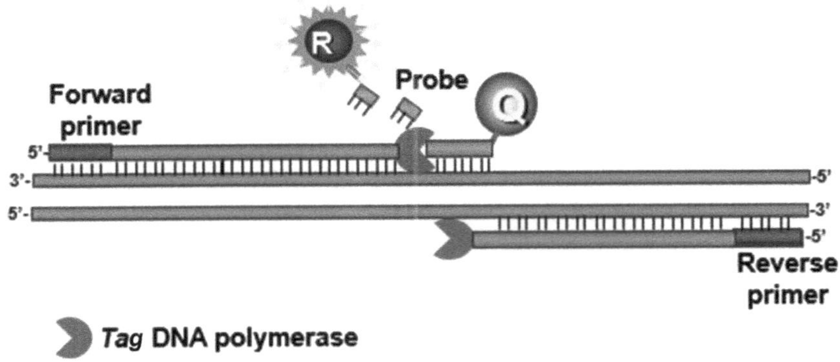

FIGURE 4.2 A schematic representation of TaqMan chemistry in real-time PCR reactions. The reporter dye (R) of the probe is being released and separated away from the quencher dye (Q) by Taq DNA polymerase during PCR amplification.

The principles of TaqMan chemistry are illustrated in Fig. 4.2. TaqMan probe is labeled at its 5′-end with a "reporter" fluorescent dye and at its 3′-end with a "quencher." As long as the two dyes are in each other's vicinity, the quencher prevents the emission of fluorescence of the reporter. The sequence of TaqMan probes is designed according to the sequence between the two primers for the PCR. During the annealing phase of PCR, the probe will anneal to the template and the specific PCR products. During the elongation phase of the real-time PCR, the Taq polymerase cleaves the annealed probe ($5' \rightarrow 3'$ exonuclease activity). The fluorescence from the freely released reporter will be detected and monitored. The fluorescence increases proportionally to the DNA quantity present in the reaction with each amplification cycle. This correlates with the increase in the original copy number of the amplified target.

The use of three oligonucleotides (two primers and one probe) allows a more specific detection. However, at the same time it reduces the flexibility of the system as it will not allow detecting some mutated sequences (e.g., caused by genetic variation of species varieties). In addition, TaqMan amplicons need to be longer as both primers and probes need to be designed in a conserved region. TaqMan real-time PCR is also a technology that allows multiplexing by using different fluorescent dyes for different targets to be simultaneously detected in one sample. However, there is a limitation in the number of fluorescent dyes that can be detected by the current real-time PCR instruments. Another potential challenge is ensuring the sensitivity or PCR efficiency of multiplex reactions is equal to singleplex reactions, if necessary.

In addition to the above-mentioned real-time PCR techniques, new alternatives and advanced technologies including high-throughput systems or platforms for multiple targets have been proposed. Some detection technologies and platforms used in GMO detection will be discussed in the next sections.

4.4.2 Ligase amplification reaction

Ligase amplification reaction (LAR) utilizes the ligation of oligonucleotide pairs that are complementary to adjacent sites on appropriate DNA templates (Wu and Wallace, 1989). Landegren et al. (1988) pioneered the oligonucleotide ligation assay to detect specific DNA sequences. The product of LAR is increased by either linear or exponential amplification using sequential rounds of template-dependent ligation. In the case of linear amplification, a single pair of adjacent oligonucleotides is ligated, the reaction is heated to dissociate the ligation product, and an additional round of ligation is performed using a thermostable ligase. Exponential amplification utilizes two pairs of adjacent oligonucleotides, complementary to each target strand. The products of the ligation reaction serve as templates for subsequent rounds of ligation. The linear and exponential amplifications of LAR are also termed as ligase detection reaction (LDR) and ligase chain reaction (LCR), respectively (Barany, 1991). Like PCR, LAR requires a thermal circler and thermostable enzyme to drive successive cycles of denaturation, annealing, and ligation, which in turn lead to a linear or exponential amplification of the desired fragment.

LCR can have greater specificity than PCR. Since a single base mismatch prevents ligation, it is possible to distinguish mutations with exquisite specificity, even at low abundance (Khanna et al., 1999). Also, LCR is easily used for multiplex reactions especially detecting products by "zip-code" hybridization microarrays (Gerry et al., 1999). One of the two oligonucleotides in LCR contains a zip-code complementary sequence in addition to the target-specific LCR sequence. The other oligonucleotide is linked with a fluorescent dye. Once ligated, the LDR products can be captured on the microarray, via hybridization between the zip codes and the complementary zip codes. However, LDR sensitivity is limited, especially when rare targets need to be detected in the presence of high wild-type DNA background. A pre-PCR amplification may be needed when detecting low-level targets with LDR detection (Morisset et al., 2008b).

Several studies have reported the use of LCR for GMO detection in food. Bordoni et al. (2004) reported a duplex LDR to detect the Cry1A(b) gene from Bt-176 transgenic maize and the reference maize gene, *zein*. Transgenic material was detected with excellent linearity within the 0.1%–2.0% range relative to wild-type maize. Peano et al. (2005) reported that the limit of detection that was estimated at 0.2 ng genomic DNA from a single GMO, corresponding to 50 copies of haploid genome of the four maize lines and 100 copies of RR (Roundup Ready) soybean haploid genome. All five transgenic lines could be detected when simultaneously present each at 0.4% relative to wild-type material by the PCR/LDR/Universal Array system. Similarly, Bordoni et al. (2005) reported a divergence in performance depending on the probe set used for each target sequence. Thus, the multiplex PCR/LDR method seems to be a promising approach for qualitative (and semiquantitative) analysis of GMO contents in a high throughput manner when associated with the hybridization of amplification products using a universal array. Mano et al. (2009) also developed a multiplex PCR-based multiplex ligase chain reaction (MPCR-MLCR) technique as an approach for the simultaneous detection of recombinant DNA segments.

4.4.3 Transcription-based amplification system

Transcription-based amplification system (TAS) uses avian myeloblastosis virus reverse transcriptase, which functions also as a DNA polymerase, and a T7 RNA polymerase to produce RNA amplified products (Kwoh et al., 1989). Every cycle of the TAS is composed of two steps. The first step is a cDNA synthesis step that produces one copy of a double-stranded DNA template for each copy of target nucleic acid (RNA or DNA). During this cDNA synthesis step, a sequence recognized by DNA-dependent RNA polymerase (e.g., T7 RNA polymerase) is inserted into the newly synthesized cDNA by bipartite primers. The second step is the amplification of the target sequence by the transcription of a double-stranded DNA template into multiple copies of RNA. Four cycles of TAS can achieve a $2-5 \times 10^6$-fold increase in the copy number of original target sequences. In the cDNA synthesis step and between the TAS cycles, heat denaturation is needed to dissociate the RNA:DNA hybrids.

The technology was later improved to an isothermal transcription-based amplification system named self-sustained sequence replication (3SR) (Guatelli et al., 1990), also known as nucleic acid sequence-based amplification (NASBA) (Compton, 1991), in which a third enzyme, RNaseH, was introduced to hydrolyze RNA from RNA:DNA hybrids. The final RNA product of 3SR/NASBA can also be used to generate a double-stranded DNA template, which is driven by a T7 promoter.

3SR/NASBA RNA product can be sequenced directly with a dideoxy method using reverse transcriptase and a labeled oligonucleotide primer. The length of the target sequence to be efficiently amplified is limited to approximately 100–250 nucleotides (Deiman et al., 2002). 3SR/NASBA amplicon detection step has significantly improved by incorporation of enzyme-linked gel assay (van der Vliet et al., 1993), electrochemiluminescent detection (van Gemen et al., 1994), molecular beacon technology (Leone et al., 1998), fluorescent correlation spectroscopy (Oehlenschläger et al., 1996), microarray (Morisset et al., 2008a) and lateral flow microarray (Carter and Cary, 2007). 3SR/NASBA has great potential for GMO identification and quantification. Morissset et al. (2008a) developed a method named NAIMA (NASBA Implemented Microarray Analysis) and applied it to GMO detection in food and feed.

specific stewardship program. As part of its product launch policy and programs, each organization must establish how it will manage products throughout its product lifecycle. Asynchronous approvals may result in several years of delay from the time of approval in the country of cultivation until identified import approvals are approved. This factor delays new innovations that would help increase their productivity and profitability while promoting enhanced environmental sustainability in reaching the hands of farmers.

Following receipt of regulatory approvals for commercialization and import into identified import markets, the stewardship program should assess what considerations impact the product, such as changing regulatory requirements. As regulatory requirements evolve, additional testing and submission for renewal may be relevant. Some countries that currently approve only single traits have pending regulatory proposals for requirements for stacked-trait products. Because most products contain multiple traits, such changes in legislation could significantly impact regulatory requirements, but should not delay the availability of new technologies to growers. Regulatory and stewardship continuously evaluate programs in place and determine what modifications, if any, should be considered to support the product life cycle.

As stated earlier, stewardship programs and quality management systems are implemented throughout the product life cycle, including through product discontinuation. Once a trait has been commercialized, especially when being grown in multiple geographies, the complexity required in the discontinuation plan to remove the product from the grain channel increases. In addition to defining the scope of the discontinuation and timelines, example decisions that must be made include disposition of existing seed and grain inventories, testing requirements, communications to stakeholders, and regulatory actions required to support the discontinuation. Often, some materials must be retained as reference materials to enable multiyear testing for low-level presence until product discontinuation is completed. Excellence Through Stewardship program's "Guide for Product Discontinuation of Biotechnology-Derived Plant Products" provides guidance on discontinuing a genetically modified trait or product (ETS, 2017).

References

Borges, B.J.P., Arantes, O.M.N., Fernandes, A.A.R., Broach, J.R., Fernandes, P.M.B., 2018. Genetically modified labeling policies: moving forward or backward? Frontiers in Bioengineering and Biotechnology 27. https://doi.org/10.3389/fbioe.2018.0018110.3389/fbioe.2018.00181. November 2018.

CAC, 2011a. Compilation of Codex Texts Relevant to Labelling of Foods Derived from Modern Biotechnology. CAC/GL 76-2011. 1. Available online at: http://www.fao.org/fao-who-codexalimentarius.

CropLife International, 2014. Compliance Management of Confined Field Trials for Biotech-Derived Plants.

ISAAA, 2019. Accomplishment Report. ISAAA, Ithaca, NY. This material is published by ISAAA. www.isaaa.org. http://www.isaaa.org/resources/publications/annualreport/2019/pdf/ISAAA-2019-Accomplishment-Report.pdf.

4.4.4 Strand displacement amplification

Strand displacement amplification (SDA) is an isothermal amplification method initially developed by Walker et al., (1992a, 1992b). It utilizes a restriction enzyme to generate primer-directed nicking and an exonuclease-deficient polymerase to initiate DNA synthesis at the nicking site and displace the downstream strand. Nicking-generated primers are composed of a nicking restriction enzyme recognition sequence part at 5′-end and a target complementation sequence part at 3′-end. The target DNA fragment is first denatured and then annealed to one side of the nicking-generated primer at 3′-end. After elongation of both strands, target DNA fragments with 5′-end restriction site modifications are produced. The nicking restriction enzyme recognizes the site and generates a nick on the primer elongated strand. Exonuclease-deficient polymerase then initiates DNA synthesis at the nicking site to displace the downstream newly synthesized strand. The displaced strand acts as the template of the other side SDA primer and subsequent SDA cycles can be carried out to achieve exponential amplification.

Generation of target fragments is an important phase in the SDA process. In the first version of the SDA process, the target fragment template was generated by a restriction enzyme (Walker et al., 1992a). In the modified version, a different pair of primers were used, which are located outside of nicking-generated primer recognition sites (Walker et al., 1992b). Nicking-generated primers elongated ssDNA were displaced by outer primers elongated DNA strand and acted as target fragment templates. The main difference between SDA and PCR is that the ssDNA template is generated by strand displacement in SDA while the PCR template is generated by denaturation of dsDNA through temperature alteration.

Optimizations of SDA include the utilization of different enzymes and adjusting the ratio of polymerase to nicking enzymes. First restriction enzymes such as HincII, BsrI, AvaI, or BsoBI, are used. They can produce nicks at hemiphosphorothioate sites generated by using a of 2′-deoxynucleoside 5’-[α-thio]triphosphate base in DNA synthesis. Lately both enzymes Nt.CviPII (Chan et al., 2004) and Nt.BstNBI (Ehses et al., 2005) have been used; these are site-specific endonucleases that cleave only one strand of DNA on a double-stranded DNA substrate. In addition to the exo-Klenow polymerase used in the first SDA study, the Klenow polymerase and the large fragment of Bst DNA polymerase have been also used in later SDA studies.

Major limitations of SDA include its inability to efficiently amplify long target sequences, and that only semiquantitation is possible by this method (Walker, 1993). Joneja et al. achieved linear amplification of a DNA template up to 5000 nucleotides in length with very little bias by using a DNA polymerase, Sequenase 2.0, which has very high processivity and strong strand-displacement capability in the presence of single-stranded binding proteins, and using nicking endonuclease, Nt.BspQI, which has a long 7-bp recognition sequence (Joneja and Huang, 2011). Nadeau et al. (1999) have described a real-time, sequence-specific DNA target detection method using the fluorogenic reporter probes and reported accurate quantification of target levels.

SDA is the basis for some commercial detection tests such as BDProbeTec (Becton Dickinson, Franklin Lakes, NJ, USA) and has been evaluated recently for the identification of *Mycobacterium tuberculosis* directly from clinical specimens (McHugh et al., 2004). Zahradnik et al. (2014) compared several isothermal amplification techniques to detect 35S promoter in GM maize.

4.5 Detection platforms

Several detection methods and platforms have been developed and used for GMO detection in the past. These methods can be generally classified as qualitative and quantitative approaches. Detection platforms may range from simple, cheap, and easy for the operation to much sophisticated and high throughput detection of biotechnology products. Each platform has its own advantages as well as limitations. Depending on the purpose, the demand for GMO detection can be usually fulfilled by more than one method and platform. Therefore, the choice of a GMO detection platform and method will largely depend on some of the following considerations:

- Purpose of the testing.
- Relevant regulatory requirements, including labeling policies and biosafety laws if applicable.
- Financial affordability.
- Capabilities to meeting the demands.
- Infrastructure and skill sets.

As mentioned previously, PCR-based target DNA detection remains the method of choice for most routine analyses of GMO contents in food and feed samples (Zel et al., 2012). Several other methods for GMO detection have been developed but have not yet been widely adopted (Randhawa et al., 2013; Milavec et al., 2014). This is not only because PCR-based methods are able to offer superior advantages such as specificity, sensitivity, reproducibility, and robustness, but also because PCR products are applicable to a wide range of detection platforms.

Once the PCR assay has been developed and validated following relevant requirements, qualitative detection for the presence/absence of a GMO target sequence by conventional PCR is fairly simple and easy in regard to operation and interpretation. The specific PCR product resulted from the target sequence can be analyzed by gel electrophoresis based on its migration rate. Both positive and negative controls are usually included in the PCR analysis and a molecular-weight size ladder is included in the gel electrophoresis. The PCR amplicon size corresponding to the target DNA can be determined under ultraviolet light upon binding to a fluorescent dye and when compared to the molecular-weight size ladder. The sensitivity of the end-point PCR for low-level GMO detection is usually adequate although the dynamic range is relatively narrower than qPCR.

Real-time PCR-based qualitative and quantitative analyses can be performed using an integrated instrument for amplification and detection. As discussed previously on Qualitative and Quantitative PCRs in Section IV (4.4.1.2), PCR products can be determined by either the use of a nonspecific fluorescent DNA binding dye or by a target sequence-specific probe in real-time during PCR cycles. Quantitative analysis of GMO is usually carried out by determination of the amount of event-specific target with respect to a taxon-specific target (i.e., endogenous gene) under the same PCR conditions. The qPCR-based GMO analysis requires a real-time PCR instrument as a detection platform and well-trained expertise for data interpretation. There are several qPCR instruments on the market to choose from. The requirements of DNA quality and concentration for qPCR analysis are usually higher than that for end-point PCR analysis. Nevertheless, accurate quantification of GMO with the qPCR platform is critical and necessary under various circumstances, although the overall applicability of the qPCR platform must be evaluated regarding most considerations mentioned above.

The main advantage of the qPCR platform is that it is versatile. It can be used to simultaneously detect single or multiple GMOs in multiple samples with either a singleplex or multiplex reaction format. It is also able to quantify the amount of each and all GMOs present in the complex sample matrix in conjunction with respective endogenous controls. On the other hand, the qPCR platform can be used as an alternative platform for the detection of products from other GMO amplification technologies, but this has not been widely adopted worldwide due to different considerations (Randhawa et al., 2013). Another platform can be a better choice than qPCR for a specific GMO detection under different considerations. As an example, microfluidic devices with miniaturization and automation such as lab-on-a-chip platform, particularly working with isothermal amplification methods, could be suitable for point-of-care applications (Zanoli and Giuseppe, 2013).

Digital PCR (dPCR) is one of the promising technical improvements in the field of quantitative GMO detection. The dPCR can quantify the absolute number of targets that are present in a sample using PCR assays along with limiting sample dilutions (Sykes et al., 1992). This is achieved by partitioning the PCR mix into a large number of individual reactions that contain about one copy of the target DNA per reaction. With the droplet digital PCR (ddPCR) platform, the absolute copy numbers of one or a few targets in a sample can be determined by counting the positive reactions after end-point PCR and using binomial Poisson statistics (Pinheiro et al., 2012). The ddPCR platform should be compatible with various amplification chemistries and is more suitable for accurate GMO quantification even without the availability of certified reference controls. Some drawbacks that may limit its application in high throughput GMO detection are low sample throughput and the need for specialized instruments and for well-trained expertise. As we will discuss in the next section, NGS technology is continually evolving, both dPCR and microarray platforms may become less popular for GMO detection.

4.6 Next-generation sequencing

Next-generation sequencing technologies, as an emerging tool and platform, can be used for molecular characterization of GM events, which have been adopted already in GM product development and for GMO detection. The utilization of the NGS platform presents clear advantages to biotechnology product developers, as it can be used to evaluate several aspects that are necessary for comprehensive molecular characterization of GM crops, thus simplifying current approaches required for this purpose. It is expected that the rapidly evolving NGS platform will become more valuable and increasingly more essential in the detection and quantification of biotechnology products, including new plants derived with innovative breeding techniques such as genome editing.

DNA sequencing is a process to determining the precise order of all nucleotides within a DNA molecule. NGS, also known as high-throughput sequencing, is a catch-all term used to represent modern sequencing technologies with massively parallel or deep sequencing capabilities (Behjati, 2013). In practice, there are generally two approaches for the use of NGS technologies. One is the whole genome sequencing (WGS) approach, where the whole DNA is extracted from a sample and analyzed without the need for any prior knowledge of the DNA sequences. On the other hand, the targeted NGS approach refers to the sequencing of DNA of interest, which could be enriched either by PCR amplification (aka, amplicon sequencing) or by hybridization methods (magnetic beads or microarrays) associated with specific probes to the DNA target(s). Therefore, this approach is also known as a targeted NGS strategy, as the sequencing requires some prior knowledge of the DNA sequences of interest (Fraiture et al., 2017a).

Several NGS technologies have been developed. Not all performed equally well for a specific application, based on users' experiences. Following are popular NGS technologies and examples of sequencing platforms:

- Sequencing-by-Synthesis technology from Illumina, which is the current leading NGS sequencer and reagents provider. Popular instruments include MiSeq, NextSeq, and NovaSeq.
- Combinatorial Probe-Anchor Synthesis and DNA NanoBalls technology from BGI-MGI. BGI-MGI offers several sequencers including DNBSEQ-T7.
- Ion Semiconductor sequencing from ThermoFisher. Instruments include Ion GeneStudio S5 System.

These sequencing technologies allow users to sequence DNA and RNA much more rapidly and cheaply than Sanger sequencing does. The NGS capability has revolutionized the research in genomics and molecular biology, as well as in product development in agricultural biotechnology. In contrast with these massive and short sequencing reads, long-range DNA sequencing (aka, third-generation sequencing) and single-molecule technologies are able to generate 10 kb or longer sequencing reads, although the sequencing error rate is relatively high. Commercially available long-range DNA sequencing technologies include:

- Single Molecule Real-Time sequencing from Pacific Biosciences. PacBio HiFi sequencing strategy can provide more accurate data with relatively long reads.
- Nanopore sequencing from Oxford Nanopore Technologies. ONT's extralong reads may be beneficial to some applications such as de novo genome assembly.

In the past, molecular characterization of GM events at the DNA level has been done mostly by Southern blot and PCR analyses in combination with Sanger sequencing, by determining the precise location of the T-DNA in the host genome and the presence of any unintended vector backbone sequences inserted anywhere on the genome. Recent advances in agricultural biotechnology have resulted in new traits that are created by new breeding techniques such as CRISPR-Cas (Chen et al., 2019; Ahmar et al., 2020). Some CRISPR plants may be subject to significant GM-type regulations in the European Union (Callaway, 2018). New requirements clearly pose challenges to current approaches including PCR-based and Southern blot analysis (Grohmann et al., 2019). Targeted (focusing on insertion and flanking region) NGS approach could provide certain answers for the molecular characterization of single GM events, breeding stacks, and new genetically edited plants at much higher efficiency. Moreover, sequencing results can be more comprehensive and accurate than hybridization or PCR results. NGS technology has become a very useful tool in molecular characterization in various laboratories (Debode et al., 2019) and holds great potential in the future as well as having limitations for GMO detection.

More specifically, the increasing sequencing capacity at decreasing costs in combination with the digital nature and the tunable resolution of NGS technologies have demonstrated its enormous value in genomic research and various applications. NGS with a genome-wide sequencing approach can facilitate GM event screening during product development and comprehensive molecular characterization of GM events more efficiently compared to Southern blot analysis (Kovalic et al., 2012). Hence the WGS, as an untargeted NGS approach, has been suggested to replace the Southern blot, PCR, and Sanger sequencing analyses all together in future GM event characterization (Kovalic et al., 2012; Wahler et al., 2013; Cade et al., 2018). On the other hand, the detection of GMOs at low levels in food samples are not practical and could not be easily achieved using the WGS approach at this time (Pauwels et al., 2015; Willems et al., 2016).

Target enrichment sequencing approach may make more sense in rare GMO detection. Besides, due to the limited capabilities of current detection systems to identify unauthorized GMOs, a new workflow based on PCR screening in combination with WGS technology has been suggested to meet different challenges (Fraiture et al., 2017a).

As NGS technology continuously evolves, several targeted sequencing approaches have been developed for the detection of authorized as well as unauthorized GMOs (Arulandhu et al., 2016). These approaches require target enrichment steps in the sequencing library preparations, either by target capture through probe hybridization or by PCR amplification, to achieve the necessary sensitivity for low-level GMO detection. The targeted NGS approaches can be much more cost- and time-effective for the detection of multiple GMOs (authorized and unauthorized) in multiple complex samples such as highly processed food samples, which can be indexed and pooled in a single sequencing run. Targeted NGS can also offer an accurate and comprehensive view of the structural information for the T-DNA, particularly in the cases of unauthorized GM events and stacked GM crops (Yang et al., 2013; Arulandhu et al., 2016). Accurate quantification of GMO contents in the complex matrix will largely rely on several factors including:

- DNA recovery and quality.
- Technology for sequencing library preparation.
- Strategy of NGS approach and sequencing data analysis.
- Statistical strategy of quantification.

Efficient DNA recovery from highly processed food or feed can be a big challenge due to DNA damage and impurities within the sample matrix but this is a common challenge to all GMO detection methods. DNA quality and quantity for sensitive GMO detection by targeted NGS approaches should be adequately high. More efforts should be made to develop suitable methods for characterization and quantification of unauthorized GMOs, stacked GM crops, as well as new crops generated by genome editing technologies (Li et al., 2017; Fraiture et al., 2017b). In the foreseeable future, NGS technologies are most likely to be commonly adopted for molecular characterization of GMOs, and eventually for detection and quantification of all kinds of GMOs and gene-edited products in complex sample matrices, once a few technical challenges have been overcome.

4.7 Closing remarks

Quantitative PCR-based methods are the current standard for GMO detection and quantification in most laboratories and will remain the top choice for GMO tracking and tracing by major authorities for some time. For rapid detection of well-known individual GMO targets, both end-point PCR and qPCR-based event-specific methods are appropriate for individual samples as well as for complex sample matrices. Other regions of T-DNA such as frequently used regulatory elements in GMOs are also suitable targets for covering multiple GMOs in a single sample matrix. Multiplex reactions for simultaneous detection of several GM targets, such as using microarrays, could be alternative approaches with better efficiency. However, the current standard GMO detection systems have limited capability for the detection of unauthorized GMOs, and some genome-edited (GE) organisms. Additional challenges include accurate quantification of all kinds of GMOs present in highly processed food and feed materials, in accordance with various biosafety laws and labeling policies.

It is reasonable to believe that only technological breakthroughs could provide ultimate solution(s) to these challenges. For example, the next generation of innovative products including GE products will require new detection methods with higher sensitivity than current WGS and higher resolution than PCR. Several exploratory studies have been done in this regard and some strategies and detection methods are promising. Particularly, the targeted NGS approach with improved throughput and sensitivity seems getting more practical for the detection of multiple GMOs in real-life food samples (Fraiture et al., 2017b). Small nucleic acid changes such as single point mutations or single allele replacements introduced through genome editing techniques or synthetic biology may only be identified and quantified through NGS technology (Wilkes et al., 2016). This application is further discussed in detail in Chapters 14 and 15. Targeted NGS technologies with increasing capacity and decreasing costs represent a major trend in biotechnology for complex GMO detection. Other technological breakthroughs beyond PCR and NGS could also be able to address some specific challenges. The CRISPR-Cas13a-based molecular detection platform, termed SHERLOCK (Specific High Sensitivity Enzymatic Reporter UnLOCKing), has been developed for rapid and sensitive nucleic acid detection (Gootenberg et al., 2017). Its sensitive DNA/RNA quantification in lieu of specific qPCR assays and the specificity to detect single-base differences opens an opportunity of using the SHERLOCK technology for rapid detection of virtually any biotechnology products in the agriculture and food industry, remotely. Several technical aspects still need to be elucidated, including sampling strategy, reference materials, DNA preparation, infrastructure, and analytical tools for data analysis and interpretation.

References

Ahmar, S., Gill, R.A., Jung, K.H., et al., 2020. Conventional and molecular techniques from simple breeding to speed breeding in crop plants: recent advances and future outlook. International Journal of Molecular Sciences 21 (7), 2590. https://doi.org/10.3390/ijms21072590.

Arulandhu, A.J., van Dijk, J.P., Dobnik, D., et al., 2016. DNA enrichment approaches to identify unauthorized genetically modified organisms (GMOs). Analytical and Bioanalytical Chemistry 408, 4575–4593. https://doi.org/10.1007/s00216-016-9513-0.

Barany, F., 1991. Genetic disease detection and DNA amplification using cloned thermostable ligase. Proceedings of the National Academy of Sciences of the United States of America 88 (1), 189–193.

Behjati, S., Tarpey, P.S., 2013. What is next generation sequencing? Archives of Disease in Childhood: Education and Practice Edition 98, 236–238.

Bordoni, R., Germini, A., Mezzelani, A., Marchelli, R., De Bellis, G., 2005. A microarray platform for parallel detection of five transgenic events in foods: a combined polymerase chain Reaction–Ligation detection Reaction–Universal array method. Journal of Agricultural and Food Chemistry 53 (4), 912–918. https://doi.org/10.1021/jf0486949.

Bordoni, R., Mezzelani, A., Consolandi, C., Frosini, A., Rizzi, E., Castiglioni, B., Salati, C., Marmiroli, N., Marchelli, R., Bernardi, L.R., Battaglia, C., De Bellis, G., 2004. Detection and quantitation of genetically modified maize (Bt-176 transgenic maize) by applying ligation detection reaction and universal array technology. Journal of Agricultural and Food Chemistry 52 (5), 1049–1054. https://doi.org/10.1021/jf034871e.

Callaway, E., 2018. CRISPR plants now subject to tough GM laws in European Union. Nature 560 (7716), 16. https://doi.org/10.1038/d41586-018-05814-6.

Cade, R., Burgin, K., et al., 2018. Evaluation of whole genome sequencing and an insertion site characterization method for molecular characterization of GM maize. Journal of Regulatory Science 6 (1), 1–14.

Carter, D.J., Cary, R.B., 2007. Lateral flow microarrays: a novel platform for rapid nucleic acid detection based on miniaturized lateral flow chromatography. Nucleic Acids Research 35 (10), e74. https://doi.org/10.1093/nar/gkm269.

Chan, S.H., Zhu, Z., Van Etten, J.L., Xu, S.Y., 2004. Cloning of CviPII nicking and modification system from chlorella virus NYs-1 and application of Nt.CviPII in random DNA amplification. Nucleic Acids Research 32 (21), 6187–6199. https://doi.org/10.1093/nar/gkh958.

Chen, K., Wang, Y., Zhang, R., Zhang, H., Gao, C., 2019. CRISPR/Cas genome editing and precision plant breeding in agriculture. Annual Review of Plant Biology 70, 667–697.

Compton, J., 1991. Nucleic acid sequence-based amplification. Nature 350 (7), 91–92.

Debode, F., Hulin, J., et al., 2019. Detection and identification of transgenic events by next generation sequencing combined with enrichment technologies. Scientific Reports 9, 15595. https://doi.org/10.1038/s41598-019-51668-x.

Deiman, B., van Aarle, P., Sillekens, P., 2002. Characteristics and applications of nucleic acid sequence-based amplification (NASBA). Molecular Biotechnology 20 (2), 163–179. https://doi.org/10.1385/mb:20:2:163.

Dellaporta, S.L., Wood, J., Hicks, J.B., 1983. A plant DNA minipreparation: version II. Plant Molecular Biology Reporter 1, 19–21. https://doi.org/10.1007/BF02712670.

Demeke, T., Jenkins, G.R., 2010. Influence of DNA extraction methods, PCR inhibitors and quantification methods on real-time PCR assay of biotechnology-derived traits. Analytical and Bioanalytical Chemistry 396, 1977–1990.

Doyle, J.J., Doyle, J.L., 1987. A rapid DNA isolation procedure for small quantities of fresh leaf tissue. Phytochemical Bulletin 19, 11–15.

Ehses, S., Ackermann, J., McCaskill, J.S., 2005. Optimization and design of oligonucleotide setup for strand displacement amplification. Journal of Biochemical and Biophysical Methods 63 (3), 170–186. https://doi.org/10.1016/j.jbbm.2005.04.005.

Esser, K., Marx, W., Lisowsky, T., 2006. maxXbond: first regeneration system for DNA binding silica matrices. Nature Methods 3, i–ii. https://doi.org/10.1038/nmeth845.

Felsenfeld, G., Miles, H.T., 1967. The physical and chemical properties of nucleic acids. Annual Review of Biochemistry 36, 407–448. https://doi.org/10.1146/annurev.bi.36.070167.002203.

Fraiture, M.A., Herman, P., Taverniers, I., De Loose, M., Deforce, D., Roosens, N.H., 2015. Current and new approaches in GMO detection: challenges and solutions. BioMed Research International 2015, 392872. https://doi.org/10.1155/2015/392872.

Fraiture, M.A., Herman, P., De Loose, M., Debode, F., Roosens, N.H., 2017a. How can we better detect unauthorized GMOs in food and feed chains? Trends in Biotechnology 35 (6), 508−517. https://doi.org/10.1016/j.tibtech.2017.03.002.

Fraiture, M.A., Herman, P., Papazova, N., et al., 2017b. An integrated strategy combining DNA walking and NGS to detect GMOs. Food Chemistry 232, 351−358. https://doi.org/10.1016/j.foodchem.2017.03.067.

Gachet, E., Martin, G.G., Vigneau, F., Meyer, G., 1999. Detection of genetically modified organisms (GMOs) by PCR: a brief review of methodologies available. Trends in Food Science & Technology 9, 380−388.

Gerry, N.P., Witowski, N.E., Day, J., Hammer, R.P., Barany, G., Barany, F., 1999. Universal DNA microarray method for multiplex detection of low abundance point mutations. Journal of Molecular Biology 292 (2), 251−262.

Gilliland, G., Perrin, S., Bunn, H.F., 1990. Analysis of cytokine mRNA and DNA: detection and quantification by competitive polymerase chain reaction. Proceedings of the National Academy of Sciences of the United States of America 87, 2725−2729.

Gootenberg, J.S., Abudayyeh, O.O., et al., 2017. Nucleic acid detection with CRISPR-Cas13a/C2c2. Science 356, 438−442. https://doi.org/10.1126/science.aam9321.

Grohmann, L., Keilwagen, J., Duensing, N., et al., 2019. Detection and identification of genome editing in plants: challenges and opportunities. Frontiers of Plant Science 10, 236. https://doi.org/10.3389/fpls.2019.00236.

Guatelli, J.C., Whitfield, K.M., Kwoh, D.Y., Barringer, K.J., Richman, D.D., Gingeras, T.R., 1990. Isothermal, in vitro amplification of nucleic acids by a multienzyme reaction modeled after retroviral replication. Proceedings of the National Academy of Sciences of the United States of America 87 (5), 1874−1878.

Haynes, E., Jimenez, E., Angel Pardo, M., Helyar, S., 2019. The future of NGS analysis in testing food authenticity. Food Control 101, 134−143. https://doi.org/10.1016/j.foodcont.2019.02.010.

Higuchi, R., Fockler, C., Dollinger, G., Watson, R., 1993. Kinetic PCR analysis: real-time monitoring of DNA amplification reactions. Biotechnology 11, 1026−1030.

Holst-Jensen, A., Rønning, S.B., Løvseth, A., Berdal, K.G., 2003. PCR technology for screening and quantification of genetically modified organisms (GMOs). Analytical and Bioanalytical Chemistry 375 (8), 985−993. https://doi.org/10.1007/s00216-003-1767-7.

Holst-Jensen, A., Bertheau, Y., de Loose, M., et al., 2012. Detecting un-authorized genetically modified organisms (GMOs) and derived materials. Biotechnology Advances 30, 1318−1335.

Jacchia, S., Bogni, A., Savini, C., Mazzara, M., Kreysa, J., 2012. Report on the Single-Laboratory Validation of a DNA Extraction Method from Carnation Leaves. European Union Reference Laboratory for Genetically Modified Food and Feed. JRC publication 77732. https://doi.org/10.2788/76844.

Jeffreys, A.J., Wilson, V., Thein, S.L., 1985. Hypervariable 'minisatellite' regions in human DNA. Nature 314, 67−73.

Ji, Y., 2017. Methods and Compositions for Preparation of Nucleic Acids. US 9556427 B2. https://patents.google.com/patent/US9556427B2/en.

Joneja, A., Huang, X., 2011. Linear nicking endonuclease-mediated strand-displacement DNA amplification. Analytical Biochemistry 414 (1), 58−69. https://doi.org/10.1016/j.ab.2011.02.025.

Khanna, M., Cao, W., Zirvi, M., Paty, P., Barany, F., 1999. Ligase detection reaction for identification of low abundance mutations. Clinical Biochemistry 32, 287–290.

Kovalic, D., Garnaat, C., Guo, L., et al., 2012. The Use of next generation sequencing and junction sequence analysis bioinformatics to achieve molecular characterization of crops improved through modern biotechnology. The Plant Genome 5, 149–163. https://doi.org/10.3835/plantgenome2012.10.0026.

Kwoh, D.Y., Davis, G.R., Whitfield, K.M., Chappelle, H.L., DiMichele, L.J., Gingeras, T.R., 1989. Transcription-based amplification system and detection of amplified human immunodeficiency virus type 1 with a bead-based sandwich hybridization format. Proceedings of the National Academy of Sciences of the United States of America 86 (4), 1173–1177. https://doi.org/10.1073/pnas.86.4.1173.

Landegren, U., Kaiser, R., Sanders, J., Hood, L., 1988. A ligase-mediated gene detection technique. Science 241 (4869), 1077–1080. https://doi.org/10.1126/science.3413476.

Leone, G., Van Schijndel, H., Van Gemen, B., Kramer, F.R., Schoen, C.D., 1998. Molecular beacon probes combined with amplification by NASBA enable homogeneous, real-time detection of RNA. Nucleic Acids Research 26 (9), 2150–2155. https://doi.org/10.1093/nar/26.9.2150.

Li, R., Quan, S., Yan, X., et al., 2017. Molecular characterization of genetically modified crops: challenges and strategies. Biotechnology Advances 35, 302–309. https://doi.org/10.1016/j.biotechadv.2017.01.005.

Mano, J., Oguchi, T., Akiyama, H., Teshima, R., Hino, A., Furui, S., Kitta, K., 2009. Simultaneous Detection of recombinant DNA segments introduced into genetically modified crops with multiplex ligase chain reaction coupled with multiplex polymerase chain reaction. Journal of Agricultural and Food Chemistry 57 (7), 2640–2646. https://doi.org/10.1021/jf803361a.

Marmiroli, N., 2008. Methods for detection of GMOs in food and feed. Anal Bioanal Chem. 392, 369–384.

Marmiroli, N., Maestri, E., Gullì, M., Malcevschi, A., Peano, C., Bordoni, R., De Bellis, G., 2008. Methods for detection of GMOs in food and feed. Analytical and Bioanalytical Chemistry 392, 369–384. https://doi.org/10.1007/s00216-008-2303-6.

Mazzara, M., Savini, C., Charles Delobel, C., Broll, H., Damant, A., Paoletti, C., Van Den Eede, G., 2015. Definition of Minimum Performance Requirements for Analytical Methods of GMO Testing. European Network of GMO Laboratories (ENGL). JRC95544. https://gmo-crl.jrc.ec.europa.eu/doc/MPR%20Report%20Application%2020_10_2015.pdf.

McHugh, T.D., Pope, C.F., Ling, C.L., Patel, S., Billington, O.J., Gosling, R.D., Lipman, M.C., Gillespie, S.H., 2004. Prospective evaluation of BDProbeTec strand displacement amplification (SDA) system for diagnosis of tuberculosis in non-respiratory and respiratory samples. Journal of Medical Microbiology 53 (12), 1215–1219. https://doi.org/10.1099/jmm.0.45780-0.

Milavec, M., Dobnik, D., Yang, L., Zhang, D., Gruden, K., Zel, J., 2014. GMO quantification: valuable experience and insights for the future. Analytical and Bioanalytical Chemistry 406 (26), 6485–6497. https://doi.org/10.1007/s00216-014-8077-0.

Morisset, D., Dobnik, D., Hamels, S., Žel, J., Gruden, K., 2008a. NAIMA: target amplification strategy allowing quantitative on-chip detection of GMOs. Nucleic Acids Research 36 (18), 1–11. https://doi.org/10.1093/nar/gkn524.

Morisset, D., Stebih, D., Cankar, K., Zel, J., Gruden, K., 2008b. Alternative DNA amplification methods to PCR and their application in GMO detection: a review. European Food Research and Technology 227, 1287–1297.

Mullis, K.B., Faloona, F.A., 1987. Specific synthesis of DNA in vitro via a polymerase-catalyzed chain reaction. Methods in Enzymology 155, 335–350.

Mullis, K.B., 1990. The unusual origin of the polymerase chain reaction. Scientific American 262, 56–61. https://doi.org/10.1038/scientificamerican0490-56, 64.

Nadeau, J.G., Pitner, J.B., Linn, C.P., Schram, J.L., Dean, C.H., Nycz, C.M., 1999. Real-time, sequence-specific detection of nucleic acids during strand displacement amplification. Analytical Biochemistry 276 (2), 177–187. https://doi.org/10.1006/abio.1999.4350.

Navarro, E., Serrano-Heras, G., Castaño, M.J., Solera, J., 2015. Real-time PCR detection chemistry. Clinica Chimica Acta 439, 231–250. https://doi.org/10.1016/j.cca.2014.10.017.

Oehlenschläger, F., Schwille, P., Eigen, M., 1996. Detection of HIV-1 RNA by nucleic acid sequence-based amplification combined with fluorescence correlation spectroscopy. Proceedings of the National Academy of Sciences of the United States of America 93 (23), 12811–12816. https://doi.org/10.1073/pnas.93.23.12811.

Pauwels, K., De Keersmaecker, S., et al., 2015. Next-generation sequencing as a tool for the molecular characterisation and risk assessment of genetically modified plants: added value or not? Trends in Food Science & Technology 45, 319–326. https://doi.org/10.1016/j.tifs.2015.07.009.

Peano, C., Bordoni, R., Gulli, M., Mezzelani, A., Samson, M.C., De Bellis, G., Marmiroli, N., 2005. Multiplex polymerase chain reaction and ligation detection reaction/universal array technology for the traceability of genetically modified organisms in foods. Analytical Biochemistry 346 (1), 90–100. https://doi.org/10.1016/j.ab.2005.08.004.

Pinheiro, L.B., Coleman, V.A., et al., 2012. Evaluation of a droplet digital polymerase chain reaction format for DNA copy number quantification. Analytical Chemistry 84, 1003–1011. https://doi.org/10.1021/ac202578x.

Porcar, M., Ramos, S., Latorre, A., et al., 2007. A simple DNA extraction method suitable for PCR detection of genetically modified maize. Journal of the Science of Food and Agriculture 87, 2728–2731. https://doi.org/10.1002/jsfa.3026.

Randhawa, G.J., Singh, M., Morisset, D., Sood, P., Zel, J., 2013. Loop-mediated isothermal amplification: rapid visual and real-time methods for detection of genetically modified crops. Journal of Agricultural and Food Chemistry 61, 11338–11346.

Safaei, P., Aghaee, E.M., Khaniki, G.J., et al., 2019. A simple and accurate PCR method for detection of genetically modified rice. Journal of Environmental Health Science and Engineering 17, 847–851. https://doi.org/10.1007/s40201-019-00401-x.

Sahu, S.K., Thangaraj, M., Kathiresan, K., 2012. DNA extraction protocol for plants with high levels of secondary metabolites and polysaccharides without using liquid nitrogen and phenol. ISRN Molecular Biology 205049. https://doi.org/10.5402/2012/205049, 2012.

Saiki, R.K., Scharf, S., Faloona, F., et al., 1985. Enzymatic amplification of beta-globin genomic sequences and restriction site analysis for diagnosis of sickle cell anemia. Science 230, 1350–1354.

Schweitzer, B., Kingsmore, S., 2001. Combining nucleic acid amplification and detection. Current Opinion in Biotechnology 12 (1), 21–27. https://doi.org/10.1016/S0958-1669 (00)00172-5.

Southern, E.M., 1975. Detection of specific sequences among DNA fragments separated by gel electrophoresis. Journal of Molecular Biology 98, 503–517.

Sykes, P.J., et al., 1992. Quantitation of targets for PCR by use of limiting dilution. Biotechniques 13, 444–449.

Turkec, A., Kazan, H., Karacanli, B., Lucas, S.J., 2015. DNA extraction techniques compared for accurate detection of genetically modified organisms (GMOs) in maize food and feed products. Journal of Food Science & Technology 52, 5164–5171.

van der Vliet, G.M., Schukkink, R.A., van Gemen, B., Schepers, P., Klatser, P.R., 1993. Nucleic acid sequence-based amplification (NASBA) for the identification of mycobacteria. Journal of General Microbiology 139 (10), 2423–2429.

van Gemen, B., van Beuningen, R., Nabbe, A., van Strijp, D., Jurriaans, S., Lens, P., Kievits, T., 1994. A one-tube quantitative HIV-1 RNA NASBA nucleic acid amplification assay using electrochemiluminescent (ECL) labelled probes. Journal of Virological Methods 49 (2), 157–167. Retrieved from. http://www.ncbi.nlm.nih.gov/pubmed/7822457.

Wahler, D., Schauser, L., Bendiek, J., et al., 2013. Next-generation sequencing as a tool for detailed molecular characterisation of genomic insertions and flanking regions in genetically modified plants: a pilot study using a rice event unauthorized in the EU. Food Analytical Methods 6, 1718–1727.

Wilkes, T., Gavin Nixona, G., Burns, M., 2016. Recent developments in DNA-based screening approaches for detection of GMO's. Journal of the Association of Public Analysts 44, 40–50.

Walker, G.T., 1993. Empirical aspects of strand displacement amplification. Genome Research 3 (1), 1–6. https://doi.org/10.1101/gr.3.1.1.

Walker, G.T., Fraiser, M.S., Schram, J.L., Little, M.C., Nadeau, J.G., Malinowski, D.P., 1992b. Strand displacement amplification–an isothermal, in vitro DNA amplification technique. Nucleic Acids Research 20 (7), 1691–1696. https://doi.org/10.1093/nar/20.7.1691.

Walker, G.T., Little, M.C., Nadeau, J.G., Shank, D.D., 1992a. Isothermal in vitro amplification of DNA by a restriction enzyme/DNA polymerase system. Proceedings of the National Academy of Sciences of the United States of America 89 (1), 392–396. https://doi.org/10.1073/pnas.89.1.392.

Willems, S., Fraiture, M.A., et al., 2016. Statistical framework for detection of genetically modified organisms based on Next Generation Sequencing. Food Chemistry 192, 788–798.

Wolcott, M.J., 1992. Advances in nucleic acid-based detection methods. Clinical Microbiology Reviews 5 (4), 370–386.

Wu, D.Y., Wallace, R.B., 1989. The ligation amplification reaction (LAR) - amplification of specific DNA sequences using sequential rounds of template-dependent ligation. Genomics 4, 560–569.

Yang, X., Staub, J., et al., 2013. Omics technologies reveal abundant natural variation in metabolites and transcripts among conventional maize hybrids. Food and Nutrition 4, 335–341.

Zahradnik, C., Kolm, C., Martzy, R., Mach, R.L., Krska, R., Farnleitner, A.H., Brunner, K., 2014. Detection of the 35S promoter in transgenic maize via various isothermal amplification techniques: a practical approach. Analytical and Bioanalytical Chemistry 406 (27), 6835–6842. https://doi.org/10.1007/s00216-014-7889-2.

Zanoli, L.M., Giuseppe, S., 2013. Isothermal amplification methods for the detection of nucleic acids in microfluidic devices. Biosensors 3, 18–43.

Zel, J., Milavec, M., et al., 2012. How to Reliably Test for GMOs. Springer Briefs in Food, Health, and Nutrition, pp. 1–95.

Chapter 5

Method validation: DNA-based detection methods

Liang Li[1], Jingwen Chen[2], Karen Moore[3] and Wujun Jin[4]
[1]*Institute of Quality Standard and Testing Technology for Agro-Products, Chinese Academy of Agricultural Sciences, Beijing, Beijing, China;* [2]*SeqMatic, Fremont, CA, United States;* [3]*Syngenta, Greensboro, NC, United States;* [4]*Biotechnology Research Institute, Chinese Academy of Agricultural Sciences, Beijing, Beijing, China*

5.1 Introduction

The number of genetically modified (GM) crops and their cultivated area are steadily increasing worldwide. Analytic methods are developed, validated, and used at several steps along the value chain including research and development programs, commercialization of genetically modified organism (GMO) products, and postcommercialization stewardship. Appropriate methods that are validated for the intended purpose should be used to ensure that the products meet quality specifications and comply with regulatory requirements such as product labeling and/or low-level presence regulations.

There are several types of analytical methods that target different analytes and use a wide variety of chemistries and strategies. For instance, DNA-based molecular methods may involve the detection of a GMO or its genetic markers by a simple polymerase chain reaction (PCR) amplification reaction, hydrolytic probes, isothermal amplification, or high-resolution melting strategies. These methods may be performed either in singleplex, arrays, or multiplex reactions. The performing laboratory may select from this wide array of test methods based on regulatory requirements, operational costs, infrastructure, and technical capability, as well as customer needs.

For a method being used to report analytical results to customers, it is important to understand the scope of the method and to be sure that the method fulfills the specified requirements. It is a requirement for testing or calibration laboratories to conform to ISO/IEC 17025, defined as: "Confirmation by examination and the provision of objective evidence that the particular requirements for a specific intended use are fulfilled." The process of ensuring these requirements are met is known as method validation, and the validation process defines a range of performance characteristics within which it has been

Application of Sampling and Detection Methods in Agricultural Plant Biotechnology
https://doi.org/10.1016/B978-0-323-99293-0.00008-8

confirmed that the method can yield acceptable results. Validation covers the entire analysis process including instrumentation as well as any computer software associated with the process. The actual procedures for method validation may vary from sector to sector. It is advisable to follow any specific sectorial guidance, if available, to ensure comparability with peer laboratories (Keer and Brich, 2008). In practice, the scope of a method refers to the types of samples that are subject to analysis (both the nature and level of target that is present, and the sample matrix).

The goal of the validation is to evaluate the performance characteristics and limitations of an analytical method. The validation process takes place either within each laboratory or, ideally, through interlaboratory comparisons to ensure that they meet minimum performance criteria and that they provide accurate results generated, preferably, following published regional, national, or international standards. During a validation process, laboratories participate in the study that is organized in accordance with internationally accepted requirements, such as the format described in ISO5725 (1994). Some of the parameters reported during the validation process are used to determine the acceptability of the method and include parameters such as applicability, specificity, dynamic range, robustness, limits of quantification (LOQ), limits of detection (LOD), and others. These elements also allow for measuring the method's level of uncertainty and determining its overall performance, for which precision and trueness complement the set of measured parameters. It is important to mention that the method must be validated with appropriate reference materials, which have been critically evaluated and for which a proper value has been assigned by metrologically sound procedures.

Among DNA-based approaches for determining GMOs and GM food/feed authenticity, qualitative and quantitative real-time PCR methods are often considered the standard methods for nucleic acid detection and quantification, respectively, and are the most commonly used techniques for analyzing the presence of nucleic acids in GM samples. Qualitative testing is used to determine the presence or absence of GM genetic elements in a given sample, that is, whether GM is detected or is not detected in that sample. Quantitative PCR (qPCR) is currently the method of choice for detection, identification, and quantification of GMOs. However, testing based on real-time PCR is time-consuming and requires a laboratory environment and is not suited for on-site rapid testing such as required at grain elevators. Recently, commercial kits based on isothermal DNA amplification have been developed for testing GMO traits in a nonlaboratory setting (http://www.envirologix.com/technology/molecular-detection/).

Both qualitative and quantitative methods, that are publicly available for commercialized GMOs developed by technology providers, have been critically assessed and the requirements for the method performance have been met. However, great care needs to be taken to address potential sources contributing to measurement inaccuracy. These may include the quantity and quality of DNA, verification of the qPCR performance parameters, characteristics of taxon-specific genes, and use of appropriate units of measurement.

qPCR is not well suited for the measurement of absolute concentrations because its accuracy is limited (standard deviation of ~20%) and it often performs poorly with low-copy number templates (Heyries et al., 2011). In addition, the use of reference materials and a standard curve often means that the evaluation of the GM content of the unknown sample is subject to "matrix effects" (Burns et al., 2010). To overcome these challenges and to cope with the continuous increase in the number of GMOs, new approaches are continuously being explored and evaluated. Digital PCR (dPCR) platforms, such as microfluidic chamber-based or droplet-based, are capable of absolute quantification of nucleic acids without the use of standard curves (Vogelstein and Kinzler, 1999). Depending on the dPCR platform, the sample is partitioned into discrete compartments (droplets or microchambers) or water-in-oil droplets that may have no target sequence or may have one or more copies of the target sequence. The presence of the target sequence is measured at the end of PCR amplification (end-point PCR), and the concentration is calculated from the fraction of positive partitions versus the total number of partitions in a statistically relevant way. Statistical strategies of quantification have already been proposed and expanded with the development of dPCR. The "Droplet" version of dPCR generally has more partitioned droplets than other platforms. This allows for very accurate copy number estimates (Pinheiro et al., 2011). New approaches are also needed for the quantification of breeding stacks of multiple GM events and for the potential quantification of organisms produced by new plant breeding techniques (Milavec et al., 2014).

5.1.1 Method validation standards and guidelines

When developing and validating in-house qualitative and qPCR methods for GMO detection, most laboratories use procedures based on the following international guidelines:

5.1.1.1 Codex alimentarius guidelines

Codex Alimentarius Commission (CCMAS, 2010) is an intergovernmental body appointed to implement the Joint FAO/WHO Food Standards Program. In their guidelines and other texts, different terminology related to GMOs is used. Some examples are "foods derived from modern biotechnology" (where the definition of modern biotechnology is taken from the Cartagena Protocol (Eggers, 2000)) or "recombinant DNA (rDNA) plants," which are defined as plants in which the genetic material has been changed through in vitro nucleic acid techniques (including rDNA and direct injection of nucleic acid into cells or organelles (Žel et al., 2012).

The Codex guideline provides information on validation criteria for methods that are applicable to the analysis of both specific DNA sequences and specific proteins of interest in food products. These two types of methods are routinely used in the analysis of materials for the detection, identification, and quantification of GMOs. The guideline emphasizes the need for method validation through a collaborative trial in conforming to an internationally accepted protocol according to ISO5725:1994 or AOAC/IUPAC Harmonized Protocol (AOAC, 2002). To evaluate a method, specific considerations must be taken into account, including method acceptance criteria for validation, applicability of the method, principle conditions, unit of measurements and measurement uncertainty, and reporting of results.

5.1.1.2 European Union guidelines

Annex I of Regulation (EC) No. 641/2004 on "method validation" (EU, 2004)) provides detailed technical provisions on the type of information that shall be provided by the applicant for detection methods and what is needed to verify the preconditions for the fitness of the method. This includes information about the method and about the method testing carried out by the applicant. Annex I of Regulation (EC) No. 641/2004 also confirms that the validation process will be carried out by the European Union Reference Laboratory for Genetically Modified Food and Feed (EU-RL GMFF) according to internationally accepted technical provisions and that all guidance documents produced by the EU-RL GMFF are to be made available (Žel et al., 2012).

In 2008, the European Union Reference Laboratory (Mazzara et al., 2008) published the "Definition of minimum performance requirements for analytical methods of GMO testing," which outlined the detailed requirements for a quantitative event-specific PCR. This document was recently updated and published with an effective date of October 2015 (EURL, 2015). The definitions and acceptance criteria are discussed in the following PCR validation section.

5.1.1.3 ISO standards

Several ISO standards, both general and GMO-specific, relate to GMO detection, and are reviewed and, if necessary, updated on a regular basis (see also chapter 12). The main technical requirements for the detection of GMOs are detailed in the following standards:

ISO 17025: General requirements for the competence of testing and calibration laboratories are the current internationally accepted standard for the accreditation of testing and calibration laboratories. ISO/IEC 17025 specifies general requirements for ISO 17025-accredited GMO Laboratories.

ISO 5725: Accuracy (trueness and precision) of measurement methods and results. During validation, a minimum number of laboratories (12–15, usually) participates in an interlaboratory study organized in accordance with internationally accepted requirements, typically the ones described in ISO 5725.

ISO 24276: **Foodstuffs**—Methods of analysis for the detection of genetically modified organisms and derived products—General requirements and definitions. The standard describes how to use other ISO standards such as sampling, DNA extraction, protein-based analysis, and qualitative and quantitative nucleic acid analysis. In addition, it provides general definitions, requirements, and guidelines for laboratory set-up, method validation requirements, description of methods, and test reports.

ISO 21571: **Foodstuffs**—Methods of analysis for the detection of genetically modified organisms and derived products—Nucleic acid extraction. This standard provides general requirements and specific methods for DNA extraction/purification and quantification. These methods are described in annexes of this ISO standard and have been established for food matrices, but could also be applicable to other matrices, such as grains and feed.

ISO 21569: **Foodstuffs**—Methods of analysis for the detection of genetically modified organisms and derived products—Qualitative nucleic acid-based methods. This standard describes the procedure to qualitatively detect GMOs and derived products by analyzing the nucleic acids extracted from the sample under study. The main focus is on PCR-based amplification methods. General requirements are given for the specific detection and identification of target sequences (DNA) and for the confirmation of the identity of the amplified DNA sequence.

ISO 21570: **Foodstuffs**—Methods of analysis for the detection of genetically modified organisms and derived products—Quantitative nucleic acid-based methods. This standard provides the overall framework around quantitative methods for the detection of GMOs in foodstuffs using PCR.

5.2 PCR validation

Screening approaches using qualitative screening methods have become an integrated part of GMO detection. Pragmatic approaches to conduct in-house and interlaboratory validation studies for GMO screening methods have been proposed (Broeders et al., 2014). The authors distilled a number of parameters from validation criteria for quantitative PCR methods and established a pragmatic validation guideline for screening methods. Such guidelines could be adapted to other areas where qualitative PCR methods are used for molecular testing allowing easier implementation of a more reliable screening phase where necessary. PCR has matured from a labor- and time-intensive, low throughput qualitative gel-based technique to an easily automated, rapid, high throughput quantitative technology.

The most reliable methods, that also enable accurate quantification of GMOs present in foodstuffs, are molecular methods based on the detection and quantitation of the DNA. The real-time PCR application allows the amounts of a target sequence specific to the GMO (GM-specific target sequence) and of a species-specific target sequence to be determined in a sample, thus leading to the determination of the percentage GMO content (Žel et al., 2008). Appropriate reference materials should be used to ensure accurate results (Trapmann and Emons, 2005). The process to evaluate PCR parameters is divided into two steps: assessment of method acceptance parameters (to be tested by the developer during in-house validation) and assessment of method performance parameters (to be evaluated via interlaboratory and collaborative trials). It should be noted that the implementation of collaborative trial validated qualitative and quantitative methods in the enforcement laboratories has not been included in this chapter. The document established by the European Network of GMO Laboratories (ENGL) (EUR 29015 EN - 2017) can hereto be used as guidance. As for many areas of molecular testing, the detection of GMO relies mainly on real-time PCR technology.

5.2.1 Qualitative and quantitative PCR method validation requirements

As discussed earlier, ISO 21569 and ISO 20570 are two international standards that provide the overall framework of PCR-based qualitative and quantitative methods for the detection of GMOs, respectively. The ENGL described the detailed quantitative PCR performance criteria for product registration in European Union in 2008 (Mazzara et al., 2008), which was updated in 2015 (ENGL, 2015). Several regulatory agencies (e.g., in Japan and China) also defined their own method performance criteria for qualitative and quantitative PCR methods. The performance parameters compiled below are mostly from ISO, Codex, and ENGL standards. For definitions refer to ISO 17025 and ISO 16577.

5.2.1.1 Specificity

Definition

The property of the method to respond exclusively to the characteristic or the analyte of interest (ISO 24276).

Acceptance criteria

For all methods: The presence or absence of the analyte has to be properly determined. The method should not produce amplification signals with target sequences different from the target sequence for which the method was developed. This should be demonstrated by similarity searches against databases (e.g., European Molecular Biology Laboratory (EMBL), Genbank, Patent, etc.) and with experimental results from testing the method with nontarget transgenic events and nontransgenic material (ENGL, 2015).

5.2.1.2 The limit of detection

Definition

The lowest amount or concentration of analyte in a sample, which can be reliably detected but not necessarily quantified (ISO 24276).

Acceptance criteria

Qualitative method: The LOD of a qualitative qPCR method should be equal or below 20 haploid genome copies (AFNOR, 2003).

Quantitative method: For combined modules, the LOD should be less than <0.045% with a level of confidence of 95%, ensuring ≤5% false-negative results. For individual modules, the LOD should be <25 copies with a confidence level of 95%, ensuring ≤5% false-negative results (ENGL, 2015).

5.2.1.3 Limit of quantification

Definition

The lowest amount or concentration of analyte in a sample, which can be reliably quantified with an acceptable level of precision and trueness (ISO 24276).

Acceptance criteria

Quantitative method: The LOQ should be ≤ the lowest amount or concentration included in the dynamic range (i.e., 0.09% or 50 copies) (ENGL, 2015).

5.2.1.4 Dynamic range

Definition

The range of concentrations over which the method performs in a linear manner with an acceptable level of trueness and precision (ENGL, 2015).

Acceptance criteria

Quantitative method The method's dynamic range should cover the values corresponding to the expected use. This can be expressed as GMO % or copy number range (ENGL, 2015).

5.2.1.5 Trueness (accuracy)

Definition

The closeness of agreement between a measured quantity value and a true quantity value of a measurand. The measure of trueness is usually expressed in terms of bias (ENGL, 2015).

Acceptance criteria

Quantitative method The trueness shall be within ±25% of the accepted reference value over the whole dynamic range of the PCR modules individually or a Z-score within the range of 2 and −2 should be obtained (ENGL, 2015; Žel et al., 2012).

5.2.1.6 Precision-relative repeatability standard deviation (RSD_r)

Definition

The precision of an analytical procedure expresses the closeness of agreement between a series of measurements obtained from multiple replicates of the same homogeneous sample under the prescribed conditions (ISO 17025).

Acceptance criteria

Quantitative method: The method is acceptable if the repeatability standard deviation (RSD_r) of the determined analyte copy number is $\leq 25\%$ over the whole dynamic range of the method (ENGL, 2015).

5.2.1.7 Precision-relative reproducibility standard deviation (RSD_R)

Definition

The relative standard deviation of test results obtained under reproducibility conditions. Reproducibility conditions are conditions where test results are obtained with the same method, on identical test items, in different laboratories, with different operators, using different equipment. Reproducibility standard deviation describes the interlaboratory variation (ENGL 2015).

Acceptance criteria

Quantitative method: The relative reproducibility standard deviation RSD_R should be $<35\%$ over the whole dynamic range. However, at relative concentrations $<0.2\%$ or at a mount <100 copies, RSD_R values of $<50\%$ are deemed acceptable (ENGL, 2015).

5.2.1.8 Robustness

Definition

The robustness of a method is a measure of its capacity to remain unaffected by small, but deliberate deviations from the experimental conditions described in the procedure (ENGL, 2015).

Acceptance criteria

Qualitative method: The method is robust if despite varied parameters, the presence of the target is still detected (Žel et al., 2012).

Quantitative method: The method should provide the expected results when small deviations are introduced from the experimental conditions described in the procedure (ENGL, 2015).

5.2.1.9 Amplification efficiency

Definition

The rate of amplification calculated from the slope of the standard curve obtained after a decadic semilogarithmic plot of cycle quantification (Cq) values over the DNA copy number/quantity. The efficiency can be calculated by the following equation:

$$\text{Efficiency} = (10\ (-1/\text{slope}) - 1) \times 100$$

Acceptance criteria

Qualitative method: The average value of the slope of the standard curve must be in the range of ($-2.9 \leq$ slope ≤ -4.1) along the dynamic range of amplification. This criterion is less stringent than for quantitative purpose because it should not affect the final result in qualitative terms (presence/absence of target) (Žel et al., 2012).

Quantitative method: The average value of the slope of the standard curve must be in the range of ($-3.1 \leq$ slope ≤ -3.6) along the dynamic range of amplification. This corresponds to an amplification efficiency of 110%–90% (ENGL, 2015).

5.2.1.10 R^2 coefficient

Definition

The R^2 coefficient is the coefficient of determination, which is calculated as the square of the correlation coefficient (between the measured Cq value and the logarithm of the copy numbers/DNA quantity of a standard curve obtained by linear regression analysis (ENGL, 2015).

Acceptance criteria

Quantitative method: The individual values of R^2 of the standard curves should be ≥ 0.98 (ENGL, 2015).

5.3 Data interpretation and sources of error

5.3.1 Data interpretation

After testing a sample for the presence of a defined number of targets, the results are compared with the information in the GMO matrix available through EU-RL-GMFF ((http://gmo-crl.jrc.ec.europa.eu/jrcgmomatrix/). A match between the results and the pattern predicted by the GMO matrix indicates that material from a particular GMO can be present in the sample. For each specific test, the result is scored according to prespecified decision criteria. Such criteria are, for example, defined in ISO 24276. To conclude that a particular GMO can be present in the sample, the following requirements must be met: The analytical results shall be positive for all tests that are applied and that according to the GMO matrix are predicted to yield a positive signal with the material of the GMO in question; The number of tests applied shall be sufficient to allow for some degree of discrimination between GMOs that can be present and GMOs that cannot be present, based on the observed results (EUR 25008, 2011).

Procedures can be regarded as a complete process from sampling to detection and analysis for data interpretation or be approached in a modular manner. While method validation has relied on considering the process as a whole for determining the global uncertainty, a current trend to adopt a modular approach allows for setting step-wise criteria that permit corrective actions before proceeding to the next step if criteria are not met. Decisions over which of these approaches is to be taken into the lab depend on the utility and particularities of the type of sample being analyzed (e.g., food matrices, commodities, vegetative tissue, environmental sample, etc.). For easiness, the following section describes procedures using a modular approach. The procedures include laboratory-associated issues related to sample handling, homogenization, and extraction for GMO detection. These steps are extremely important in the detection and identification of GMOs as significant analytical errors can be introduced at these stages. This section does not involve a description of sampling, which is an extremely important step of an analytical procedure but is beyond the scope of this document (CBD, 2014).

5.3.2 Source of error

5.3.2.1 General considerations

Biological sources of errors

In determining the percent GM value for an unknown sample, the laboratory must convert the analytical result (copies of the GM gene/copies of the endogenous gene) into a percent GM value (weight to weight if required by the customer). This conversion assumes there is a direct 1:1 relationship between the endogenous control gene and the GM gene. However, there are many biological factors that can affect this 1:1 relationship and, as such, this basic assumption is not valid in many circumstances. Of most significance is the effect of biological factors on the 1:1 relationship. This effect is most pronounced in maize and wheat grains and grain products, but soybeans and cotton are not exempt from the basic physiological issues discussed below. Major factors that can impact the 1:1 ratio assumption include hybrid status, copy number, DNA degradation, DNA endoreduplication, outcrossing versus inbreeding, and effects of grain processing and variability in the genome.

Analytical/instrumental sources of errors

Total analytical error (or measurement error) refers to assay errors from all sources derived from a data collection experiment. The accuracy and precision of a PCR method for GM detection or quantification are subject to influences of total analytical error. Total analytical error is of paramount importance in judging the acceptability of PCR-based GM detection or quantification methods. Errors in PCR assays can be classified and are discussed as random (indeterminate) error and systematic (determinate) error (Lipp et al., 2005).

As mentioned in the Codex Guideline on Measurement Uncertainty (CAC, 2004), laboratories are required to estimate the uncertainty of their quantitative measurements. Sample preparation and analytical methods are two significant sources of error that should be considered when evaluating an analytical measurement. Analysts using methods that have been validated according to these guidelines should have sufficient information to allow them to estimate the uncertainty of their results. For details, refer to the Codex Guideline on Measurement Uncertainty (CAC, 2004), the section entitled "The Use of Analytical Results: Sampling Plans, Relationship between the Analytical Results, the Measurement Uncertainty, Recovery Factors and Provisions in Codex Standard" from the Codex Procedural Manual (CAC, 2004).

Measurement Uncertainty (MU) is an important element in the assessment of validated methods. MU is commonly applied to quantitative measurements (the estimation of a target concentration), but the concept will also apply to qualitative methods (i.e., confirmation of the presence/absence of a target). The MU, which should take account of all effects on a measurement process, is the most important single parameter that describes the quality of measurements. Accreditation standard ISO 17025 requires that a testing laboratory shall have a procedure to estimate MU; the estimation of MU is considerably facilitated, and more rigorous, when well-recognized, validated methods specify the values of the major sources of uncertainty of measurement; in the case of methods validated in collaborative trials, the value of reproducibility standard deviation (RSD_R), reflecting the interlaboratory variability, is a useful indicator of the upper overall uncertainty expected for that particular measurement. Validated methods reporting an estimation of the method variability (i.e., precision) are therefore a useful tool for the establishment of MU in each laboratory (Van den Eede, 2010).

5.3.2.2 False positive and false negative

False-positive and false-negative rates are the probability that, respectively, a negative sample would be classified positive or a positive one would be regarded as negative (CCMAS, 2010). If DNA purification is inefficient or DNA is extracted from highly processed food materials, the DNA may be degraded or contaminated with compounds that interfere with the PCR process. With such DNA preparations, the performance of PCR analysis will not be optimal. Also, DNA degradation will reduce the sensitivity of PCR amplification. Some contaminants may reduce the efficiency of PCR amplification, while some contaminants may alter the specificity with which primers interact with the DNA template, resulting in the production of artifactual PCR products that greatly confuse the interpretation of results. Contaminants that lead to artifactual PCR products can lead to false positives (classification of a sample as containing GM soy/corn/canola when in fact it does not), while contaminants or degradation problems that reduce PCR efficiency can lead to false negatives (classification of a sample as free of GM material when in fact it contains such material). Either of these incorrect results can create economic hardship for the buyer or seller of a food product.

False positives

The probability that a known negative sample is classified as positive, for convenience this rate is expressed as a percentage.

%False positive results = 100 × number of misclassified known negative samples/total number of known negative samples.

The main criteria to be considered for a GMO detection method are its specificity (selectivity) and its sensitivity. In GM testing, a false positive occurs if the test result is positive (GM target or a specific GMO is detected) when the actual condition is negative (GM target or the specific GMO is absent). Laboratories under accreditation are obliged to take the necessary precautions to minimize the occurrence of false positives. In the light of the matrix approach, wherein the likely presence of a GMO is deduced from the combined presence of particular targets within the GMO, the evidence of linkage of the positive signals needs to be carefully weighted. Indeed, the presence of some quantity of GM material from an undeclared species is often a source of positive analytical results that do not correspond to the results predicted from a declared list of ingredients.

False negatives

The probability that a known positive sample is classified as negative, for convenience this rate is expressed as a percentage.

%False negative results = 100 × number of misclassified known positive samples/total number of known positive samples.

False-negative results occur when no nuclear acid amplification is detected, yet the target sequence is present. This can be caused by poor primers, suboptimal thermocycling parameters, or insufficient starting DNA concentration in the reaction due to degradation or poor extraction. To reduce the occurrence of false negatives, optimization of PCR assay design and cycling conditions is necessary. This includes the selection of quality primers, optimization of temperatures during cycling, and the use of quality template nucleic acid. For highly processed materials where the presence of DNA is limited or highly degraded, special care should be devoted to the quantity and quality of DNA purified. In GM testing, a false negative occurs if the test result is negative (GM target or a specific GMO is not detected) when the actual condition is positive (the GM target or the specific GMO is present).

5.4 Reference materials

Material, sufficiently homogeneous and stable with reference to specified properties, which has been established to be fit for its intended use in measurement or in the examination of nominal properties (ISO Guide 99).

Certified reference material (CRM) is reference material accompanied by a certificate issued by a recognized body indicating the value of one or more properties and their uncertainty (see also Chapter 8).

GM reference materials for commercialized GMOs are available from Joint Research Center–Institute of Reference Materials and Measurements (JRC-IRMM: https://ec.europa.eu/jrc/en/research-topic/reference-materials-gmo-analysis?search) and/or from American Oil Chemist's Society (AOCS: http://www.aocs.org/LabServices/content.cfm?ItemNumber=19248). Such materials are provided by trait providers, and the qualified organizations mentioned above certify the values and their quality through a lengthy certification process.

JRC-IRMM reference materials are processed at different concentrations. Some of the JRC-IRMM reference materials are available from Sigma–Aldrich as well. For the implementation of the European legislation that regulates the authorization and the labeling of GMOs, the JRC-IRMM develops, produces, and distributes CRMs. These CRMs are used for the calibration or quality control of GMO quantification measurements, typically carried out by qPCR. Numerous sets of reference materials for different GM events in maize, soybean, canola, rice potato, sugar beet, and cotton are offered to testing and control laboratories worldwide. Research activities are ongoing to develop further GMO CRMs, including CRMs for new plant species. The majority of CRMs intended for the quantification of GMOs are powders produced from seeds or other tissues. These so-called matrix materials are mixtures of non-GMO materials with GMO materials, which have gravimetrically been certified for their mass fraction (expressed in g/kg) of a specific GMO event. The available concentrations differ for the individual GMO event (and the set of CRM) and range from nominal 0 g/kg to 1000 g/kg.

The AOCS reference materials are available as powder or leaf tissue DNA and also as seeds in the case of canola. They are usually provided as 100% GM materials and non-GM negative controls, so the users have the flexibility to make their own calibration standards by mixing GM and non-GM materials together. In addition, testing laboratories can extract DNA from the nominal 100% powders, or use available DNA nominal 100% CRM, to make serial dilutions starting with the 100% DNA sample. This allows the operator to obtain Cq values for both endogenous and trait targets from the same diluted DNA sample. The benefits of this approach are the elimination of (1) uncertainty due to variation in DNA extraction efficiencies between materials and (2) uncertainty of the DNA concentration measurements. The only remaining uncertainty comes from the dilution step and, if done properly, dilution introduces only a small uncertainty. Therefore, the total uncertainty using this approach could be lower than using a premixed CRM sample such as 1% GM CRM.

There are growing interests in using plasmid reference materials as calibrators as they are relatively easy to produce in large quantities with high purity. Several laboratories investigated the use of plasmid RM in PCR and demonstrated that they are appropriate for use as calibrators (Burns et al.,

2006; Toyota et al., 2006; Caprioara-Buda et al., 2012). These studies indicated that exhaustive validation is required to compare quantification values generated from using plasmid RM with data generated using plan-derived genomic RM. Correction factors may be incorporated for accurate quantification (Allnuttet al., 2005, Meng et al., 2012).

Additionally, the National Institute of Metrology, China (http://www.ncrm.org.cn), provides several CRMs including matrix and plasmid CRMs for GMO quantification.

5.5 Assay transfer and proficiency testing

5.5.1 Interlab verification

Data from the interlaboratory validation process are evaluated according to the document "Definition of Minimum Performance Requirements for Analytical Methods of GMO Testing" (Mazzara et al., 2008; ENGL, 2008, 2015). When an interlaboratory validated method is used by an accredited laboratory, the laboratory must, before its use, ensure that the chosen method shows performance characteristics similar to those assessed in the interlaboratory study. The verification process must be documented and recorded in the quality system. The laboratory must record the procedure used, the results obtained, and a statement as to whether the method is fit for the intended use. The process includes the following:

- Design and planning of the verification;
- Description of the method;
- Acceptance criteria and performance requirements, as decided by the laboratory;
- Test records;
- Approval of the method.

Interlaboratory validated methods are assessed according to the acceptance criteria and performance requirements described in the document "Definition of Minimum Performance Requirements for Analytical Methods of GMO Testing." During the verification process, a laboratory should ensure compliance with the requirements described in the following standards: ISO 21570, ISO 24276, ISO 21568, ISO 21569, and ISO 21571.

5.5.2 Proficiency testing

Proficiency testing (PT) is crucial for an independent assessment of laboratory performance. Participation in proficiency tests depends on the methods introduced in the laboratory and the type of samples analyzed. Laboratories can use many methods for testing; however, not all of them can be assessed as each proficiency test covers only a limited number of GMOs. Therefore, it is recommended to prepare a plan for participation in proficiency tests and to assess individual methods periodically. It is also important to cover different types of matrices that are subject to testing during routine analyses (CBD, 2014).

According to ISO/IEC 17025, quality control procedures should be planned activities. Once the level and frequency of participation have been established, laboratories should develop a PT strategy for which the content and extent of this strategy will depend on the circumstances and scope of the individual laboratory and the availability of PT schemes. The strategy should be a part of the laboratory's overall quality control strategy. The document on the level and frequency of PT participation recommends that the strategy covers at least one accreditation cycle. Furthermore, the laboratory should review the strategy annually and evaluate its appropriateness.

The advisory documents guidance on the level and frequency of PT participation (ILAC-P9:06/2014) and ILAC Policy for Participation in Proficiency Testing Activities (ILAC, 2010) provided guidance on the level and frequency of PT participation. The document does not state a fixed number of tests to be performed within a specified period of time, which still is a requirement from accreditation bodies (EA members) in some countries. Rather, the Guidance states that it is up to individual laboratories to define their level and frequency of participation after careful consideration of their other quality management measures. Laboratory PT is an essential element of quality assurance. Participation in relevant PT schemes is a requirement of ISO/IEC 17025 or other accreditation standards such as those provided by International Seed Testing Association (ISTA).

Proficiency testing guidance and programs for GMO detection are provided by different institutions, some of which are listed in the following paragraphs.

5.5.2.1 Technical guidance for PT

- UKAS TPS 47-UKAS Policy on Participation in Proficiency Testing Schemes
- EA document EA-04/18 Guidance on the level and frequency of proficiency testing participation.
- ISO/IEC 17043:2010 Conformity assessment—General requirements for proficiency testing.

5.5.2.2 Programs for PT

- Grain Inspection, Packers and Stockyards Administration of United States Department of Agriculture
- EURL GMFF
- ISTA
- Laboratory Proficiency Program of AOCS

5.6 Summary

DNA methods for the detection and quantification of GM content in food or feed samples are developed and validated following international standards and national and regional guidelines that include validation parameters and their acceptance criteria. Method validation may also include interlaboratory transferability, where method performance is assessed across participating laboratories in what is known as a collaborative trial. These detection methods are based on PCR amplification of intended GM targets and can be quantitative or qualitative. These analytical assays are subject to errors, whose source should be appropriately identified and taken into consideration when reporting GM content in a sample (e.g., as measurement uncertainty). In addition, for qPCR, the availability of CRMs with certified amounts of GM content is a prerequisite because these materials are used as calibrators for the accurate quantification of unknown GM samples.

Finally, before using DNA-based detection methods for estimating content in unknown GM samples, testing laboratories need to obtain the proper accreditations and can choose to participate in proficiency testing programs to monitor their performance as it pertains to the specific methods that they are accredited for.

References

AFNOR XP V03-020–022, 2003. Produits alimentaires. Détection et quantification des organismes végétaux génétiquement modifies et produits dérivés. Partie 2: méthodes basées sur la réaction de polymérisation en chaîne. Norme expérimentale.

Allnutt, T.R., Chisholm, J., Hired, H., Oehlschlager, S., Henry, C.M., 2005. Plasmid Standards for Real Time PCR and GM Enforcement Testing accessed Janury 26th 2021. http://randd.defra.gov.uk/Default.aspx?
Menu=Menu&Module=More&Location=None&Completed=0&ProjectID=13930.

AOAC, 2002. Appendix D: Guidelines for Collaborative Study Procedures to Validate Characteristics of a Method of Analysis, AOAC OFFICIALMETHODS OFANALYSIS available at: http://www.aoac.org/aoac_prod_imis/AOAC_Docs/StandardsDevelopment/Collaborative_Study_Validation_Guidelines.pdf. (Accessed 22 February 2021).

Broeders, S., Huber, I., Grohmann, L., Berben, G., Taverniers, I., Mazzara, M., Morisset, D., 2014. Guidelines for validation of qualitative real-time PCR methods. Trends in Food Science & Technology 37 (2), 115–126.

Burns, M., Corbisier, P., Wiseman, G., Valdivia, H., McDonald, P., Bowler, P., Ohara, K., Schimmel, H., Charels, D., Damant, A., Harris, N., 2006. Comparison of plasmid and genomic DNA calibrants for the quantification of genetically modified ingredients. European Food Research and Technology 224, 249–258.

Burns, M.J., Burrell, A.M., Foy, C.A., 2010. The applicability of digital PCR for the assessment of detection limits in GMO analysis. European Food Research and Technology 231 (3), 353–362.

Caprioara-Buda, M., Meyer, W., Jeynov, B., Corbisier, P., Trapmann, S., Emons, H., 2012. Evaluation of plasmid and genomic DNA calibrants used for the quantification of genetically modified organisms. Analytical and Bioanalytical Chemistry 404, 29–42.

CBD, 2014. Secretariat of the Convention on Biological Diversity (CBD). Technical Tools and Guidance for the Detection and Identification of LMOs. https://bch.cbd.int/protocol/cpb_detection/toolsandguidance/topic1.shtml.

Codex Alimentarius Commission (CAC), 2004. Guidelines on Measurement Uncertainty - CAC/GL 54–2004. FAO/WHO Joint Publications, Roma. Report No.: CAC/GL 54–2004.

Codex Committee On Methods Of Analysis And Sampling (CCMAS), 2010. Guidelines on performance criteria and validation of methods for detection, identification and quantification of specific DNA sequences and specific proteins. In: Foods. CAC/GL 74–2010. Rome: Codex Alimentarius Commission- World Health Organization.

Eggers, B., Mackenzie, R., 2000. The Cartagena protocol on biosafety. Journal of International Economic Law 3 (3), 525–543.

Essentials of nucleic acid analysis: a robust approach. In: Keer, J.T., Birch, L. (Eds.), 2008. International Organization for Standardization. Royal Society of Chemistry, Geneva, Switzerland.

EU, 2004. Official Journal of the European Union L102.14 available at: http://eur-lex.europa.eu/LexUriServ/LexUriServ.do?uri=OJ: L:2004:102:0014:0025:EN: PDF. (Accessed 22 February 2021).

European Network of GMO Laboratories (ENGL), 2011. Guidance document from the European Network of GMO Laboratories: Overview on the Detection, Interpretation and Reporting on the Presence of Unauthorised Genetically Modified Materials. Document No EUR 25008 EN-2011. https://doi.org/10.2788/89665.

European Network of GMO Laboratories (ENGL), 2008. Definition of Minimum Performance Requirements for Analytical Methods of GMO Testing.

European Network of GMO Laboratories (ENGL), 2015. Definition of Minimum Performance Requirements for Analytical Methods of GMO Testing. Publications Office of the European Union. November 2009. http://gmo-crl.jrc.ec.europa.eu/doc/MPRReportApplication20_10_2015.pdf.

European Network of GMO Laboratories (ENGL), 2017. Verification of Analytical Methods for GMO Testing when Implementing Interlaboratory Validated Methods DocumentNo. EUR 29015-EN. https://doi.org/10.2760/645114.

Heyries, K.A., Tropini, C., VanInsberghe, M., Doolin, C., Petriv, I., Singhal, et al., 2011. Megapixel digital PCR. Nature Methods 8 (8), 649–651.

ILAC-P9:11/2010, 2010. ILAC policy for participation in proficiency testing activities. http://gac.gov.ge/files/ILAC.pdf. Accessed 6 June 2022.

ILAC-P9:06/2014 ILAC Policy for Participation in Proficiency Testing Activities (https://ilac.org/?ddownload=3259) Accessed January 26, 2021.

ISO/IEC 17043. Conformity Assessment – General Requirements for Proficiency Testing.

ISO 5725. Accuracy (Trueness and Precision) of Measurement Methods and Results. International Organization for Standardization, Geneva, Switzerland.

ISO 24276. Foodstuffs. Methods of Analysis for the Detection of Genetically Modified Organisms and Derived Products – General Requirements. International Organisation for Standardisation, Geneva, Switzerland.

ISO 16577, Foodstuffs — Molecular Biomarker Analysis – Terms and Definitions. International Organisation for Standardisation, Geneva, Switzerland.

ISO 17025. General Requirements for the Competence of Testing and Calibration Laboratories. International Organisation for Standardisation, Geneva, Switzerland.

ISO 21569. Methods of Analysis for the Detection of Genetically Modified Organisms and Derived Products – Qualitative Nucleic Acid Based Methods. International Organisation for Standardisation, Geneva, Switzerland.

ISO 21570. Methods of Analysis for the Detection of Genetically Modified Organisms and Derived Products – Quantitative Nucleic Acid Based Methods. International Organisation for Standardisation, Geneva, Switzerland.

ISO 21571. Methods of Analysis for the Detection of Genetically Modified Organisms and Derived Products – Nucleic Acid Extraction. International Organisation for Standardisation, Geneva, Switzerland.

Lipp, M., Shillito, R., Giroux, R., Spiegelhalter, F., Charlton, S., Pinero, D., Song, P., 2005. Polymerase chain reaction technology as analytical tool in agricultural biotechnology. Journal of AOAC International 88, 136–155.

Mazzara, M., Savini, C., Delobel, C., Broll, H., Damant, A., Paoletti, C., Van den Eede, G., 2008. Definition of minimum performance requirements for analytical methods of GMO Testing European Network of GMO Laboratories (ENGL). Joint Research Centre and Office for Official Publications of the European Communities. https://publications.jrc.ec.europa.eu/repository/bitstream/JRC49650/jrc49650 - final version - min perf req.pdf.

Meng, Y., Lu, X., Shu, L., Wang, S., Zhang, D., Yang, L., 2012. Applicability of plasmid calibrant pTC1507 in quantification of TC1507 maize: an interlaboratory study. Journal of Agricultural and Food Chemistry 60 (1), 23–28.

Milavec, M., Dobnik, D., Yang, L., Zhang, D., Gruden, K., Žel, J., 2014. GMO quantification: valuable experience and insights for the future. Analytical and Bioanalytical Chemistry 406 (26), 6485–6497.

Pinheiro, L.B., Coleman, V.A., Hindson, C.M., Herrmann, J., Hindson, B.J., Bhat, S., Emslie, K.R., 2011. Evaluation of a droplet digital polymerase chain reaction format for DNA copy number quantification. Analytical Chemistry 84 (2), 1003–1011.

Toyota, et al., 2006. Rapid Quantification Methods for Genetically Modified Maize Contents Using Genomic DNAs Pretreated by Sonication and Restriction Endonuclease Digestion for a Capillary-type Real-Time PCR System with a Plasmid Reference Standard.

Trapmann, S., Emons, H., 2005. Reliable GMO analysis. Analytical and Bioanalytical Chemistry 381 (1), 72–74.

Van den Eede, G., 2010. Compendium of reference methods for GMO analyses. Publications Office of the European Union, Luxembourg, Luxembourg. http://publications.jrc.ec.europa.eu/repository/handle/JRC60583.

Vogelstein, B., Kinzler, K., 1999. Digital PCR. Proceedings of the National Academy of Sciences of the United States of America 96, 9236–9241.

Žel, J., Mazzara, M., Savini, C., Cordeil, S., Camloh, M., Štebih, et al., 2008. Method validation and quality management in the flexible scope of accreditation: an example of laboratories testing for genetically modified organisms. Food Analytical Methods 1 (2), 61–72.

Žel, J., Milavec, M., Morisset, D., Plan, D., Van den Eede, G., Gruden, K., 2012. How to Reliably Test for GMOs. Springer US, pp. 1–95.

Chapter 6

Protein-based detection methods

Ray Shillito[1], Rong Guo[1] and Tom Currier[2]
[1]*BASF Corporation, Morrisville, NC, United States;* [2]*Bayer CropScience, Morrisville, NC, United States*

6.1 Introduction

Detecting and measuring specific proteins is an important activity in the development and use of genetically engineered (GE) crops. During product development, thousands of samples are routinely analyzed at multiple stages from the characterization of the initial transgenic plants to measuring the introduced proteins either qualitatively or quantitatively (Privalle et al., 2012) for regulatory purposes. Protein-based detection methods are then used in the breeding process and may be used for purity control during the production of commercial seed. In addition, protein-based methods are used for quality control of both GE and non-GE grain that enters the commodity markets.

It is important to select the most appropriate protein detection techniques for each of these uses. Considerations that influence the selection of a protein detection technology are protein levels in the sample, the complexity and number of matrices, and the number of samples to be analyzed. These factors directly affect the resources that must be expended for analysis and whether it requires highly specialized core facilities and personnel.

Proteins can be detected and quantified using nonspecific physical methods (Bradford, 1976; Lowry et al., 1951; Smith et al., 1985) or using antibodies to specifically detect them. This chapter focuses on methods used to detect specific proteins as used in the Agricultural Biotechnology Industry. It will focus primarily on immunoassays—the most widely used method for detection of the proteins in plants, seed, and grain. Prior to the routine use of immunoassays, assay of these proteins was difficult (Kim et al., 2006; Ocaña et al., 2009). Mass-spectrometry has been used more recently to characterize the expression of some proteins for regulatory and other purposes (Hill et al., 2015; Chang et al., 2014), but the cost and time taken for mass spectrometry render it less useful for large-scale deployment.

Application of Sampling and Detection Methods in Agricultural Plant Biotechnology
https://doi.org/10.1016/B978-0-323-99293-0.00006-4

Definitions

- Antibody: protein produced by B lymphocytes that recognize a foreign "antigen," and thus triggers an immune response[1]
- Antigen: substance that is recognized as foreign by the immune system and elicits an immune response through stimulating antibody production[1]
- Immunogen: a specific type of antigen that can elicit an immune response.
- Monoclonal antibody: antibody produced from a single hybridoma clone and directed to a single antigen determinant[1]
- Polyclonal antibody: mixture of antibodies capable of reacting specifically with a certain immunogenic substance[1]

1 Definitions from ISO 16577:2016 Molecular biomarker analysis—Terms and definitions.

This chapter focuses on the applicability of The Enzyme-Linked ImmunoSorbent Assay (ELISA) and Lateral Flow Strip (LFS) immunoassay formats in agricultural biotechnology, the primary formats, production of antibodies, and the design and use of the assays.

6.2 History

Immunoassays exploit the unique ability of antibodies to recognize and bind to a specific protein. This affinity provides exceptional properties of specificity, stability, and robustness for the recognition of biomolecules and particularly proteins.

Immunological techniques developed over many decades from the 1959 seminal work published by Berson and Yalow (1959), to the demonstration by several groups, among them Nakane and Pierce (1966), that enzymes can be coupled with antibodies to allow detection, and then the 1971 publication on enzyme immunoassays (Engvall and Pearlman, 1971). The modern version of today's ELISA came about by the evolution of the microtiter plate format (Voller et al., 1974) and commercially available microplate readers. Application of novel solid phases for easy and rapid separation of bound and unbound antibodies, along with the development of the sandwich immunoassay or "two-site" immunoassays led to the widespread adoption of immunodiagnostic techniques (Asensio et al., 2008; Morisset et al., 2009; Kamle and Ali, 2013) for detection and quantification of proteins. A review of the history of immunoassays has been published by Hosseini et al. (2018).

Western blotting (Towbin et al., 1979), a protein detection technique that involves combining size-based separation of proteins via SDS-PAGE with blotting and subsequent probing by antibodies to identify specific proteins, is not a tool that is used beyond the research phase and so will not be described here.

The fast, solid-phase format known as lateral flow immunochromatography (also called LFS) has widespread use in the field and supply chain (Grothaus et al., 2006). The application of these LFS devices and liquid-based multi-well ELISA methods are the focus of this chapter.

6.3 Principle

Antibodies are immunoglobulins secreted by B lymphocytes in vertebrates as a response to an invading foreign substance. Antibodies have a main stem and two arms; their structure is like the letter Y. Their task is to recognize and direct an immune attack on foreign objects in the animal body. The immune system develops thousands of different antibodies where the Y tips in each have a unique molecular structure that binds specifically to a targeted antigen. This ability to recognize and bind to a specific target provides a wide application of antibodies in the diagnostic and analytical fields.

The development and production of antibodies for diagnostic purposes make use of the fact that an intentionally introduced substance (immunogen) can cause the production of antibodies targeted to that immunogen. Antibodies can be produced to recognize peptides, proteins, hormones, and in general any molecule that is >1000 Da in size and are ways to make antibodies against smaller molecules such as pesticides and drugs. An animal that has been immunized will produce a mixture of different antibodies (termed polyclonal antibodies) that each recognizes their own target site (epitope) on the immunogen. Some antibodies may also cross-react with similar structures or protein sequences on molecules other than the immunogen, in many cases with lower affinity. It is also possible to isolate a single B lymphocyte and then immortalize and culture it as a hybridoma cell line to produce a single type of antibody, called a monoclonal antibody, which recognizes only a single epitope (and antigen). Monoclonal and polyclonal antibodies have their advantages and disadvantages in protein immunoassays, as described in Section 6.7.

In Agricultural Biotechnology applications, the target antigen is usually a protein. The goal of a protein assay is to have consistent, reliable, and accurate method that is specific to the target protein. Successful immunoassays depend on having an effective immunogen for developing antibodies with suitable binding specificity and affinity to allow the target protein to be separated from other molecules in extracts of the plant matrices.

6.4 Application of immunoassays to GE crops

6.4.1 Why use immunoassays

Almost all GE crop varieties on the market express one or more proteins in planta to impart trait characteristics. Immunoassay-based detection techniques are a tool used extensively to characterize and track these plants and their seeds and grain. The type of assay to be used will depend on whether the objective is to detect the presence of the protein or to quantify the amount of protein and also on whether the assay is to be carried out in the field or in a laboratory setting.

Several characteristics make immunoassays appropriate for these purposes. They have excellent sensitivity and specificity and are highly reproducible. They can detect and quantify single components in complex matrices without additional purification steps. There is versatility in the type of formats, and they are relatively easy to use. Some can be automated, leading to favorable economies of scale.

Immunoassays are used at almost all stages of product development and commercialization. They are developed in-house by trait developers or under contract with a kit developer, as the proteins expressed in GE crops are generally proprietary. Most trait developers make the tests readily available once the related GE crop becomes commercial.

As well as the benefits, certain limitations to immunoassay technologies need to be considered when applied to agriculture. As mentioned, antibodies may have cross-reactivity with proteins that share similar protein sequences or structural components. This is more often seen when using polyclonal antibodies to detect proteins that are highly similar to each other. Examples of these are Cry1Ab and Cry1Ac proteins in YieldGard[2] (Mon810) and Agrisure[3] (Bt11) maize, and in Bollgard,[4] Widestrike,[5] and TwinLink[6] Cotton. This characteristic can also be used to advantage for detecting related proteins from multiple products, for example in a screening program. If the same protein is expressed in multiple products (e.g., CP4 EPSPS in Roundup Ready (RR1) and Roundup Ready 2 Yield (RR2)) or the same protein is expressed in different species (e.g., Cry1A in cotton, maize, and soybeans or PAT protein in LibertyLink soybeans and maize), then an immunoassay would not be an appropriate way to differentiate between these products.

Variability of protein expression during the plant lifecycle or between plant tissues may lead to difficulties in using immunoassays for detection in seed or grain. An example of this was the tissue-specific expression of Cry1A (Fearing et al., 1997) in NaturGard KnockOut[7] (Bt176); where the protein was expressed in leaf tissue but not in seed.

---

2. Yieldgard is a Trademark of Bayer Ag.
3. Agrisure is a Trademark of Syngenta Inc.
4. Bollgard, Roundup Ready and Roundup Ready two Yield are Registered Trademarks of Bayer Ag.
5. Widestrike is a Trademark of Corteva Inc.
6. TwinLink and LibertyLink are Registered Trademarks of BASF.
7. NaturGard KnockOut is a Trademark of Syngenta Inc.

Applications developed for raw commodities or other plant parts like leaves or grain do not translate well into applications for food or foodstuffs (Bogani et al., 2008; De Luis et al., 2009; Margarit et al., 2006), as most proteins are denatured in many foods by processing at high heat, pressure, and shearing. However, such immunoassays for denatured proteins have been developed (Stave et al., 2004, cited in (Grothaus et al., 2006)

It is important to note that immunoassays will not necessarily indicate whether a protein is functionally active.

6.4.2 Immunoassay formats

The most common ELISA format used in connection with GE crops is known as a sandwich ELISA. The ELISA is based on multi-well plates.

Whereas there are many formats of ELISA design, the primary ones used in Agricultural Biotechnology applications are the sandwich ELISA. In the case of a protein being expressed in plants for the first time, a simple ELISA assay using polyclonal antibodies can be quickly developed to confirm expression. This ELISA is qualitative or at best only semiquantitative.

The amount of a trait protein expressed in a transgenic plant is important for the regulatory process. Detailed measurement of the amount of the trait protein expressed in multiple plant tissues, including seed, roots, stems, leaves, and in some cases, pollen are required by some governments. Thus, a robust, widely applicable, stable, repeatable quantitative ELISA assay is required. Such assays typically take 12–24 months to develop and validate.

The other commonly used assay is the Lateral Flow Assay (known as Lateral Flow Device, LFS or Immunostrip©[8]). For the purpose of consistency, we will use LFS to describe lateral flow methods in this chapter. An LFS typically takes from 9 to 18 months to develop and will typically be needed for breeding and quality control purposes, as well as assay of low levels of GE in grain.

6.4.3 Sandwich ELISA

The ELISA is the most common immunoassay format used for the detection of proteins in GE crops. The Sandwich assay is so called because the analyte being measured is sandwiched between two antibodies. In a sandwich ELISA (Fig. 6.1), the process consists of a number of steps. (1) Capture antibodies are bound to microtiter plates using the ability of polystyrene or, more recently, proprietary coatings to irreversibly bind proteins. Plates with 96 wells are most common although 384 well plates are sometimes used. (2) After the antibody is bound, additional binding sites on the plate are blocked with a nonspecific protein such as bovine serum albumin. (3) The sample is added, and the

8. Immunostrip is a Copyright of Agdia, Inc.

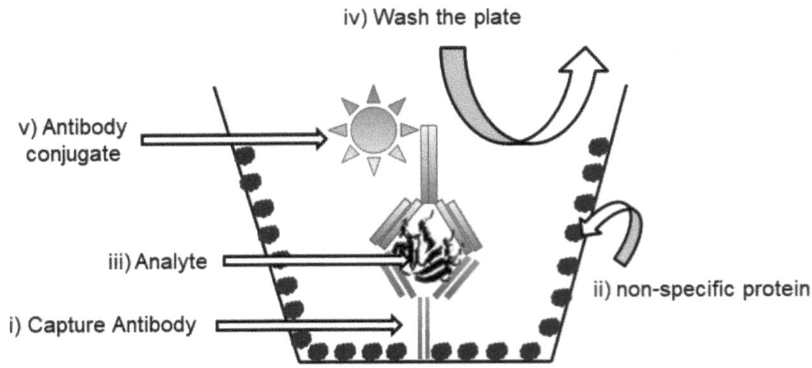

FIGURE 6.1 The sandwich ELISA process.

analyte is captured from the solution by the antibody on the plate. (4) Uncaptured solution and any excess protein is removed by washing the plate and (5) an antibody linked to an enzyme (the conjugate) or other means to develop a colored product that can be read in a plate reader, is added, and binds to the captured protein. Two common conjugates are an antibody linked to reporting enzymes such as horseradish peroxidase or alkaline phosphatase or an antibody linked to biotin.

The sandwich ELISA can be highly sensitive, it works well in crude extracts and can have low background values. It is often used for quantitation and is the format that is generally available from commercial vendors.

In a variation of this ELISA, the second antibody is not labeled with a detection system. Instead, a third antispecies antibody is added that is reactive with the second antibody and it is labeled with a detection molecule. In this form, the capture antibody and the second antibody are generally made in different species. Such a procedure is intended to increase sensitivity but is rarely needed for assays used on GE crops.

6.4.4 Lateral flow strip assay

A LFS assay (Fig. 6.2) is a solid-phase assay that is fast and can easily be used outside of a laboratory setting. The most recognizable types of this device might be the home pregnancy tests, rapid tests used in doctor's offices to test for influenza, or in the recently popular Covid-19 home tests. The strip consists of an inert support backing, which is attached to a wicking material (typically nitrocellulose) that measures about 0.5×10 cm. Each end of the strip contains an adsorbent pad. The lower adsorbent pad—the sample pad—contains an antibody (detection antibody) specific for the analyte being tested. The detection antibody is most commonly linked to a visible label such as colloidal gold (Ngom et al., 2010) or colored nanoparticles. Many antibody molecules

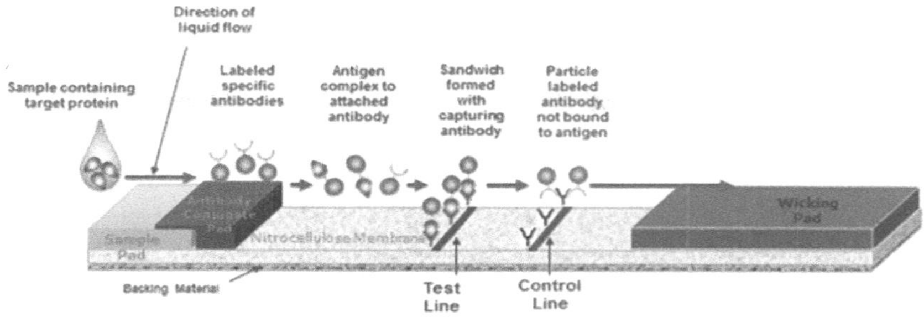

FIGURE 6.2 The lateral flow assay.

are attached to each particle. The upper adsorbent pad does not contain reactants. It acts as a fluid reservoir and keeps liquid flowing through the strip by capillary action.

The wicking material between the two pads has at least two transverse bands of immobilized capture antibody. The antibody in the lower band (closest to the extract) has an affinity for the analyte being tested. The upper antibody band is an antispecies antibody directed to the antibody in the lower pad and is used as a control band. The purpose of the upper control band is to validate that the strip has not been compromised in a way that makes it inoperable. LFS must not be exposed to high temperatures or humidity, as these lead to denaturation of the antibodies, thus making the strip unable to perform as designed.

To run the assay, the bottom of the strip is placed in a small amount of sample extract in an upright position with the sample pad in the test sample. Alternatively, the test sample can be applied to the sample pad. As the extract enters the strip, the antibody conjugate in the lower pad binds to the target analyte if it is present. The antibody-conjugate-analyte complex and unreacted antibody-conjugate migrate along the strip by capillary action. The remaining free analyte may react with the antibody conjugate during this migration period. The analyte complex is captured by the first band. Thus, a sandwich is formed between the conjugated antibody—analyte complex traveling up the strip and the capture antibody in the band. This results in an accumulation of the directly visible conjugate in the lower band. As the extract continues to flow up the strip, any antibody conjugate that did not react with the analyte is captured by the antibody in the second band.

After an LFS test is run there are three types of results as illustrated in Fig. 6.3. (1) A signal is visible at both bands. This indicates the test analyte was present in the sample and the strip ran properly. (2) Only the second band is visible. This indicates that the analyte was not present in the sample and the strip ran properly. (3) No bands are visible. This result indicates the strips are not functioning properly and no conclusion can be made concerning the presence of the analyte in the sample. In some rare cases, the lower band is

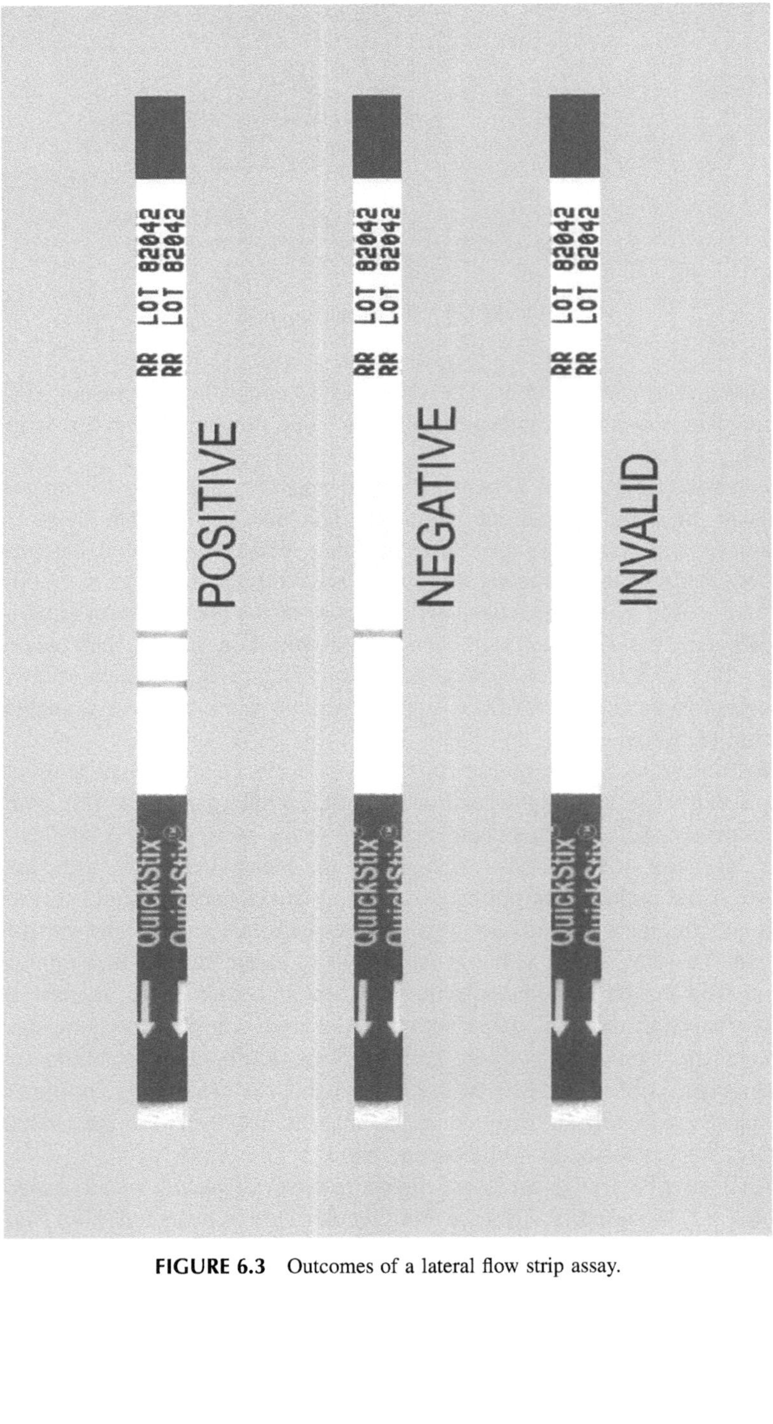

FIGURE 6.3 Outcomes of a lateral flow strip assay.

visible and very intense but there is no upper band (not shown). This condition may result from so much analyte in the extract so that all the available conjugate reacts with the analyte and is captured by the lower pad. Thus, all the antibodies will be captured in the lower band and there will be no antibody conjugate available to be captured in the upper band. This may also occur if the liquid did not reach the upper band. If this condition is observed and is not due to poor liquid flow, the extract can be diluted until a band is observed at both the upper and lower bands. Strips may also contain a small amount of nonspecific antibody that does not recognize any expected analyte, to always give a signal in the upper band if the strip is functional.

While most LFS is designed to be highly sensitive and contain only one analyte-specific band for detection of a single protein, strips with multiple analyte bands, for use in assessing single plants or seeds, are increasingly becoming available. Each band on the strip is specific for a single protein analyte. To date, strips with up to four protein detection bands plus a control band are available.

6.4.5 Other formats

In addition to the sandwich ELISA, an ELISA can be performed as a direct ELISA, an indirect ELISA, or a competitive ELISA (Sakamoto et al., 2018). These are not generally used to analyze GE crops.

There are other immuno-based formats that are similar to ELISAs. These are typically produced by dedicated high technology analytics companies and require specialized instruments for interpretation of the results. These assay formats are marketed as having some advantages over traditional ELISA; they allow assay of multiple analytes in the same extract at the same time and/or they have a much greater dynamic range in which the assay can be performed. The ability to assay multiple traits in a single extract may be an effective way to reduce the cost of materials and time required to perform individual separate assays. A traditional ELISA has a working range of two orders of magnitude at best and while some have only a 10-fold range newer methods area improving on this. Certain more advanced multiplex formats have working ranges of three or four orders of magnitude depending on the assay. On the other hand, due to the simultaneous assay of multiple analytes, these technologies can be complex to validate and are limited to the combination of analytes in the specific product and can thus be relatively expensive compared to traditional ELISAs when used in limited volumes.

6.5 Production of antibodies

Antibodies for use in the ELISA or LFS are the most important components of the assays. The first step in designing an immunoassay is to select antibodies that will recognize the analyte. For an ELISA or LFS, monoclonal and polyclonal antibodies can be used. ELISA's were historically made with a polyclonal capture antibody and a polyclonal antibody conjugate. However, the use of at least one monoclonal antibody component has increased because of its specificity and sensitivity. In most cases, one monoclonal antibody is used for capture and a different monoclonal antibody is used as a conjugate. This makes it possible to consistently produce the ELISA as long as the cell lines producing the monoclonal antibodies are maintained. In cases where it is not possible to have a monoclonal antibody as capture and conjugate, it is desirable to have the polyclonal antibody as the conjugate because it requires much less antibody as a conjugate than it does for capture. If an ELISA with a monoclonal antibody component is going to be used to detect a protein expressed in a plant and if a reference standard for the ELISA was produced in a microbial system, it is important to make sure that the ELISA reacts equally well with the proteins from both sources because some monoclonal antibodies can recognize slight conformational differences in proteins produced in different expression systems.

Factors that are important in choosing an antibody production strategy are the format of the desired assay (e.g., ELISA, LFS, etc.), the number of assays needed (is this an assay that will be used for a limited number of tests or for a large number of tests over many years), whether the assay will be qualitative or quantitative and the required specificity. The availability or access to other proteins or potential matrices against which the antibodies will need to be screened for reactivity or cross-reactivity is also important.

There are a number of excellent references published (e.g., Tijssen, 1985) for the generation of antibodies including Harlow and Lane (1988).

6.5.1 Choice of immunogen

The development of any assay relies on the availability of a well-characterized high-quality antigen/immunogen, in sufficient quantities to support whatever scale of antibody production is required. The immunogen used to elicit the antibody response is typically a purified protein, produced in bacteria, using baculoviruses to infect insect cells, or in plants. However, peptides and partial proteins can also be used. Peptide or partial protein immunogens are often used to generate polyclonal antibodies that target unique difficult-to-reach epitopes or where production of the target protein in sufficient quantities is a challenge. The purity of the immunogen should ideally be 80% or above for antibody production, but lower purities are less of an issue for monoclonal antibody production, as the clonal lines are selected for their ability to produce monoclonal antibodies of high quality. Certain other immunogen characteristics also determine the likelihood of successful antibody production such as size, hydrophobicity, and tertiary structure. The quality of the immunogen strongly influences the quality of the antiserum produced and thus has a direct impact on the quality of the data generated using the respective immunoassay.

The physicochemical and functional characteristics of the protein (e.g., concentration, composition, lipophilic) to be used as the immunogen should be known before undertaking any large-scale antibody production. Specific functional activity of the protein may be a factor in assuring that the immunogen comprises mostly functional protein and therefore generate high-quality antibodies, especially in the case of a polyclonal production. In addition, larger proteins (immunogens) generally give rise to a better immune response.

The longevity and reliability of the future assay are highly dependent on the strategy selected to produce antibodies and the availability of an appropriate immunogen.

6.5.2 Polyclonal antibodies

Production of large quantities of polyclonal antibodies is relatively quick and inexpensive compared to monoclonal antibodies (Fig. 6.4). Polyclonal antibodies are antibodies produced by B cells within the animal. A collection of immunoglobulin (Ig) molecules are produced and react against a specific antigen; each of them identifies a different epitope with different specificities and epitope affinities. Depending on the amount of antibody needed, polyclonal antibodies are commonly produced from rabbits; however, other species, such as mouse, rat, and chicken can be used, or goats, and many other animals. Typically, multiple animals that are used as antibody response will vary between individuals.

The chosen immunogen is injected into the mammal, typically in a series of booster injections over several weeks, along with an adjuvant to promote antibody production, inducing the B-lymphocytes to produce

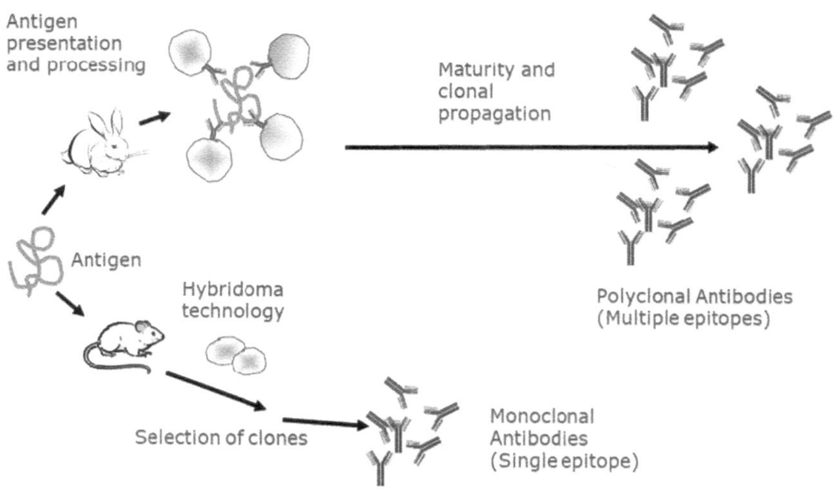

FIGURE 6.4 Basic steps of monoclonal and polyclonal antibody production.

immunoglobulins (IgG) specific for the antigen. This polyclonal IgG is purified from the mammal's serum.

Selection of the quantity of immunogen varies with the properties of the immunogen and the adjuvant selected. In general, micrograms to milligram quantities of protein in adjuvant are necessary to elicit high titer antibodies. Antigen dosage is also species, rather than body weight, associated. The so-called "window" of immunogenicity in each species is broad, but too much or too little antigen can induce tolerance, suppression, or immune deviation toward cellular immunity.

For example, for a rabbit, 50–100 μg of purified immunogen is typically injected and followed by several injections to boost the response. The initial immune response will be assessed 21–28 days after the first injection; test bleeds will be monitored at 14 days intervals to determine the antibody titer. Once the antibody titer is high enough, serum can be collected throughout the life of the animal, possibly with some booster injections. Larger mammals may be preferred for some applications, as the amount of serum that can be collected is greater, but the amount of immunogen required is likely to be much greater.

Due to phylogenetic distance, chickens are used as a source of antibodies with immunogenicity to and less cross-reactivity with mammalian proteins. IgY antibodies are avian equivalents of mammalian IgG antibodies. Chicken antibodies are produced by immunization of laying hens and subsequent purification of IgY from the egg yolk. Very low quantities of immunogen are required to obtain a high and long-lasting IgY titer in the egg yolk and chickens lay eggs regularly, providing a continual source of IgY antibody.

The fact that polyclonal antibodies are a mixture of antibodies capable of recognizing multiple epitopes on any one antigen increases the capacity to bind the target protein and offers a more robust detection tool (Table 6.1). However, there can be variability from batch to batch, the profile will vary from one animal to another, and the supply from each animal is limited. Thus manufacturers may combine batches to produce large numbers of consistent tests over multiple years.

Polyclonal antibodies are used extensively for research purposes in many areas of biology, such as immunoprecipitation, histochemistry, ELISA, diagnosis of disease, immunoturbidimetric methods, western blots, and Biochip technology. Polyclonal antibodies are ideally suited for use in sandwich assays as second-stage antigen detectors.

6.5.3 Monoclonal antibodies

Monoclonal antibodies come from a single B lymphocyte generating antibodies to one specific epitope on the antigen (Fig. 6.4). The production of monoclonal antibodies is more complex and takes longer than polyclonal antibodies.

The key to the production of monoclonal antibodies is hybridoma technology developed by Köhler and Milstein (1975). In this process, the immunogen is injected into a mouse, and following antibody formation B-cells are isolated. Cells producing antibody that binds to the antigen are selected for hybridizing with myeloma cells. Clones of the resulting immortalized hybridoma cells are screened using ELISA; single clones that produce an antibody of interest are selected for further proliferation. The selected clones are usually propagated in vitro in tissue culture. The antibody production rate in culture is typically 10–100 mg per mL of culture. Alternatively, antibody can be produced in vivo, where the clone is injected into the peritoneal cavity of histocompatible mice and allowed to multiply and produce antibody. The production is higher on a per mL basis but requires live mice and so the in vitro process is more routinely used. The monoclonal antibodies can be purified as needed by affinity chromatography.

Antibody-producing cell lines can be stored indefinitely under liquid nitrogen. Thus, in theory, infinite amounts of the selected antibody with stable characteristics can be produced. If experimental conditions are kept constant, results from monoclonal antibodies are highly reproducible.

TABLE 6.1 Comparison of the characteristics of polyclonal and monoclonal antibodies.

	Polyclonal	Monoclonal
Time to develop	Relatively quick	Relatively long development time due to need to select clones
Homogeneity	Variation between different animals if production is continuous	Very homogeneous with repeatable affinity and sensitivity over time
Availability (quantity)	Limited supply	Unlimited quantities
Purity of immunogen	Purity of immunogen must be higher than for monoclonal production	Purity of immunogen is not as important
Ease of labeling	Stable and easily labeled in various assay formats	Can be rendered nonfunctional by some labels
Sensitivity	Usually has high sensitivity	Sensitivity can be a challenge
Specificity	Likely to cross-react with similar proteins	High specificity
Cross-reactivity	Cross-reactivity can be addressed, and/or is an advantage	Little or no cross-reactivity (can be selected)

Monoclonal antibodies are an excellent choice for protein detection tools such as quantitative ELISA, which are required to produce consistent and standardized data (Table 6.1). However, two antibodies binding to different sites on the target protein are needed. Monoclonal antibodies are also often used in LFS applications as they are available in unlimited amounts and produce consistent results.

6.5.4 Recombinant antibodies

Synthetic or genetically engineered antibodies called recombinant antibodies (rAbs) are a recent development. These antibodies can be created using antibody genes made in a laboratory, completely eliminating animals from the antibody-production process. rAbs can be used in all applications in which traditional monoclonal antibodies are used.

In rAB development, bacteria and bacterial viruses (phages) are used to express and select recombinant antibodies that have all the target recognition qualities of natural antibodies (Barbas et al., 1991). The phages are genetically engineered so that a particular antibody is fused to a protein on the phage's envelope, and the gene encoding the displayed antibody is contained inside the phage particle. Phage libraries typically contain a billion different antibodies and desired antibodies are selected by "screening" the library against the target molecule. Phages with antibodies that recognize the target molecule bind tightly, whereas others are washed away. Bacteria, yeasts, or vertebrate cells are then transfected with the DNA contained within the bound phage and produce the selected antibody.

Using a recombinant antibody has significant advantages compared with conventional monoclonal antibodies. The fact that no animals are needed, and the reduced development time is an advantage. The quality of the final antibody product can be higher than that of the traditional nonrecombinant approach.

6.6 Assay design and operation

To assay the protein in a plant-derived sample, several considerations must be explored:

- What proteins of interest are to be measured? A product may express a single protein or several proteins that need to be studied simultaneously
- Is there a specific tissue that must be analyzed?
- When is the protein expressed during the plant's lifecycle?
- What is the most effective assay buffer system that can be used—can we avoid the use of multiple buffer systems required for multiple proteins.?

6.6.1 Microtiter plate ELISAs

6.6.1.1 ELISA creation and quality control

After the antibody pairs for an ELISA have been developed and selected, the next steps in preparing an ELISA are coating of the plate with a capture molecule (usually antibody), blocking the plate with a nonreactive protein to prevent nonspecific adsorption of proteins added in subsequent steps and storage of the ELISA plates. For a detailed description of how to optimize each of these steps, the reader is referred to the chapter by Brown (2010).

Though ELISA plates can be prepared as needed, this is rarely done, except in a research environment, due to the need for comparability of results over the long term. Thus, batches of plates are typically prepared and dried for longer-term storage at refrigerated or ambient temperatures. Plate to plate reproducibility is an important requirement as is storage stability. Documentation of these factors requires numerous repetitions of assays over time particularly if the ELISA is going to be used for long term and particularly when used for studies required for submission to regulatory bodies. While trait developers may develop ELISAs in-house, development through a commercial vendor with this expertise is a convenient solution.

Quality control measures need to be applied at all stages of manufacture to ensure the ELISA plates produce reproducible results. Commercially manufactured plates are typically produced using automated equipment and under conditions that ensure plates are consistent within and between batches.

Performing quality control measures is also important when the plates are made "in house," and will consist of a series of tests where replicate samples and replicate standard curves are tested across the entire plate to ensure well to well consistency, plate to plate consistency, and to check for plate edge effects.

Although commercial plates may come with a set of instructions for a specific application, the procedure for running the assay can be altered to fit the user's needs, as long as the use is properly validated. For example, a plate sold for qualitative purposes can often be adapted for use in a quantitative assay if a reliable protein standard is available.

For a detailed discussion of ELISA assay development, see Keith et al. (2) and Shan et al. (3).

6.6.1.2 Grinding/tissue disruption

To assay the amount of a protein in a matrix, it must first be extracted. Extraction efficiency depends upon the protein characteristics and subcellular location, the extraction buffer, how the extraction is performed, and the extent to which the cells in the tissue matrix have been disrupted. These parameters are all interrelated and affect the quantification of the protein.

The most critical step in achieving high extraction efficiency is the first step of grinding the matrix into a fine powder. Several methods have been used to achieve a high degree of cell breakage. Frozen or lyophilized tissue can be pulverized with dry ice in a blender or ground with a mortar and pestle cooled with liquid nitrogen. Several grinding mills can be used including those that grind the tissue at the temperature of liquid nitrogen. Liquid nitrogen grinders are effective, but sample throughput is low. A good, simple test for efficient cell disruption is microscopic examination of the ground tissue. If large pieces of tissue are present, the extraction process will not likely be highly efficient.

6.6.1.3 Choice of buffer

For many proteins, Phosphate Buffered Saline plus Tween is a good starting place. Polyvinyl pyrrolidone 40 at a concentration of 1%–2% is commonly used in extraction buffers to increase extraction efficiency. Some proteins extract much more efficiently with more acid or alkaline buffers (citation). If the analyte is present in the chloroplasts or is membrane bound, the inclusion of mild nonionic detergents such as Triton ×100 may be useful to achieve efficient extraction. There are many good buffers available from manufacturers of ELISA kits, but the exact composition of the buffer is not usually provided. However, there has been a movement among kit developers and users to try and standardize on a single buffer system where possible to allow assay of multiple proteins in a single extract.

6.6.1.4 Protein extraction

A common way to test extraction efficiency is to perform multiple extractions on the same sample. At least three extractions of the pellet obtained by centrifugation of the extract will indicate the extraction efficiency. If the extraction is good, 80% or more of the analyte should be present in the supernatant of the first extract. However, this does not mean that most of the analyte has been extracted. One way to test for the residual analyte is to perform a Western blot on the pellet of the final extraction. The extraction process in a Western blot is highly denaturing due to high extraction temperature and the presence of the detergent sodium dodecyl sulfate, so an antibody that can detect the denatured protein is required. If little of the analyte is detected in the pellet from an extraction, there is a high degree of confidence that most of the analyte was extracted.

The extraction ratio can be expressed in units of g/mL. It represents a dilution of the analyte that must be taken into account when calculating the concentration of the analyte in the matrix. The magnitude of this dilution can have a dramatic effect on the extraction efficiency. As a general rule, the extraction ratio should be no more concentrated than 1 part matrix to 10 parts of extraction buffer. Extraction at this ratio might be necessary if the level of expression of the analyte in the matrix was low. However, such a concentrated extract may not pass a spike and recovery validation due to matrix effects (see Section d below). Extraction with a low volume of extraction buffer for a highly expressed protein may result in poor extraction efficiency due to the saturation of buffer with the analyte. Thus, the level of expression of an analyte determines how much dilution can be done at the extraction stage. There is a practical limit to the amount of dilution at the extraction step, which is between 1:100 and 1:400. It is critical to be able to accurately weigh out the matrix to be extracted, so this will rarely be less than 30–100 mg. An amount of extraction buffer of the 3–40 mLs can then be used. As the extraction ration increases, larger volumes of buffer are required, which can be cumbersome and expensive. The background signal of a nontransgenic matrix control at these dilutions is expected to be low because there would be little matrix in the extract. Additional dilution of the extract after extraction may be required for highly expressed proteins, in which case any background signal from the matrix would be even less significant.

The procedure used for extraction also affects the efficiency. Most extractions are done in the cold. The procedure can be as simple as rotary shaking or continual inversion of the extract in a tube. Other more vigorous procedures can be used. These include vigorous shaking of the extract with glass or steel beads as well as continual homogenization with rotating steel blades immersed in the extract. The more vigorous extraction procedures may help increase extraction efficiency, but many proteins can be denatured by these treatments and lose immunoreactivity.

6.6.1.5 Standard curves

For qualitative assays, a standard curve may not be used, and a signal above a set background level is considered positive. Alternatively, positive and negative samples can be used to determine the level at which a sample is considered positive.

For an assay to give quantitative results, a standard curve must be available. A standard curve requires the availability of a standard reference protein with a known concentration. As most proteins are either expressed at a low concentration in plants or difficult to purify enough to use as a standard, or both, the reference protein is usually produced in a microbial system. Ideally, the same protein source used in the development of the antibodies (the immunogen) will be used as the reference. Microbially produced proteins may not contain posttranslational modifications found in some plant proteins, and these factors must be controlled for (citation). The important factor is that antibodies in the assay must react equally to the plant and microbially produced standard protein.

Because the analytical samples are usually a crude extract or a dilution of a crude extract, some analysts prepare a standard curve by spiking the standard reference protein into an undiluted extract of a control matrix lacking the analyte—thus mimicking a matrix-matched sample. Matrix effects can be determined by comparing a standard curve prepared in an extraction buffer with a standard curve prepared in the extract of a control matrix. If the matrix does not result in unacceptable changes in the standard curve, a standard curve of the analyte in the extraction buffer can be used, thus eliminating the need to prepare a different standard curve for each matrix.

To determine the error in measurement for each level of the standard curve, the percent difference in absorbance (coefficient of variation) for replicate samples is determined. A difference between duplicates or a %CV of $\leq 25\%$ is usually acceptable. Fitting a curve to the measured OD values of a standard is done through software that may be supplied by the manufacturer of the plate reader or other means. The measured values for a standard curve can be fit with a linear, quadratic, 4-parameter logistic, or 5-parameter logistic curves. Higher-order equations generally give a better fit over the entire range of the standard curve such that the standard curve is linear when transformed by the curve fitting, and the useable range is defined by this range. After an ELISA has been validated and considered "fit for purpose," the reproducibility of the standard curve is monitored to assess day-to-day and operator-to-operator variability. The R^2-value of the standard curve, its slope, and intercept with the axis are useful parameters for historical reference and quality control performance criteria, except that this may be less useful when using 4- or 5-parameter curve fitting. Changes in these parameters over time can be used to determine if a problem might be developing that needs to be addressed.

6.6.1.6 Cross-reactivity

As described in Section 6.4.1, an antibody may show cross-reactivities to other analytes. It is therefore essential that the cross-reactivity of the assay for other proteins is assessed, and this is usually evaluated during the antibody production process. Polyclonal antibodies are more prone to cross-reactivity. Cross-reactivity issues can be overcome by the use of affinity-purified antibodies or utilizing monoclonal antibodies. Rigorous screening of the antibodies against possible cross-reaction during development is used to avoid cross-reactivity in the final assay.

An apparent cross-reactivity of an antibody to two proteins is sometimes observed regardless of their apparent lack of homology. This apparent cross-reactivity may come from the presence of small amounts of contaminating proteins from the organism producing the immunogen. If these contaminating proteins are highly immunogenic, the antibodies produced will cross-react with another protein sample containing the same contaminating proteins. This may occur if the protein used as the immunogen and protein being tested for cross-reactivity were produced in the same expression system and contain the same contaminating proteins. For example, if a Cry1 and a Cry2 protein were both produced in *Escherichia coli*, there would be low concentrations of *E. coli* protein in both of the Cry protein preparations. Antibodies made against the contaminating proteins in the Cry1 antibody sample would cross-react with the same set of contaminating proteins in the Cry2 protein preparation. This apparent cross-reactivity would not be observed when analyzing the plant-produced protein because the contaminating *E. coli* proteins would not be present in the plant. This apparent cross-reactivity is usually not seen if the immunogens were produced in different organisms, for example, *E. coli* and *B. thuringiensis*, or if monoclonal antibodies are used.

6.6.1.7 Determination of the useful range of the assay

ELISAs have a range in which they are validated for accurate quantification of an analyte. This range is defined by a limit of detection (LOD), a limit of quantitation (LOQ), an upper limit of quantitation (ULOQ), and an LOQ, which is sometimes referred to as a lower limit of quantitation. These parameters define the limits of the usefulness of an assay and are determined by measuring the accuracy and precision of the assay in spike and recovery experiments where known concentrations of the analyte are added to the extraction. Typically, acceptable recoveries range from about 60% to 130% but other ranges can be used. The precision of the assay should also be determined by calculation of a %CV (coefficient of variance) from the percent recovery data from multiple spikes and recovery determinations. Values of the %CV of $\leq$25% are common.

There are two common ways of performing spike and recovery studies. In the first approach, the nontransgenic matrix is extracted with a buffer spiked with the analyte and then clarified by centrifugation. Alternatively, the matrix is extracted with a buffer not containing the analyte and a sample of the clarified extract obtained after centrifugation is spiked with the analyte. Spiking the buffer enables the analyte to interact with and possibly adsorb to insoluble materials in the matrix. This can be important in matrices that contain components that adsorb proteins. If the experiment is done by spiking only the clarified extract, the analyte does not have the opportunity to interact with insoluble materials in the matrix. However, it requires much less analyte to perform because only a subsample of the clarified extract is spiked. In addition, a determination of the degree of interaction of the analyte with insoluble materials in the extract can be determined on a small scale and a correction can be made if deemed necessary. Spiking the extraction at high levels of analyte can consume large quantities of the purified analyte, which may be difficult to produce.

Because large amounts of purified analyte are needed to perform spike and recovery experiments at high analyte concentrations, the ULOQ may be set at a level where the acceptance criteria are met but higher levels of spike and recovery were not tested. This is acceptable as long as the ULOQ is not exceeded when analyzing samples. Conversely, determination of the LOQ usually involves spiking at lower and lower levels of the analyte until the acceptance criteria are not met. When this is observed, the lowest level of spiking that meets the acceptance criteria (e.g., <25% CV) is set as the LOQ.

The LOD is usually defined as the amount of analyte that is statistically different from zero or from the background level of the signal observed in matrix blanks. The LOD was often traditionally set to equal 30% of the LOQ (4) although replicates of matrix blank can be used to establish the experimental LOD.

6.6.1.8 Other factors

Other factors that affect the performance of the assay include incubation temperatures and incubation times for the various steps. In addition, assays can be run with or without shaking. Shaking on an orbital shaker generally shortens the process as it allows more efficient capture of the antigen by the antibodies. It should be defined in the method as shaking too fast can prevent this binding. Plates are normally incubated at ambient temperature—ambient being generally understood as a temperature between 68 and 77°F (20 and 25°C).

6.6.2 Lateral flow strips

6.6.2.1 LFS creation and quality control

Lateral flow strips are commonly designed and produced by commercial suppliers that specialize in their design and manufacture. The design process is similar to that of the ELISA and involves selecting a pair of antibodies that will perform well in an LFS format. Because of the physical and biochemical differences between LFS assays and microtiter ELISAs, antibodies that perform well in an ELISA may not work in an LFS format. ELISAs are, however, used in the screening process. As for an ELISA, all available antibodies need to be screened in all possible combinations to select the best performing combinations.

LFSs designs used in agricultural biotechnology applications generally use a nano-bead of colloidal gold conjugated to the antibody that can be detected by visual inspection. The gold can be attached to specific regions of the antibody so that the reactive site of the antibody is not masked. This colloidal gold can be detected by visual inspection, as well as by readers that measure the intensity.

For agricultural use, two types of lateral flow strips are generally available. One type of strip is made for the detection of an analyte in single seeds (SS) or in a single leaf (SL). The other type of strip is primarily used for measuring low amounts of a GMO event in samples consisting of samples from bulk seeds or grains (BG) or samples consisting of multiple leaf (or another plant part) samples bulk leaf (BL). The difference in these two applications relates to the sensitivity required.

6.6.2.2 *LFS for testing single seed or single leaf*

SS/SL strips are designed to give a simple, yes/no nonquantitative result, where it is not necessary to obtain maximum extraction efficiency and sensitivity. These SL strips are used in plant breeding to monitor the segregation of a trait after making crosses or for detecting the presence of the trait in regenerated plants following the initial transformations. The SS LFS is commonly used to determine seed purity by testing individual seeds for the presence of the trait (most often used in cotton). A percent purity and confidence interval can be calculated based on the number of seeds tested and the number expressing the trait. The number of seeds to test and results can be estimated statistically. A convenient way to do these calculations is using the free Excel application SeedCalc that can be downloaded from the International Seed Testing Association.

SS/SL LFS for single or multiple traits are available as single strips or in a comb format. The latter combines 8 strips attached at the upper end and spaced so that each strip will fit into eight wells of a 48-well microtiter plate. This format allows 48 seeds to be simultaneously crushed, extracted, and tested with six combs of 8 strips per comb.

6.6.2.3 *LFS for testing bulk grain or bulk leaf samples*

LFS BG/BL strips detect only a single trait; they are designed to be as sensitive as possible because they are used for the detection of a low level of the protein.

BL testing is most commonly used in plant breeding or testing a small plot of plants for purity, to remove plants containing a trait that is not desired. For example, leaf punches from 10 to 20 plants are combined (see Chapter 10), extracted, and tested for the presence of a plant with an unwanted trait. If the unwanted trait is detected, each plant that made up the bulk is tested, the unwanted plant is removed, and the remaining plants retested. Given similar levels of expression of the trait in leaves and grain, an LFS for BL used for testing a 20 leaf discs bulk does not need to be as sensitive as an LFS used for BG testing a 200 seed bulk.

There are two main applications for the BG LFS—testing for low levels of an undesired trait in seeds, and testing bulk grain samples. For BG, a desirable sensitivity is the detection of 1 transgenic seed/grain in 200, or in transgenic where the undesired trait needs to be excluded. Some products may only detect 1 in 150. The strip is used to test for the presence of an undesired trait (such as from a regulated event) or to determine whether the level of an undesired (but allowed) trait is below a desired threshold. The result of testing multiple pools of seeds is used to determine the impurity of a seed lot, that is, it can estimate how much of the undesired seed/grain is present in a seed lot. The pools typically consist of 150–200 seed/grains, limited by the sensitivity of the strip and the testing strategy, and the data are analyzed using SeedCalc. There are a large number of commercially available LFSs that can detect a trait at this level and higher; the sensitivity of a strip is dependent on the level of protein expressed in the seed/grain of the specific event. It is important to have the strip as sensitive as possible and it is also helpful if the level of the analyte is high in the seed.

6.6.2.4 *Evolution of the LFS design*

LFSs for SS/SL are available for the detection of up to four traits on a single strip, with each trait being detected by a single line. This is convenient in that it allows for multiple traits to be assayed in a single extract (e.g., of a seed). It may theoretically be possible to detect more traits on a single strip, but this has not yet been demonstrated. These types of LFS are used primarily for testing seeds (usually cotton) for purity.

In the past, strips from different (or even the same) manufacturers required different extraction buffers, necessitating multiple extractions to test for multiple traits. Almost all of the BG/LG strips now use water as the extraction buffer and the buffer is supplied in the lower pad of the strip. The water used can influence the strip and cause false negatives or false positives if it contains a high level of some contaminants such as ferrous iron, so should be at least of drinkable quality.

Combs consisting of multiple single BG strips are sometimes used where each strip in the comb tests for a specific trait. This comb can be inserted in a single water extract to test for multiple traits as a single operation.

Several manufacturers have produced strip readers for quantitative analyses. The strips contain information about the standard curve for the trait on the strip as printed QR-like code. The strips may be supplied in comb format to simplify the operation. This approach can give an estimate for the amount of analyte in an extract from which the %GM in the sample is estimated. However, care must be taken to use a standardized extraction protocol to minimize any sample to sample differences in results.

6.6.2.5 *Validation of a lateral flow assay*

Performance optimization of LFS assays primarily involves ensuring sufficient sensitivity of the assay and quick response. The manufacturer performs extensive quality control of each component that makes up the final strip. Samples of the final product are taken at intervals during a production run. Many strips are now made using a continuous auto-laminator rather than in small batches. Each batch is validated using the test samples withdrawn during manufacture, typically using 300 strips. Cross-reactivity to other traits is also tested. In addition, trait providers that contract for the production of an LFS may have their own internal validation criteria that have to be met before the product can be released. More information on validation of immunoassays can be found in Chapter 7 (Shan and Schneider, 2022).

6.7 Conclusions

Antibodies are the workhorses in detection and quantification of proteins throughout the development, characterization and commercialization of crops using Agricultural biotechnology. Whether done in-house by developers, or by contractors, the development of such assays is fairly routine but not without challenges. The choice of polyclonal or monoclonal systems, and a plate or lateral flow approach is decided according to the need. The assays used are almost exclusively the ELISA and lateral flow device. Systems in which multiple proteins are assayed simultaneously are used less often in Agricultural biotechnology as assays are less routine than for example in medical applications. Spectrophotometric methods are gaining ground as the technology improves, and especially for intractable (Bushey et al., 2014) proteins.

Acknowledgments

The authors thank Yelena Dudin (Bayer Crop Science US) for her extensive contributions and review during the preparation of this chapter.

References

Asensio, L., Gonzalez, I., Garcia, T., Martin, R., 2008. Determination of food authenticity by enzyme-linked immunosorbent assay (ELISA). Food Control 19, 1–8.

Barbas 3rd, C.F., Kang, A.S., Lerner, R.A., Benkovic, S.J., 1991. Assembly of combinatorial antibody libraries on phage surfaces: the gene III site. Proceedings of the National Academy of Sciences of the United States of America 88, 7978–7982.

Berson, S.A., Yalow, R.S., 1959. Quantitaitve aspects of the reaction between insulin and insulin binding antibody. Journal of Clinical Investigation 38, 1996–2016.

Bogani, P., Minunni, M., Spiriti, M., Zavaglia, M., Tombelli, S., 2008. Transgenes monitoring in an industrial soybean processing chain by DNA-based conventional approaches and biosensors. Food Chemistry 113, 658–664.

Bradford, M.M., 1976. A rapid and sensitive method for the quantitation of microgram quantities of protein utilizing the principle of protein-dye binding Anal. Biochem 72, 248–254.

Brown, 2010. microtiter plate elisa in: immunoassays in agricultural biotechnology pp: 59–89 Guomin Shan. In: 2010 John Wiley & Sons, Inc.

Bushey, D.F., Bannon, G.A., Delaney, B.F., Graser, G., Hefford, M., Jiang, X., Lee, T.C., Madduri, K.M., Privalle, L.P., Ranjan, R., Saab-Rincon, G., Schaffer, B., Thelen, J.J., Zhang, J.X.Q., Harper, M.S., 2014. Characteristics and safety assessment of intractable proteins in genetically modified crops. Regulatory Toxicology and Pharmacology 69 (2), 154–170. https://doi.org/10.1016/j.yrtph.2014.03.003.

Chang, P.-C., Reddy, P.M., Ho, Y.-P., 2014. Quantification of genetically modified soya using strong anion exchange chromatography and time-of-flight mass spectrometry. Analytical and Bioanalytical Chemistry 406, 5339–5346.

de Luis, R., Lavilla, M., Sánchez, L., Calvo, M., Pérez, D.M., 2009. Immunochemical detection of Cry1A(b) protein in model processed foods made with transgenic maize. European Food Research and Technology 229, 15–19.

Engvall, E., Perlmann, P., 1971. Enzyme-linked immunosorbent assay (ELISA). Quantitative assay of immunoglobulin G. Immunochemistry 8 (9), 871–874.

Fearing, P.L., Brown, D., Vlachos, D., Meghji, M., Privalle, L., 1997. Quantitative analysis of CryIA(b) expression in Bt maize plants, tissues, and silage and stability of expression over successive generations. Molecular Breeding 3, 169–176.

Grothaus, G.D., Bandla, M., Currier, T., Giroux, R., Jenkins, G.R., Lipp, M., Shan, G., Stave, J.W., Pamntella, V., 2006. Immunoassay as an analytical tool in agricultural biotechnology. Journal of AOAC International 89 (4), 913–928.

Harlow, E., Lane, D., 1988. Antibodies. A Laboratory Manual. Cold Spring Harbor Laboratory Press, New York.

Hill, R.C., Oman, T.J., Shan, G., Schafer, B., Eble, J., Chen, C., 2015. Development and validation of a multiplexed protein quantitation assay for the determination of three recombinant proteins in soybean tissues by liquid chromatography with tandem mass spectrometry. Journal of Agricultural and Food Chemistry 63 (33), 7450–7461.

Hosseini, S., Vazquez, P., Rito-Palomares, M., Martinez-Chapa, S.O., 2018. Fundamentals and history of ELISA: the evolution of the immunoassays until invention of ELISA. 10.1007/978-981-10-6766-2_1. In: Enzyme-linked Immunosorbent Assay (ELISA), SpringerBriefs in Applied Sciences and Technology, pp. 1–18. January 2018.

ISO 16577, 2016. Molecular Biomarker Analysis — Terms and Definitions.

Kamle, S., Ali, S., 2013. Genetically modified crops: detection strategies and biosafety issues. Gene 522, 123–132.

Kim, Y.H., Choi, S.J., Lee, H.A., Moon, T.W., 2006. Quantitation of CP4 5-Enolpyruvylshikimate-3-Phosphate synthase in soybean by two-dimensional gel electrophoresis. Journal of Microbiology Biotechnology 16 (1), 25–31.

Köhler, G., Milstein, C., 1975. Continuous cultures of fused cells secreting antibody of predefined specificity. Nature 256, 495–497.

Lowry, O.H., Rosebrough, N.J., Farr, A.L., Randall, R.J., 1951. Protein measurement with the folin pheno reagent. Journal of Biological Chemistry 193, 265–275.

Margarit, E., Reggiardo, M.I., Vallejos, R.H., Permingeat, H.R., 2006. Detection of BT transgenic maize in foodstuffs. Food Research International 39, 250–255.

Morisset, D., Demsar, T., Gruden, K., Vojvoda, J., Stebih, D., Zel, J., 2009. Detection of genetically modified organisms-closing the gaps. Nature Biotechnology 27, 700–701.

Nakane, P.K., Pierce Jr., G.B., 1966. Enzyme-labeled antibodies: preparation and application for the localization of antigens. The Journal of Histochemistry and Cytochemistry 14 (12), 929–931.

Ngom, B., Guo, Y., Wang, X., et al., 2010. Development and application of lateral flow test strip technology for detection of infectious agents and chemical contaminants: a review. Analytical and Bioanalytical Chemistry 397, 1113–1135. https://doi.org/10.1007/s00216-010-3661-4.

Ocaña, M.F., Fraser, P.D., Patel, R.K.P., Halket, J.M., Bramley, P.M., 2009. Evaluation of stable isotope labeling strategies for the quantitation of CP4 EPSPS in genetically modified soya. Analytica Chimica Acta 634, 75–82.

Privalle, L.S., Chen, J., Clapper, G., Hunst, P., Spiegelhalter, F., Zhong, C.X., 2012. Development of an agricultural biotechnology crop product: testing from discovery to commercialization. J. Agric. Food Chem. 60 (41), 10179–10187. https://doi.org/10.1021/jf302706e.

Sakamoto, M.F., Putalun, W., Vimolmangkang, S., Phoolcharoen, W., Shoyama, Y., Tanaka, H., Morimoto, S., 2018. Enzyme-linked immunosorbent assay for the quantitative/qualitative analysis of plant secondary metabolites (2018). Journal of Natural Medicines 72 (1), 32–42. https://doi.org/10.1007/s11418-017-1144-z.

Shan, G., Schneider, M., 2022. Method validation: protein detection methods. In: Shillito, R., Shan, G. (Eds.), Application of Sampling and Detection Methods in Agricultural Plant Biotechnology. Elsevier. Chapter 7 Submitted for publication.

Stave, J.W., Brown, M.C., Chen, J., McQuillin, A.B., Onisk, D.V., 2004. ACS symp. Ser 866, 69–82.

Smith, P.K., Krohn, R.I., Hermanson, G.T., Mallia, A.K., Gartner, F.H., Provenzano, M.D., Fujimoto, E.K., Geoke, N.M., Olson, B.J., Klenk, D.C., 1985. Measurement of protein using bicinchoninic acid. Analytical Biochemistry 150, 76–85.

Tijssen, P., 1985. Practice and theory of immunoassays. In: Laboratory Techniques in Biochemistry and Molecular Biology. (Burdon, RH and Van Knippenberg, PH, , first ed.vol. 15. Elsevier Science, New York, USA.

Towbin, H., Staehelin, T., Gordon, J., 1979. Electrophoretic transfer of proteins from polyacrylamide gels to nitrocellulose sheets: procedure and some applications. Proceedings of the National Academy of Sciences 76 (9), 4350–4354.

Voller, A., Bidwell, D., Huldt, G., Engvall, E., 1974. A microplate method of enzyme-linked immunosorbent assay and its application to malaria. Bulletin of the World Health Organization 51, 209–211.

Chapter 7

Protein methods: antibody-based protein method validation and assay verification

Guomin Shan and Michael Schneider
Corteva Agriscience, Indianapolis, IN, United States

7.1 Introduction

To ensure an analytical method is suitable for its purpose, it is important to conduct a proper validation before implementation. Method validation is a process to establish documented evidence through laboratory studies and to provide a high degree of assurance that the performance parameters of the method meet the requirements for the intended analytical applications.

Immunoassay has been the method of choice for protein detection in agricultural biotechnology since the introduction of commercial genetically engineered (GE) crops, and it will continue to be a predominant technology in foreseeable future. Antibody-based protein detection technology will be our focus in this chapter. Chapter 2 in this book provides the general principles and considerations related to the development and use of immunoassay for products of agricultural biotechnology. This chapter will cover the validation of developed methods before implementing as an analytical tool in the laboratory. In the agricultural biotechnology field, this process usually includes method validation, interlaboratory validation, and assay verification.

Method validation generally refers to in-house validation and consists of a series of experiments that evaluate and document the performance of an analytical method including sensitivity, accuracy, specificity, precision, range, and limit of quantitation. For GE protein detection in plant samples, matrix effect and extraction efficiency are required as part of method validation. Due to the difference in assay scope and intended use, quantitative and qualitative methods differ in assay parameters during method validation, which is discussed later in the chapter.

Application of Sampling and Detection Methods in Agricultural Plant Biotechnology
https://doi.org/10.1016/B978-0-323-99293-0.00005-2

Inter-laboratory validation or collaborative validation trial is a process to further evaluate the repeatability and reproducibility of an assay after the completion of in-house method validation and assess the transferability of methods between laboratories (AOAC Official Methods of Analysis, 2005). For novel proteins in insect-resistant traits, US Environmental Protection Agency (EPA) requires an independent laboratory validation (ILV) under Good Laboratory Practice (40CFR 160). In GE crop protein detection, test kits are usually obtained from and manufactured by kit companies, where a good quality control system is required for kit production. Therefore, it is uncommon to conduct a formal interlaboratory validation for GE protein quantification. In this chapter, we will discuss independent laboratory validation in Section 4.

Finally, assay verification refers to the process required when adding a new test method or when an alternative kit is required for a given analyte, as well as bridging between lots of internal assays or kits from vendors where it is needed to ensure consistent assay performance. More details will be discussed in Section 5.

7.1.1 Standards and validation parameters

Method validation is a well-established scientific discipline, and international standards and guidelines have been developed for various technologies and applications (ISO 5725, 1993; ICH, 2005; FDA, 2013). The Codex Alimentarius guidelines provide specific method validation requirements for the detection and quantification of GE proteins in foods (CAC/GL 74–2010), and the International Organization for Standardization (ISO) also provides guidance for immunoassay-based methods for testing foodstuff (ISO 21572). Draft Guidance for bioanalytical method validation published by the US Food and Drug Administration (FDA, 2013) includes some specific recommendations for validation of immunoassays or ligand binding assays. In addition, numerous technical and review articles have been published, which provide vast field-specific knowledge and practical information in GE protein detection application (Lipton et al., 2000; Grothaus et al., 2006; Schmidt and Alarcon 2011). Validation parameters for qualitative and quantitative methods of analysis are summarized in Table 7.1. Among these assay parameters, ruggedness and robustness are not routinely evaluated for immunoassays. In this chapter, parameters are discussed.

Limit of detection (LOD) is defined as the lowest amount of analyte in a sample which can be detected but not necessarily quantitated with precision.

Limit of quantitation (LOQ) is the lowest amount of analyte in a sample that can be quantitatively determined with suitable precision and accuracy. LOQ is a parameter of quantitative assays for low levels of protein of interest in sample matrices.

TABLE 7.1 Summary of method validation parameters.

Quantitative method	Qualitative method
Sensitivity	Sensitivity
Limit of detection	
Limit of quantitation	
Linearity and range	Hook effect
Precision	Precision/repeatability
Accuracy	False positive and false negative rate
Specificity (matrix effects, protein cross-reactivity)	Specificity
Ruggedness	Ruggedness
Robustness	Robustness

Linearity is its ability to obtain consistent test results across a given range that are directly proportional to the concentration of analyte in the sample. Range is the interval between the upper and lower concentration of analyte in the sample for which it has been demonstrated to have a suitable level of precision, accuracy, and linearity.

Precision is the closeness of agreement between a series of measurements obtained from multiple sampling of the homogeneous sample under the prescribed conditions. Precision may be considered at three different levels: intraassay precision (or repeatability such as dilution agreement/parallelism), intralaboratory precision, and interlaboratory precision (or reproducibility).

Accuracy is the closeness of agreement between the value that is accepted either as a conventional true value or an accepted reference value, and the value obtained by the assay being evaluated. Accuracy may be measured by extraction efficiency and/or spike recovery, depending on the assay, although the true value of a protein in the target tissue may not be known.

Specificity is the ability to assess unequivocally the protein of interest in the presence of components that may be expected to be present. Typically, these include matrix and cross-reactivity to the protein of interest.

Ruggedness is the reproducibility of the assay under a variety of normal, but variable, test conditions. Variable conditions might include different instruments, operators, and reagent lots. Ruggedness is a measure of experimental reproducibility with unavoidable variability in assay conduct.

Robustness is the capacity of an assay to remain unaffected by small but deliberate changes in test conditions. Robustness provides an indication of the ability of the assay to perform under normal usage. Robustness is a measure of the effect of deliberate changes such as incubation time, temperature, sample preparation, and buffer pH that can be controlled through specifications in the assay protocol and is usually examined during the development of methods before multi-laboratory validations.

Hook effect is a phenomenon that causes a false negative reading when a high level of the target analyte is present. Hook effects are commonly observed with lateral flow devices and need to be assessed during the validation.

The false-positive rate is the percent of times a known negative sample is detected by the assay as a positive result. Conversely, the false-negative rate is the percent of times a known positive sample, which usually contains a level of target protein close to the LOD, results in a negative result.

7.2 Quantitative ELISA method validation

Quantitative ELISA method validation includes sensitivity, specificity, accuracy, and precision. Extraction efficiency and dilution agreement are also evaluated for plant matrix samples.

7.2.1 Sensitivity

During method development and optimization, a standard curve is usually established and tested as described in Chapter 2, which is normally used to define method lower limit of quantitation (LLOQ or LOQ) and upper limit of quantitation (ULOQ). In prevalidation studies, each of the points along an extended curve including LLOQ and ULOQ is assessed for its accuracy by evaluating the interpolated values in the assay buffer system. There is no required criteria or guidance for determining the quantitative curve range for matrix-matched or buffer-based standard curve. However, guidance has been proposed as a result of an American Association of Pharmaceutical Scientists (AAPS)/FDA Bioanalytical workshop (Viswanathan et al., 2007). This guidance recommends a matrix-matched standard curve must be able to predict at least 75% of the concentration points in the quantitative range within ±20% of the theoretical value except at the LLOQ and ULOQ, which should be within ±25% of the theoretical value. A standard curve based on buffers can also be suitable.

LOD is used for a qualitative assay to define the lower concentration limit for which a positive sample can be reliably differentiated from a negative sample. It is a function of the variability of the blank sample and the sensitivity of the assay. The LOD may be included in quantitative assays as the lowest standard curve point, although it is not expected to meet the accuracy criteria for the LOQ. In practice, the LOD can be determined by several approaches. One common approach is to calculate the concentration based on the mean value of the absorbance reading for a negative control sample plus two or three standard deviations. To validate method LOD and LOQ, one common practice is to test control samples fortified with analyte at a concentration at or near the targeted LOQ following the method of Keith et al. (1983). The standard deviation is then calculated based on the results of a minimum of five fortified samples. In this approach, the target LOD and LOQ may be considered if the LOD and LOQ values are equal or less than three times or 10 times the standard deviation. An example based on a Cry1F ELISA method in corn grain is shown in Table 7.2. The calculated 3xSD and 10xSD values support the method LOD and LOQ (0.25 and 0.5 ng/mL, respectively) for Cry1F protein with corn grain tissue. Use of 3 standard deviations to determine LOD and 6 or 10 Standard deviations for LOQ is not the only approach to determining these characteristics of the assay. Alternatively, the LOD and LOQ can be determined experimentally, which may provide a better estimate of the real performance of the method.

7.2.2 Specificity

Due to the nature of immunoassay, the antibodies used in the assay may cross-react with other transgenic or endogenous proteins in plant matrices. It is important that the method specificity is validated thoroughly before its application. Evaluation of assay specificity during assay development includes cross-reactivity testing with other known GE proteins that it might be encountered during use and the assessment of matrix effect. Matrix effects may be determined using matrix-matched extracts, which can be used to demonstrate that there is acceptable accuracy across the range when subtracting matrix blanks (using spiked samples).

7.2.2.1 Cross-reactivity with purified proteins

Many commercialized GE crops express multiple transgenic proteins. Some stacked (coexpressed) GE products may contain up to 8 GE proteins (https://www3.epa.gov/pesticides/chem_search/reg_actions/pip/smartstax-factsheet.pdf). It is important to ensure the assay has a demonstrated specificity prior to the implementation. Access to pure protein or reference samples from multiple GE crops will rarely be possible, and this should be part of the design and manufacturing process. In addition, in many situations (e.g., regulatory studies) where a seed or plant of a known origin is being assayed, any presence of other interfering proteins is unlikely. Where necessary, a cross-reactivity experiment designed to demonstrate the specificity of an assay should include all other relevant GE proteins that may be present in samples. For proteins labile to proteases, multiple forms may coexist in the sample, these forms should be evaluated in a cross-reactivity study. For example, full-length Cry1Ac protein and truncated Cry1Ac core toxin coexist in WideStrike cotton

TABLE 7.2 Cry1F ELISA method LOD and LOQ validation summary.

Tissue	Spiked Cry1F (ng/mL)	Average recovery (ng/mL)	SD	3× SD	Target LOD ng/mL	10× SD	Target LOQ ng/mL
Corn grain	0.50	0.51	0.035	0.11	0.25	0.35	0.5

Note: Five samples were tested for validation each containing 15 mg grain tissue, which was spiked with 1.5 mL of 0.5 ng/mL Cry1F in assay buffer.

tissue samples, both forms were included in method validation experiments for evaluating assay specificity (Shan et al., 2007).

One way to test cross-reactivity is to assay the target protein and each nontarget protein at concentrations from 0 ng/mL to 10,000 ng/mL serially diluted in the assay buffer. A dose-response curve for each protein is generated and compared with the target protein standard curve and the cross-reactivity is calculated as follows (Shan et al., 2007).

First, the concentration-response curves are obtained by plotting the absorbance against the logarithm of protein concentration, for example using a four-parameter logistic equation:

$$y = \{(A - D) / [1 + (x/C)^B]\} + D$$

where A is the maximum absorbance at infinite concentration, B is the curve slope at the inflection point, C is the concentration of analyte giving 50% responses (RC_{50}), and D is the minimum absorbance for no analyte.

Then, the cross-reactivity (CR) values are calculated based on obtained RC_{50} using the following formula:

$CR\% = (RC_{50}$ of nontarget protein/RC_{50} of target protein$) \times 100$.

Based on the assay, a different curve fit may be appropriate, for example, a 5-parameter curve fit. Once a nontarget protein is determined as cross-reactive, its potential interference with the quantification of the target protein needs to be assessed in the intended sample system. The significance of interference depends on two key factors: the degree of cross-reactivity and nontarget protein level in the sample. Schmidt and Alarcon (2011) suggested to assess the assay accuracy at LLOQ and ULOQ of the assay with specified amounts of the nontarget protein. If the concentration of the target protein in the presence of the nontarget protein differs by $\geq$25% from the concentration of the target protein alone, the interference by this nontarget protein is considered as significant. Once a significant cross-reactivity by a coexisting nontarget protein is found, the assay may still be used for quantification by applying a calibration factor if the dilutional agreement of assay in precision test with positive samples meets the validation criteria. The CV of the adjusted results from several dilutions of a single sample extract is $\leq$20%. In many applications (such as studies on known materials), cross-reactivity is not an issue, as the cross-reactive protein would not be present in the assayed materials.

7.2.2.2 Matrix effects/interferences

Due to the nature of plant samples, compounds in the tissue extracts/matrices may interfere with immunoassays used for protein quantitation. This needs to be properly assessed during method validation. Endogenous proteins present and other compounds in plant matrices and nonspecific binding of plant matrix components to the target protein or antibodies are two main causes for such effects. To assess the matrix effects in an assay, standard curves are prepared in assay buffers with different concentrations (e.g., 100%, 33%, and 11% or 1X, 3X, and 9X dilutions) of negative control sample extracts (matrix) for the plant tissue of interest. The matrix-containing standard concentrations were interpolated from a nonmatrix standard curve run on the same ELISA microtiter plate. A difference of greater than, for example, 15% between the measured (a nonspiked standard curve used to interpolate the matrix-spiked standard concentrations) and theoretical values for all standard concentration levels might be considered indicative of a potential matrix effect (Shan et al., 2007); this value will depend on the variability of the method as uncertainty effects must be considered. If a matrix effect is observed only at lower dilutions, a minimum dilution of the tissue extract that does not have matrix effects should be specified. For example, tissue-dependent matrix effects were found in cottonseed at 3X dilutions and in leaf, bolls, and whole plant at 1X dilutions. Therefore, a minimum 2X dilution is recommended for tissues of leaf, bolls, and whole plant to minimize the matrix effects, while a minimum 4X dilution is applied for a cottonseed matrix (Shan et al., 2007).

One other approach to manage matrix effects is to use matrix-matched standard curves for quantitation. In the example described above, instead of applying a minimum dilution of 2X or 3X to cotton leaf extracts, the standard curve could be prepared in undiluted leaf extract. This option would require each tissue type having its own matrix-matched curve for analysis, which can be challenging when multiple tissues need to be tested or when the negative sample supply is limited. In addition, matrix matching means that dilutions of the matrix would be necessary for each level of dilution of the extract, and this can add complexity and add to the uncertainty budget of the assay.

7.2.3 Extraction efficiency

To accurately quantify transgenic protein levels in a sample, one needs to ensure the target protein is properly extracted in a form suitable for measurement from plant tissues. Some GE proteins may not be fully extracted in single extractions. Thus, during the method validation, the extraction efficiency of a method needs to be determined to ensure it meets the quantification needs.

To measure extraction efficiency, typically three to five extractions are performed on transgenic plant tissues known to express the target protein (Shan et al., 2007; Schmidt and Alarcon, 2011). First, the extraction buffer is added to the tissue sample and extracted with the buffer. Following extraction and centrifugation, the extracted solution supernatant is removed from the sample container, another volume of buffer is then added to the tissue, and the extraction process is repeated. This procedure is repeated multiple times until five consecutive extractions are completed. The concentration of GE protein in each extraction is then determined. The final pellet can be extracted with Laemmli sample buffer and evaluated using Western blot to confirm only negligible protein remains unextracted. The apparent efficiency of the tissue extraction process is determined by a comparison of the target protein in the first extract with the total target protein in all five extracts. As a common practice, a 70% or greater of extraction efficiency is desired; however, a lower efficiency may be considered acceptable if the relative protein amount extracted by the method is consistent (≤20% CV).

7.2.4 Accuracy

To determine the method accuracy, ideally known concentrations of purified standard reference protein are fortified into negative plant matrices, and then these samples are extracted and measured following the procedure of the method. However, proteins may be denatured by this process, and an alternative process is to spike plant extracts. In general, at least three fortification concentrations are used for the testing of each tissue type: at or near the LLOQ and ULOQ, and the mid-point of the standard curve with minimum recommended dilutions. The fortified concentrations are served as theoretical value, and the method accuracy is determined by comparing the measured value with the fortified value, which is expressed as percent recovery. At least three biological replicates for each fortification level should be performed (ICH, 2005). The mean recovery value at each fortification level should be between 70% and 120% with a CV less than 20% (CAC/GL 74, 2010; Lipton et al., 2000). However, lower recovery values (i.e., below 70%) may be considered acceptable if they are consistent (with a CV<15%). By way of example, for the Cry1Ac method validation (Shan et al., 2007), an aliquot of negative control leaf tissue was fortified with Cry1Ac protein solution (at 0.1, 0.25, and 0.8 μg/g dry weight tissue) and then extracted with assay buffer. Five sets of experiments were performed and analyzed. A 2X dilution was applied for leaf extract analysis. The percent recovery was calculated as the average of all replicate concentrations divided by the fortification concentration. The fortification recoveries are summarized in Table 7.3.

TABLE 7.3 Summary of accuracy results for cotton leaf tissues.

Fortification level		Recovery rate (%)			
ng/mg	ng/mL	Mean	Range	CV%	*n*
0.8	8	81	74–92	8.1	7
0.25	2.5	77	65–89	13.0	7
0.1	1 (LOQ)	84	77–93	4.9	10
0.1–0.8	1.0–8.0	81	65–93	9.0	24

7.2.5 Precision

To validate the precision of a method, generally two approaches are used: (1) testing a fortified negative tissue matrix by different analysts on multiple days or (2) testing the agreement of a series of dilutions of extract (*Dilution Linearity*) from positive samples on the same plate.

7.2.5.1 Fortified sample approach

Like accuracy validation fortification samples, the negative sample extracts are fortified with known concentrations of standard protein at LLOQ, ULOQ, and mid-point of the standard curve. Each level of the fortified extract is analyzed by different analysts on the same day. If the fortified extract samples are stable, the same samples can be analyzed on different days by the same or different analysts. The mean measured concentration, standard deviation (SD), and percent coefficient of variation (% CV) are calculated for each sample on each day by different analysts or on different days by the same analyst.

The acceptance criteria for the precision across days and analysts should be within 20%–25% CV. Table 7.4 is an example of precision summary results for Cry1Ac ELISA using cotton whole plant tissue in our laboratory. The overall CVs represent interday and interanalyst precisions.

7.2.5.2 Dilution linearity

The second approach to evaluate precision is the dilution linearity test, which is designed to demonstrate whether the method gives equivalent results regardless of sample extract dilutions (if the OD value of samples falls within the standard curve range). Positive tissues are required for the test, and serial dilutions of extracts from each positive tissue are assayed on the same plate. Ideally, a minimum of three dilutions resulting in values that fall within the quantitative range of the standard curve are required to assess the dilution agreement. If applicable, at least one dilution above the ULOQ of the assay and one below the LLOQ of the assay are desired. If a minimum dilution is recommended due to matrix effects, any dilution that is less than the minimum dilution should be excluded from the study. Table 7.5 is an example of good dilution agreement results of maize grain, leaf, and root samples using a Cry35Ab1 ELISA.

7.3 Independent laboratory validations

In the United States, plant-incorporated protectants (PIPs) are regulated as pesticides by the EPA. GE crops expressing insect-resistance traits are considered PIPs and methods for detection and quantification are required to comply with EPA guidelines for pesticide residue analytical methods, including an ILV (US EPA OPPTS 860.1340). The requirements for an ILV are described in Pesticide Registration Notice 96-1: Tolerance Enforcement

TABLE 7.4 Summary of assay precision results in cotton whole plant samples.

Fortified protein (µg/g DW)	Day 1			Day 2			Mean (µg/g)	SD	CV %
	Analyst 1	Analyst 2	Analyst 3	Analyst 1	Analyst 2	Analyst 3			
0.8	0.60	0.75	0.75	0.90	0.59	0.90	0.74	0.094	12.6
0.4	0.36	0.43	0.41	0.43	0.33	0.43	0.38	0.045	12.0
0.1	0.08	0.096	0.079	0.09	0.11	0.09	0.091	0.009	10.4

TABLE 7.5 Dilution linearity study summary of Cry35Ab1 ELISA for maize samples.

Tissue	Number of quantifiable dilutions	Mean ng/mg	Stdev ng/mg	% CV	Range ng/mg
Grain	5	6.79	0.70	10.3	5.73–7.45
Leaf	5	93.6	6.58	7.04	84.3–100
Root	4	22.7	2.40	10.6	19.8–25.0

Methods—Independent Laboratory Validation by Petitioner (EPA, 1996), which includes specific requirements for the laboratory performing ILV trial and how to conduct the study including the equipment, instruments, and personnel specifications. In general, an ILV study includes at least one set of samples or one trial, which consists of two control samples, two control samples fortified at the LLOQ, and two control samples fortified at another level proposed by the developer such as 2X the LLOQ. Specific requirements to judge pass or fail the ILV are defined in the guideline.

7.4 Qualitative method validation

Qualitative analytical methods are routinely used in the agricultural industry to indicate whether the analyte response in a sample can be interpreted as present or absent. Qualitative methods for plant tissue testing are generally preferred for routine testing because in most cases they can be more easily adapted to a high throughput testing workflow. Like quantitative analytical methods, a qualitative method also requires validation (Macarthur and von Holst, 2012). ISO Technical Specification 16393 gives guidance on characterization and validation of qualitative (binary) methods (ISO/TS 16393, 2019).

The performance characteristics requiring validation for a qualitative method are very similar to a quantitative assay (Table 7.1). Varying circumstances will often dictate what resources are available for validating a method, so an attempt is made in the following sections to provide alternative but practical approaches to conducting these validation experiments.

7.4.1 Specificity/selectivity

Specificity and selectivity assessments evaluate if the qualitative method is specific for the intended analyte and can detect the analyte within a complex matrix without positive or negative interference.

7.4.1.1 Specificity

When testing for method specificity transgenic proteins other than the targeted protein are spiked at high concentrations in extract buffer, and the samples are observed for a signal greater than expected for extract buffer alone. Other transgenic traits commonly combined or stacked with the targeted trait protein should be evaluated for potential cross-reactivity. This is normally carried out by the manufacturer, and so is not a concern in routine use. If such a study is required, protein samples at the upper limit of detection (e.g., 10 mg/L) are prepared in extraction buffer, and then the spiked extraction buffer and non-spiked extraction buffer are dispensed in triplicate. For a plate-based method, one calculates and compares the mean optical density (OD) responses for both spiked and nonspiked samples. The results of spiked samples can be calculated as % cross-reactivity. If cross-reactivity is detected, the severity of impact must be evaluated. If cross-reactivity is $\leq$110% of the nonspiked extraction buffer, then the method can be considered specific for the target protein.

7.4.1.2 Selectivity

A selectivity experiment for validating a qualitative method should involve the evaluation of matrix interference. Matrix effects within a plant extract usually cause signal suppression as compared with the response observed in extraction buffer, but occasionally enhancement is observed as well. Therefore, a common approach for selectivity assessment is to use a target protein reference standard, if available, and create serial dilutions curves in extraction buffer and plant tissue extract. Such studies are not usually required in routine use but are routinely performed during the manufacture of qualitative assays.

If it is desired to do so, it is recommended that at least 12 calibration levels be prepared ranging from 10 mg/L to 0 mg/L depending on the starting concentration of the reference standard. This assay range may be further optimized but must include the expected range for typical target protein expression. The extracts should be prepared by blending pooled seed or plant tissue from a nontransgenic source. For a more robust experiment, additional pooled extracts can be prepared each from at least two different sources of transgenic seed not containing the target protein. A total of four calibration curves (one in extract buffer and three in plant extract) are prepared and dispensed in duplicate for testing.

For a visual comparison plot the results of all calibration curves using an appropriate software package. Ideally, all plant extract curves will overlay on the spiked extraction buffer curve indicating no matrix effect. If this is not the case, then the impact of the matrix effect will need to be determined. One simple way to evaluate the extent of the matrix interference is to simply calculate the % bias of the OD response by calibration level between the plant extract and extraction buffer curves. If the calculated % bias is within $\pm$20%

TABLE 7.6 Summary of OD values for a matrix effect with spiked corn seed extract.

Conc. (ng/mL)	Extraction buffer	Seed extract	% Bias
0	0.067	0.068	NA
1	0.125	0.12	−4.0%
5	0.31	0.25	−19.4%
10	0.508	0.45	−11.4%
25	1.281	1.05	−18.0%
50	1.699	1.45	−14.7%
100	2.018	1.8	−10.8%
250	2.281	2	−12.3%
500	2.39	2.1	−12.1%
1000	2.45	2.16	−11.8%
5000	2.5	2.19	−12.4%
10,000	2.52	2.2	−12.7%

for seed extracts, then the extent of the matrix effect is deemed insignificant (Table 7.6; Fig. 7.1).

More significant matrix effects, especially in ELISA, can be minimized by using a greater ratio of extraction buffer to plant tissue material. However, this will reduce the overall sensitivity of the method. Minimizing matrix effects can also be accomplished through optimization of the extraction buffer by increasing detergent, salt content, or addition of protein like bovine serum albumin to mitigate nonspecific binding of extract components with the immobilized capture antibody and/or target protein. In the case of an ELISA, one option could be to increase the sample extraction incubation time or consider redesigning the ELISA assay.

Of importance is whether the qualitative method is suitable for its intended purpose. If method performance is good enough for determining the presence or absence of targeted transgenic protein with low false-positive or/and false-negative rates, then observed matrix interference may be justified and easily accommodated during routine sample testing.

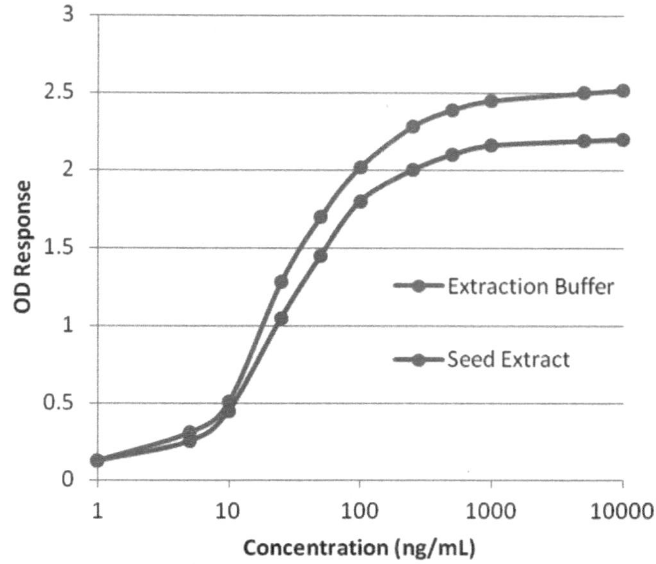

FIGURE 7.1 Example of matrix effects.

7.4.2 Sensitivity, establishing the cut-off value (fixed vs. floating)

7.4.2.1 Sensitivity

The sensitivity of a qualitative ELISA method can be investigated by using the following or similar protocol. For seed testing, it is recommended that at least 30 single seeds are selected from at least three different negative seed lots so that a total of 90 single seeds are tested. Typically, testing of nontransgenic seed lots is also done, but a more robust approach may be to test a combination of sources including nontransgenic seed and transgenic seed not containing the target protein. These sample sources are considered as negative control samples for validation.

A total of at least 90 OD values will be generated from the selected negative-control seed batches and used to calculate the mean OD response and SD. Typically, the LOD is calculated as the mean OD response + 3X SD (and up to + 5X SD in some cases). However, a more practical approach may be to define a threshold or cutoff value above the LOD so that any result equal to or above this threshold will be considered as a positive result while a result below the threshold is a negative result.

An alternative approach to establishing sensitivity is to use a protein reference standard and perform a serial dilution in an assay buffer or matrix. The concentration or response value of the highest diluted level that can be clearly distinguished from the background instrument response of a nonspiked assay buffer or matrix is a more practical estimation of sensitivity. This approach should be considered when validating lateral flow strips.

If no protein reference standard is available, then a series of extracts can be prepared by diluting transgenic seed/tissue with nontransgenic seed/tissue. In this case the actual target protein concentration is unknown, so the ratio of transgenic seed/tissue to nontransgenic seed/tissue should be reported. Once the limit of detection is identified, several replicates of extract should be tested to confirm whether the lateral flow strips consistently report positive results. Ideally, at least three lots of lateral flow strips should be used.

7.4.2.2 Defining the threshold value

For a qualitative method, it is important to establish a threshold or cutoff value to determine whether a sample result is considered as positive for the targeted protein analyte or whether it is considered a negative result. There are more rigorous statistical procedures that can be used for establishing this threshold value, but these will not be discussed. This section will focus on simple approaches that can be used for estimating the threshold value, especially when there are limited lots of seed or tissue initially available for the method validation effort.

For trait purity testing of individual seed/tissue the integrity of the result will be based on the acceptable false-positive and false-negative rates for the analytical method. There are several ways to define the initial threshold value, and they can all be based on establishing sensitivity data. A few options are presented here, and each has its advantages and disadvantages that should be considered when determining whether an analytical method is fit for purpose.

- The threshold value can be based on the calculated LOD (mean OD of negative controls + 3SD). While false-negative occurrences will be minimized, there is a greater probability for generating false-positive results, especially when there is an unanticipated increase in background response, for example, due to lot-to-lot kit variability, and unexpected matrix interference).
- The threshold value is established using the practical estimation of sensitivity as described previously. The threshold is set at the detection signal derived from the highest diluted sample of a serial dilution curve, in which a detection response is clearly distinguishable from the negative control signal. In this case, the serial dilution should be prepared in seed or tissue extract to accommodate any matrix effects. This approach relies on the sensitivity of the test kit, but not on the typical expression range expected for transgenic seed or tissue.
- The threshold value can be based on the mean OD response calculated from the negative control samples and multiplied by a factor of ≥ 3. In this case, it is expected that there will be a significant difference between a detection signal from a negative control and that observed from typical

expression levels of the targeted protein in transgenic seed/tissue. Otherwise, setting the threshold value in this fashion will increase the frequency of false-negative results and could cause a considerable amount of confirmatory testing or retesting during routine sample analysis.

Whatever approach is used to establish the threshold value, it is important to understand both the test kit sensitivity and the typical expression range expected for transgenic tissue that confers relevant trait efficacy in the field. Obviously, a test kit must be sensitive enough to reliably detect true positive results, so that the detection threshold value can be ideally set to capture most, or all positive results expected from the normal expression range. Furthermore, the typical expression range for a given crop and tissue may not be immediately known when multiple lots are unavailable during the method validation stage. In these cases, the comparison of results to another established and validated test method is useful. In addition, it is highly recommended that relevant performance trending be monitored when routine sample testing commences with the newly validated method. This can be considered as an in-study validation phase and should be helpful to determine whether the initial threshold value may need to be adjusted as more data are analyzed.

One additional consideration should be whether a fixed or floating threshold value should be implemented for a qualitative method during routine testing. A floating threshold value is a plate-based cut-off and can be advantageous for normalizing with-in lot variability (e.g., differences from one ELISA plate to the next) and lot-to-lot variability in test kits and supporting raw materials required for routine testing. It can also aid in normalizing variability among analytical instruments. A floating threshold value is calculated for each ELISA plate, and it is usually dependent on the mean OD value derived from the negative control samples. For example, the mean OD response of three negative control wells is 0.05 for ELISA plate #1, and so the corresponding threshold value is 0.15 OD (0.05 OD X 3). The negative control mean response is 0.10 for ELISA plate #2, so the threshold value is 0.30 OD (0.10 OD X 3).

7.4.3 Hook effect

During method validation of plate-based immunoassay kit, it is important to determine if a hook effect can be expected. A hook effect can occur in sandwich ELISA methods where both the sample extract and antibody conjugate are added simultaneously. Extremely high concentrations of the targeted protein essentially saturate the immobilized capture antibody sites on the solid phase as well as exhaust all available antibody conjugate in solution, and the sandwich is prevented from forming. This situation can cause ambiguity in results interpretation because extremely high levels of a target protein are easily interpreted as a low-level response. Hook effects are determined when ELISA assays are designed (e.g., by the manufacturer).

Many qualitative methods for plant tissue testing are designed for high throughput analysis and may use simultaneous addition of sample extract and antibody conjugate to reduce the total assay time. This is often a validation component that is investigated by a kit manufacturer, so it may be possible to have this information provided. However, if this information is unavailable, then it is best to conduct an experiment for potential hook effects and evaluate the potential impact on routine testing.

Selectivity data may be used to determine if a hook effect exists, if high enough transgenic protein concentrations were used. If conducting a separate experiment, then one needs to spike either assay buffer or nontransgenic seed or leaf extract with a high concentration of the protein reference standard (e.g., 10 mg/L). If purified reference protein is unavailable, one needs to identify a sample lot with the highest expression level available and prepare an extract. Performing a serial dilution in extraction buffer or plant extract is useful so that at least 8–12 calibrator levels are created. Plotting the results (OD) can be helpful for a visual assessment. For a normal curve, the high calibrator levels should continue to increase in OD response (highest calibrator leads to highest OD response) or a plateau be observed. With respect to the entire curve, a hook

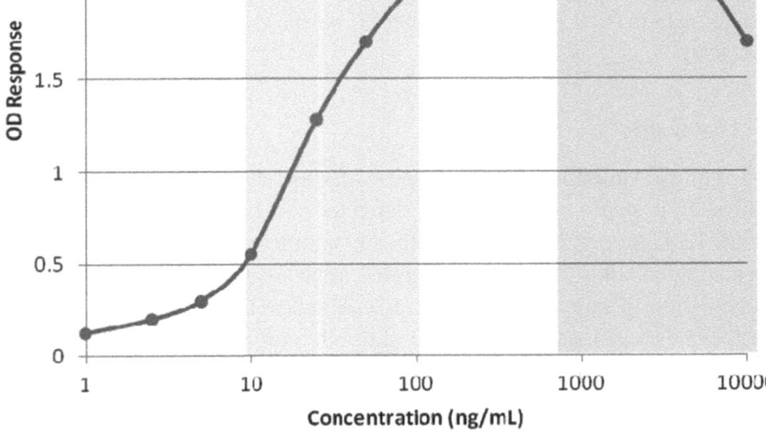

FIGURE 7.2 Example of hook effect.

effect is recognized by a continual and significant decrease in OD response as the concentration level increases at the high end of the curve (Fig. 7.2).

If a hook effect is prominent, the impact on eventual routine testing should be evaluated. For a qualitative method, the impact on data integrity may be less significant than for quantitative methods. If the hook occurs at detection levels far above what is expected for normal expression levels, justification exists that the current methodology is most likely fit for purpose with no alteration needed. Even if an unusually high expressing sample is encountered, the response will most likely be well above the threshold value. If the hook effect appears within the necessary working range of the method, special consideration should be made for high response samples, or the method should be altered to reduce or eliminate the hook effect.

The typical solution to eliminating a hook effect is by increasing the dilution of the sample extract, but this is at the expense of assay sensitivity. Hook effects are rare in commercially produced assays, but if two dilutions are used it can be assessed whether the higher dilution level gives a lower signal level, which would indicate such an effect. Another possibility is to redesign the method. One example would be to eliminate the simultaneous addition of sample extract and antibody conjugate and use a sequential format. A separate sample extract incubation time can be introduced with a subsequent washing step followed by antibody conjugate addition.

7.4.4 Accuracy

For a qualitative analytical method, accuracy is usually represented by the rate of false-positive and false-negative results, which is determined based on the established threshold value. For determining method accuracy, transgenic seed lots containing the targeted protein will be used as positive control samples, and seed lots (nontransgenic and/or transgenic) free of the targeted protein analyte will be used as negative control samples. A false-positive result is documented when a detection signal is observed at or above the threshold value for an individual negative control seed/tissue. A false-negative result is recorded when the detection signal falls below the threshold value for an individual positive control seed/plant tissue.

Establishing reliable analytical method accuracy during the validation is highly dependent on the purity of the seed lots or plant tissue used as the positive and negative control, especially in methods used in supporting individual seed/plant tissue testing. It is recommended that the highest quality seed or tissue be used. If possible, one should use high purity transgenic lots containing the targeted protein that has been established by other methods (DNA testing, herbicide bioassay, or immunoassay).

Some of the wells in a 96-well ELISA plate layout can be used to include positive and negative controls, or in the case of a quantitative assay, standard curves. An assessment of ability to detect a positive sample can be done by including nontransgenic seed extract spiked with a reference standard to a concentration 1 or 2X of the defined threshold (or LOD) value. In addition, seed extract can be spiked at the middle and upper portion of the expected working range for a total of three levels to be evaluated. The results should be reviewed for potential false-positive and false-negative results.

If false-positive and/or false-negative results are indicated with a plate-based method, it is prudent to confirm these values using an alternative validated lateral-flow strip method if available, and this should be performed immediately with the same fresh extracts.

If a high purity seed containing the transgenic protein (or protein reference standard) is unavailable, it is still possible to evaluate method accuracy for an ELISA plate-based method by using a previously verified or validated lateral flow strip (LFS) as a confirmatory reference. A minimum of three seed/tissue lots should be used with 90 individual seeds/leaf punches per lot to be tested. The trait purity score should be calculated for all three lots.

$$\frac{\#\text{ of Positive Wells}}{\#\text{ of Tested Wells}} \times 100 = \%\text{ Trait Purity Score}$$

The same extracts should be tested by LFS, and the results from both plate and LFS can be compared for discrepancies. Ideally, results that fall below the threshold value should be confirmed by a negative result on a corresponding LFS test. If a higher trait purity score is observed using the LFS, the initial threshold value may be set too high and could be lowered to better match the LFS trait purity score. If no LFS method is available, other testing technologies (e.g., DNA, herbicide bioassay) or comparison with a third-party lab can be considered.

Establishing acceptance criteria for rates of false-positive and false-negative results is subjective and will depend on what is considered an acceptable risk. Acceptable rates of false-positive and false-negative results need to be determined on a case-by-case basis as there are several factors that need to be considered, including but not limited to the intended purpose of the method, defined crop purity specifications followed by a specific organization, and assumed risk to the business and the stakeholder or customer. False-negative and false-positive rates may need to be monitored for trends, and the qualitative method may need to be revalidated if a change is deemed necessary. It is the opinion of the authors that a more conservative approach should be followed that would allow higher rates of false-positive results than false-negative results. A higher number of false-positive results will cause more retesting or confirmatory testing to be conducted while a higher number of false-negative may pose higher stewardship challenges for products potentially entering process streams desiring exclusion of the GE event.

7.4.5 Precision

Precision evaluations provide information on expected method variability due to random error and are usually expressed as coefficient of variation (%CV) or standard deviation. Depending on the situation, there are different aspects of method performance that may require investigation by precision-based validation experiments. These are typically categorized as repeatability and reproducibility.

Repeatability refers to the use of fixed testing conditions in which the same sample(s) is tested independently. For example, a repeatability experiment is normally conducted within the same laboratory under the same conditions, and often testing is performed by the same analyst. Repeatability experiments are usually conducted within the same day but can extend over the course of several days if desired. Repeatability can be demonstrated by analyzing, on the same day, replicates of a common extract in which the volume is sufficient to dispense across a total of three plates. All other variables should be uniform, including reagent lots, equipment used, and the analyst. Intraplate and interplate precision can be investigated by using a pooled extract of a source containing the transgenic target protein dispensed in 90 wells per plate and across three plates. The mean OD response and corresponding %CV can be calculated per plate for intraplate precision. The mean OD responses per plate can be calculated to determine the interplate %CV. The qualitative method is most often deemed acceptable if the %CV is $\leq$20%.

For individual seed or plant tissue testing methods using LFS, it may be more meaningful to calculate the % trait purity for each plate. The mean % trait purity score is determined, and the CV can be calculated. In this case, a different acceptance criterion should be used, which defines the maximum variability allowed in the % trait purity score. For example, a difference of 1 seed in 90 seeds results in a 1.11% change in the trait purity score, so a potential acceptance criterion could be $\leq$1.11% in the trait purity score. For three plates of 90 replicates, each test would allow for three failing extracts among the 270 replicates tested.

Reproducibility primarily refers to measuring precision using the same method but comparing results across different laboratories, and it is often used to measure method variability associated with different analysts and equipment. Reproducibility can be conducted in a similar fashion as described for the repeatability experiment.

7.4.6 Ruggedness

Testing the ruggedness of a qualitative method is important in determining the reliability of generated test results and should be considered especially for immunoassay methods.

Validating the ruggedness of an analytical method is performed by making small but planned deviations to the experimental conditions stated in the procedure. Changes to experimental conditions should be treated one parameter at a time, and the generation of results should be repeated each time a planned change has occurred. It is generally good practice to test multiple kit lots during a validation. For kit lot variability spanning multiple days, 90 individual seed/tissue samples can be used from a high purity transgenic lot to generate trait purity scores for the targeted protein analyte. The same transgenic lot is used for different kit lots. The trait purity scores will be compared for agreement within a specified tolerance of $\pm 1.11\%$ (i.e., the difference of a single seed result among 90 seeds), and the ruggedness validation will be deemed successful if this acceptance criterion is met. If multiple kit lots are to be tested simultaneously or within a short period of time, sufficient extract can be prepared from a blended seed pool. Aliquots of the same seed extract will be dispensed across all kit lots and the mean OD response can be compared to determine lot-to-lot variability within the kit.

7.5 Assay verification

Internal assay verification (lot to lot verification; kit to kit verification).

Method verification is often used to confirm that an analytical method continues to perform as determined during the validation stage. Verification of the analytical method demonstrates whether the assay performs as expected in the hands of the end user and/or if laboratory competency has been established to perform the method as intended. Typically, assay verification consists of evaluating a subset of key analytical performance characteristics to generate relevant data and avoid the time, labor, and consumables necessary for repeating the validation process.

In general, method verification is conducted when implementing a method that has been previously validated internally within the laboratory or when a validated kit (e.g., lateral flow strip or ELISA plate) has been purchased from a vendor/manufacturer for the first time. In addition, method verification might be performed when a new lot of raw material is used. For example, a new lot of lateral flow strips or a different lot of vendor-supplied antibody conjugate shipped with an ELISA kit might require verification that the method performance characteristics can be reproduced as established during the validation stage, if experience shows there is lot to lot variation. Method verification may be an integral part of the assay, as when continued verification of the assay is part of the process. Method verification may be warranted under the following conditions as well:

1. When there is a new source of raw material(s) or supporting consumables required to perform the method.

2. When establishing a new laboratory facility or reestablishing laboratory operations after moving equipment to a different facility.

While there will be situations for which it is unclear whether to perform assay verification or validation, in this section the focus will be on verifying the method performance of new kits and to manage lot-to-lot variations observed for critical reagents and consumables.

7.5.1 Kit verification

A kit verification would be performed when an end user has purchased a manufacturer's testing kit. In these situations, the kit manufacturer will usually have validated the method, and the end user will follow the manufacturer's kit protocol during the verification process. Deviations to directly following the protocol are most likely outside of the scope of the manufacturer's validation and may require revalidation instead of assay verification. For example, if a quantitative kit is to be used qualitatively, then an in-house validation process would be performed.

Useful guidance documents are available and can be used to develop a method verification procedure (USDA/GIPSA 2004). The GIPSA verification methodology can be easily adapted and applied to plant tissue testing when using either plate-based ELISA kits or lateral flow strips. Kit verification should include data generation using the following method performance characteristics: testing for false-positive and false-negative results; repeatability, and possibly detection of a hook effect or detection range.

It is recommended that the end user should review the validation report (if available) provided by the manufacturer for the intended application. Typical performance specifications and acceptance criteria should be included for method sensitivity, assay format, assay time, accuracy/precision, and any relevant stability measurements for the protein analyte. Due to the proprietary nature of some kits, the validation report may not be readily available from the manufacturer. In these cases, it may be necessary to glean as much information from the kit protocol, obtain the validation information by special request, or search for available literature associated with the protein analyte.

Before presenting procedures for verifying kits, terminology should be briefly presented. For any method verification, a set of control samples are required for data generation. In general, this will include the following:

1. Negative Control Samples: Samples of seed or plant tissue that do not contain genetically modified seed and plant tissue.
2. Positive Control Samples: Samples that are supplied by the vendor or from seed/plant tissue lots containing the target protein of interest.
3. Fortified Samples: Samples of seed or plant tissues that contain a predetermined concentration of genetically modified plant material or purified protein reference material.

In addition, the following concepts are important for assessing method verification when using a manufacturer's kit:

Detection Threshold: The lowest concentration of a genetic protein that can be reliably detected by a LFS or ELISA plate assay. This concentration will be specified by the manufacturer in product insert.

Hook Effect: A false-negative result with an LFS or ELISA plate assay due to very high concentrations of protein analyte.

7.5.1.1 Verification of plate-based ELISA kits

The following verification procedure may be used for plate-based ELISA kits, and all experimental processes including sample preparation and processing of sample extracts will follow manufacturer's instruction from the test kit.

1. False-positive testing

   At least 30 independent analyses will be conducted using at least three different ELISA kit lots with at least 10 wells containing negative control samples for each plate. At least two wells shall contain a positive control sample. All test results must be negative for the protein of interest. If there is a positive result (refer to vendor's protocol), a repeat study shall be performed.
2. False-negative testing

   At least 30 independent analyses will be conducted using at least three different ELISA kit lots with at least 10 wells containing fortified samples for each plate. Fortified samples should be prepared so that they are 1–2X greater than the claimed or intended detection threshold. All test results must be positive for the protein of interest. A fortified sample may be created from a purified protein reference standard and/or an extract from a blended pool of positive seed or plant tissue. If these materials are not available, a positive control from the vendor may be used.

   If there is a negative result, a repeat study must be performed. Use of a protein reference standard spiked into a conventional seed extract is recommended. If the repeat study fails, the kit manufacturer should be contacted if the use of this vendor's ELISA kit is necessary.
3. Repeatability

   Inter-plate precision is determined using at least 30 analyses of an extract composed of a 100% positive seed or tissue. It is recommended that a blended pool be used to create a sufficient volume of extract for this study. Alternatively, a sample extract may be prepared using a negative control fortified to a concentration expected of positive seed or tissue. Three ELISA plates are used for this experiment and contain the identical extract in at least 10 wells using at least three test lots. The extract results from all three ELISA plates will be combined, and the mean OD response and standard deviation will be calculated. The CV will be calculated to evaluate precision. When a plate reader is not available, the repeatability experiment cannot be conducted.

4. Hook effect or detection range

If a hook effect is noticed or documented by the manufacturer for a given trait protein, then the extent of the hook effect should be evaluated, and the detection threshold should be known or established. The detection threshold is the lowest concentration of a genetically modified protein that can be reliably detected by ELISA. This concentration will typically be specified by the manufacturer in the product insert or can be independently established if the protein reference material is available.

It is recommended that a serial dilution (see note b below) be prepared in which at least six replicates at three different concentration levels should be evaluated. For example, concentrations at or near the detection threshold, approximate middle of detection range, and upper portion of detection range should be used.

a) Use suggested upper limit of protein of interest, if available;
b) Use purified protein and spike into control matrix extract for determining the potential hook effect;
c) Confirm with positive samples to determine whether the hook effect will be a factor for real sample analysis.

5. Acceptance criteria

The following acceptance criteria should be followed when assessing the method verification data.

a) False-positive testing: 100% negative
b) False-negative testing: 100% positive
c) Repeatability testing: %CV $\leq$20% using plate reader.

At times, it may be difficult to procure three different ELISA kit lots at one time. When three kit lots are unavailable, different shipments of the same ELISA kit lot may be considered as different test lots. Otherwise, individual ELISA plates from the same lot should be tested but from different packets/boxes, if possible.

7.5.1.2 Verification of lateral flow test strips

The following verification procedure may be used for lateral flow test strips, and all experimental processes including sample preparation and processing of sample extracts will follow manufacturer's instructions from the test kit. As indicated below, it is recommended that three different lots of lateral flow strips should be used for data generation. When procurement of three different LFS lots is not possible then strips from three different packages may be used.

1. False-positive testing.

Thirty independent analyses are conducted using three different test lots with 10 negative control samples for each lot. All test results must be negative for the protein of interest. If there is a positive result, a repeat study is warranted.

2. False-negative testing.

   Thirty independent analyses are conducted using three different test lots with 10 fortified samples for each lot, at the claimed or intended detection threshold. All test results must be positive for the protein of interest. Both purified protein standard and positive plant tissue or either of them will be used for fortification testing. If there is a negative result, a repeat study is warranted.

3. Hook effect or detection range assessment

   If a hook effect is observed or documented by the manufacturer then the extent of the hook effect for the trait protein should be evaluated and the detection threshold should be known or established.
   - Use suggested upper limit of protein of interest if available;
   - Use purified protein and spike into control matrix extract to determine the potential hook effect;
   - Confirm with positive samples to determine whether the hook effect will be a factor for real sample analysis.

4. Acceptance criteria

   The following acceptance criteria should be followed when assessing the method verification data for lateral flow strips.

   a) False-positive testing: 100% negative;

   b) False-negative testing: 100% positive

   c) Hook effect should not be a factor for real sample testing.

7.5.1.3 Critical consumable and reagent verification

It is good practice to ensure that raw materials identified as quality critical consumables or critical reagents are verified as conforming to predetermined requirements before use in sample testing. One of the key pitfalls in immunoassay-based test methods is the loss of consistent analytical method performance due to lot-to-lot variability as kits are consumed during routine testing and different kit lots are introduced into the testing workflow. Therefore, kit lot verifications should be conducted and documented to track any variability observed during sample testing. The following section describes examples of simple procedures for verifying a new lot of ELISA kits or lateral flow test strips and can be easily adapted to best fit the needs of the end user.

a) Lot verification of ELISA kits

A proprietary or commercially available ELISA kit lot is often composed of several consumables and reagents and individual lot numbers will be associated with each. Therefore, it is acceptable that an ELISA kit may be considered as a complete and integrated unit for lot testing and verification. It is also acceptable that lot testing can be conducted on the individual ELISA plate and/or critical reagents. A critical reagent consists of any antibody-derived reagent that is essential for binding and immobilizing the protein analyte of interest (e.g., detection antibody conjugate). The simplest approach in lot verification is to monitor the performance of the integrated ELISA kit and compare new kit lots with a verified lot that is currently being used. When using either approach for lot verification, a certain number of previously verified kits or critical reagents will need to be held in reserve for results compared with the new lot.

1. ELISA kit or plate evaluation.

   One ELISA plate from the new lot is compared with one ELISA plate from a previous lot that was verified for use. A seed or plant tissue sample is selected, which expresses the trait(s) of interest. Ninety individual seed or plant tissue extracts are prepared. In addition, six extracts are prepared, which represent the control samples (two positive control extracts, two negative control extracts, two extracts containing only assay buffer). The first aliquot of each extract is dispensed into the ELISA plate representing the new lot. The second aliquot of each extract is dispensed into the ELISA plate representing the established or previously verified lot.

   The trait purity percentage results from both of the ELISA plates are compared for agreement within a specified tolerance of 1/(total # seeds) *100. For example, a 90 seed test is conducted, and the trait purity score is 97.78% using the established ELISA plate lot. Therefore, the trait purity score of the new ELISA plate lot must agree within ±1.11% (i.e., 100*[1/90] = 1.11%). If the new ELISA plate lot result is out of tolerance, the new ELISA plate lot is deemed not equivalent and the vendor should be contacted for further instruction.

   When the new ELISA plate lot is out of tolerance, retesting is an option; however, it is recommended that the plate comparison should be repeated using twice the number of seeds used for testing in the original evaluation. Ultimately, each instance will be handled on a case-by-case basis, and the decision and justification to use or reject the new kit lot should be documented.
2. Critical reagent evaluation.

   Critical reagents are a component of validated kits and do not normally need reevaluation. A bridging verification would be performed when a critical reagent (e.g., detection antibody conjugate) lot is required that was not originally matched by the vendor with a specific ELISA plate lot. This should be considered the development of a new integrated ELISA kit lot and evaluation should be conducted as described in the ELISA plate or kit evaluation section. For example, a new ELISA kit lot could be created when a lab site has a surplus of ELISA kits in which the antibody conjugate lot has expired. In this case, the extra ELISA plates from the kit could be matched with a new lot of antibody conjugate.

b) Lot verification of lateral flow strips

Lot verification of LFS is very similar to what was described for integrated ELISA kit testing. The approach in lot verification is to monitor the performance of the LFS and compare new lots with a verified lot that is currently being used. When using this approach, a certain number of LFS from a previously verified lot (if available) will need to be held in reserve for results compared with the new lot.

It is recommended that at least six LFS from both a previously verified lot and a new lot be used for comparison. If using six strips from each lot, six seeds or individual plant tissue samples should be prepared and extracted. Three of these extracts will be from conventional stock (absence of the trait protein) and the other three extracts will be considered as positive controls that contain the trait protein of interest. Ensure that an adequate volume of extract is prepared to accommodate two LFS per tube or plate well. Place one strip from the LFS lot to be verified and one strip from a previously accepted LFS lot in each of the six prepared extracts. After LFS results develop, evaluate the strips. For a new LFS lot to be considered as verified the test results from the new lot must match the results from the accepted lot within each extract. The test observations and results should be verified. If LFS results do not match, retesting may be performed and the decision and justification to use or reject the new LFS lot should be documented. In addition, the vendor should be contacted for further instructions.

References

CAC/GL 74, 2010. Guidelines on Performance Criteria and Validation of Methods for Detection, Identification and Quantification of Specific DNA Sequences and Specific Proteins in Foods. http://www.codexalimentarius.org/standards/list-of-standards.

Environmental Protection Agency, 1996. Pesticide Registration (PR) Notice 96-1: Tolerance Enforcement Methods – Independent Laboratory Validation by Petitioner. http://www.epa.gov/PR_Notices/pr96-1.html. http://www.epa.gov/opptsfrs/publications/OPPT_Harmonized/860_Residue_Chemistry_Test_Guidelines/index.html. U.S. Environmental Protection Agency, OPPTS 860.1340.

FDA, 2013. Guidance for Industry: Bioanalytical Method Validation, 2013. http://www.fda.gov/downloads/drugs/guidancecomplianceregulatoryinformation/guidances/ucm368107.pdf. (Accessed 25 March 2015).

Grothaus, G.D., Bandla, M., Currier, T., Giroux, R., Jenkins, G.R., Lipp, M., Shan, G., Stave, J.W., Pantellam, V., 2006. Immunoassay as an analytical tool in agricultural biotechnology. Journal of AOAC International 89 (4), 913–928.

ICH Harmonised Tripartite Guideline, 2005. Validation of Analytical Procedures: Text and Methodology Q2(R1). http://www.ich.org.

ISO 5725, 1993. The International Vocabulary of Basic and General Terms in Metrology, ISO 5725 Accuracy (Trueness and Precision) of Measurment Methods and Results. ISO, Geneva, Switzerland.

ISO 21572, 2013. Foodstuffs–Methods for the Detection of Genetically Modified Organisms and Derived Products–Protein Based Methods. ISO, Geneva, Switzerland.

ISO/TS 16393, 2019. Molecular biomarker analysis – Determination of the performance characteristics of qualitative measurement methods and validation of methods. ISO, Geneva, Switzerland.

Keith, L.H., Crummett, W., Deegan Jr., J., Libby, R.A., Taylor, J.K., Wentler, G., 1983. Principles of environmental analysis. Analytical Chemistry 55, 2210–2218.

Lipton, C.R., Dautlick, J.X., Grothaus, G.D., Hunst, P.L., Magin, K.M., Mihaliak, C.A., Rubio, F.M., Stave, J.W., 2000. Guidelines for the validation and use of immunoassays for determination of introduced proteins in biotechnology enhanced crops and derived food ingredients. Food and Agricultural Immunology 12 (2), 153–164.

Macarthur, R., von Holst, C., 2012. A protocol for the validation of qualitative methods of detection. Analytical Methods 4, 2744–2754.

Official Methods of Analysis, 2005. AOAC INTERNATIONAL, Gaithersburg, MD, Appendix D: Guidelines for Collaborative Study Procedures to Validate Characteristics of a Method of Analysis, eighteenth ed.

Schmidt, J., Alarcon, C., 2011. Immunoassay method validation. In: Shan, G. (Ed.), Immunoassays in Agricultural Biotechnology. Jon Wiley & Sons, Hoboken, New Jersey, pp. 115–138.

Shan, G., Embrey, S.K., Schafer, B.W., 2007. A highly specific Enzyme-linked immunosorbent assay for the detection of Cry1Ac insecticidal crystal protein in transgenic WideStrike cotton. Journal of Agricultural and Food Chemistry 55, 5974-5959.

United States Department of Agriculture/Grain Inspection, Packers and Stockyards Administration, 2004. Performance Verification of Qualitative Mycotoxin and Biotech Rapid Test Kits, Directive 9181.2. http://archive.gipsa.usda.gov/reference-library/directives/9181-2.pdf.

Viswanathan, C.T., Bansal, S., Booth, B., DeStefano, A.J., Rose, M.J., Sailstad, J., Shah, V.P., Skelly, J.P., Swann, P.G., Weiner, R., 2007. Workshop/conference report—quantitative bioanalytical methods validation and implementation: best practices for chromatographic and ligand binding assays. The AAPS Journal 9 (1). http://www.aapsj.org.

Chapter 8

Reference materials and working standards

Ray Shillito[1] and Nancy Brooker[2,3]
[1]*BASF Corporation, Morrisville, NC, United States;* [2]*Recursion Pharmaceuticals, York, Alabama, United States;* [3]*Abbott Toxicology Laboratories, Gretna, LA, United States*

8.1 Background

Many aspects of society are supported in some way by analytical measurements. For example, import testing and export testing are vital to commerce and trade, as are conformity to regulations and health and safety standards. An analytical method used in one location (e.g., export country) must give the same result as one used in another location (e.g., import country) to avoid the potential for litigation and trade disruption. To produce dependable results, a laboratory needs equipment, trained personnel, protocols, and some form of reference material. Reference materials (RMs) have been developed to act as a standard, much as all meter rules are in some way traceable back to a single standard meter. Use of RM brings confidence and repeatability to analytical measurements.

ISO Guide 30:2006 states a Reference Material is defined as, "material or substance, one or more of whose property values are sufficiently homogeneous and well established to be used for the calibration of an apparatus, the assessment of a measurement method, or for assigning values to materials." Additionally, RM is a generic term. Depending on the measurement process where the RM is used, the RM properties can be defined as quantitative (with an assigned value) and/or qualitative, for example, identity of substances or species. Uses of RM range from calibration of a measurement system, assessment of a measurement procedure, assigning values to other materials, and quality control. However, a single RM cannot be used for both calibration and validation of results in the same measurement procedure.

8.2 Uses of reference materials

The use of an RM contributes greatly to the reliability of measurements, and their use is required when following many laboratory quality standards such as

Application of Sampling and Detection Methods in Agricultural Plant Biotechnology
https://doi.org/10.1016/B978-0-323-99293-0.00012-X

ISO 17025. Reference materials are used to calibrate analytical methods, either by comparison of a result to a standard curve produced using an RM, or by calibration of equipment. RMs ensure reliable traceability of measurement results. This is important for the sharing and equivalence of values of measured or assigned properties between testing and measurement/calibration laboratories. They are thus important for method validation, calibration, estimation of measurement uncertainty, training, internal quality control, and external quality assurance purposes (Fig. 8.1). In the area of agricultural biotechnology, RMs are used as a standard to confirm the identity, and in some cases, to quantify transgenic/genetically modified (GM) and nontransgenic (conventional) organisms, specific genetic constructs, and expressed proteins and specific genetic species. One such application is ensuring the accuracy of testing and tolerance thresholds of transgenics in products due to mandatory labeling requirements within some countries. These labeling requirements are primarily based on the detection of DNA and are complex, varying widely from country to country. Quantitative threshold requirements have been put in place in some countries that require quantitative measurements for testing compliance with the regulations.

For these reasons, RM availability has become that much more critical and is required to meet the ever-changing global regulatory requirements for testing, detection, and traceability applications in feed and food. In some cases, a Certified Reference Material (CRM: certified according to ISO guides 30, 31, 33, and 35 and produced in accordance with ISO 17034) is required (European Commission (EC) 2011). CRMs have a very high level of

Significance of Reference Materials (RMs)

Reliability and Repeatability of Analytical Measurement Data ➢ Confidence in decision-making process is altered	Instrumental techniques using modern methods of analysis ➢ Analytical results are directly related to reference materials
National and International Requirements ➢ Comparability & commutability of test results produced by different accredited labs	Purity and Quality of Reference Materials ➢ Measurement reliability, testing accuracy & confidence in analytical results are impacted

FIGURE 8.1 Relation between the attributes of reference material and the impact on measurements and credibility of test results.

metrological characterization and traceability and are offered by accredited organizations.

8.3 What constitutes a reference material?

An RM has specific characteristics with regards to its purity, uniformity, and stability (Linsinger et al., 2001). RMs cover a vast array of substances and analyses, ranging from oils, to metals, pesticides, and other chemicals, fabrics, and in the context of this paper, seed, ground seed, other plant materials, proteins, and DNA.

Different classes of RM and examples are described in Box 8.1. CRM is the best-characterized material and is the most traceable back to International Standard units. In some cases, a material that is well characterized but not certified to ISO standards is sufficient for the purpose, "fit for purpose," provided that the source of the material and its characterization is known. These are termed reference materials, working standards, or control materials. In a quantitative analysis, the CRM is the best characterized and has the lowest uncertainty as to any quantitative measurement. However, not all applications require the use of a CRM. For example, in-house RM are suitable for calibration of assays (ISO Guide 80), as long as they are calibrated against a CRM, and the increased uncertainty inherent in their use is recognized.

8.4 Reference materials in agricultural biotechnology

An appropriate RM should be available in testing laboratories using detection methods for agricultural transgenic material, both for the validation of the detection method and as an internal positive and negative (conventional RM)

BOX 8.1 Types of reference materials (RM) and their uses

Working standard (WS) or control material—this is material available in the laboratory, which is well characterized, and is used in the day-to-day operations of the laboratory. Such materials are used if a CRM or commercial RM is not available, or available in limited quantities, in which case the WS should be calibrated back to the RM or CRM.

Reference material (RM)—"material or substance, one or more of whose property values are sufficiently homogeneous and well established to be used for the calibration of an apparatus, the assessment of a measurement method, or for assigning values to materials." ISO Guide 30.

Certified reference material (CRM)—"reference material characterized by a metrologically valid procedure for one or more specified properties, accompanied by an RM certificate that provides the value of the specified property, its associated uncertainty, and a statement of metrological traceability" ISO Guide 30.

for the testing materials. As with any RM, it must have sufficient information on its quality and be fit for purpose (ISO Guides 30, 31, 33 and 35). Some regulatory systems (e.g., the European Union (EU)) require that the RM be suitable for polymerase chain reaction (PCR) measurements and be certified (a CRM). The EU specifically requires information concerning homogeneity and storage stability (Annex II of Regulation (EC) 641/2004) characteristics that are typically supplied for a CRM.

RM for each commercially available transgenic event in biotechnology-derived agricultural products are made available as single events (Chapters 2, 5). Material containing low-level mixtures of the event with nontransgenic crop materials (<1%) are also made available in some cases. Construction of calibration curves or instrument calibration for purposes of quantitative estimation of low levels of GM materials can be achieved by the use of these low-level CRM or by dilution of a high (typically 100%) concentration CRM. The CRM can also be used to establish identity and method sensitivity. CRMs are made available through two primary sources: The American Oil Chemists' Society (AOCS) and the European Commission, Joint Research Center (JRC) Directorate F - Health, Consumers and Reference Materials Unit (formerly Institute of Reference Materials and Methods, IRMM).

8.5 Practical limitations on reference material

Ideally, the matrix of the RM should be identical to the test substance being evaluated. However, this would require that thousands of CRM are made for all the products and matrices that may be tested. The materials (matrices) that can be analyzed in a laboratory are highly diverse, ranging from seed to plant parts, processed materials, partial products, and finished food products such as the thousands of products found in a typical grocery store. For any target analyte, it is not possible to provide RM in every matrix that might be analyzed. The ideal situation is to provide a single reference material that acts in the same manner as all the samples that are analyzed. The reality is that it is feasible to produce reference materials either as plant DNA or from which plant DNA can be isolated, for the primary use of analyzing seeds and grain.

Commercially available RM are almost universally produced as single GM events. However, laboratories often use methods that screen for genetic elements present in multiple GM events (products). There are no RM materials at present that are certified for this purpose, so laboratories generally use one event that contains the targeted element as a proxy. These reference material samples are then used to test the sensitivity and specificity of the detection methods. Quantification via this approach may be complicated by the presence of multiple copies of the target element in some events, however that is not the focus of this chapter. Adding to this complexity in testing is the ever-expanding set of traits and crops with more than one transgenic trait (stacks). The stacking of traits increases the number of target assays that must be used for detection and identification. Stacking also increases the complexity of interpreting test results and the costs associated with testing. The issue of RM in relation to stacks is discussed in Section 8.8.3.

RMs for GM analysis are produced and commercially available as single events. Current detection methods are not able to differentiate between a bulk lot of grain containing a mixture of seed, each containing one event, from a bulk lot of grain with individual seeds containing multiple events. Therefore, producing RM containing multiple events is not necessary. Those wishing to calibrate assays for the presence of multiple events can easily do so by combining (spiking) multiple RMs in a single sample.

8.6 Sources of reference materials

RM other than CRM may be in use in a laboratory. According to ISO 17025, which is the accepted standard for food analytical laboratories, the reference standard used must be of the highest possible quality. Thus, CRMs are used where available. The RM can be used directly as a calibrant to measure the concentration of the GM, such as in real-time PCR reactions. However, if the cost of the CRM is a concern, a CRM can be used to calibrate a working RM (ISO Guide 80). Quantitative measurements generated using the working RM will inherently have a higher uncertainty than measurements using a CRM, due to the uncertainty of the method used to calibrate the working standard. When used in this way to generate a secondary calibrant, less of the primary CRM is needed, which is economically beneficial. The working standard can be created by mixing materials of suitable concentration, but many laboratories that use this approach repurpose naturally occurring samples that they encounter in the course of their analytical work.

Proficiency testing (PT) in laboratories can use RM if that laboratory needs to demonstrate proficiency in a method with traceability back to an assigned value as required by quality management systems such as those based on ISO 9001:2015, ISO 17025:2005, ISO 17043:2010, or ISTA[1] certification. RMs used in PT programs provide information about origin, metrological traceability, and measurement uncertainty of any assigned value. RM qualities include demonstrated stability, homogeneity, and purity for the measurand (target analyte). However, "*the required metrological traceability chain can differ depending on the type of proficiency test item, the measure and or characteristic, and the availability of traceable calibrations and reference materials*" (ISO 17043 4.4.5.3 note).

Proficiency samples obtained by participation in programs such as those offered by the USDA GIPSA,[2] ISTA, FOFSA,[3] and AOCS[4] are often used as

1. ISTA: International Seed Testing Association.
2. USDA GIPSA: United States Department of Agriculture Grain Inspection, Packers and Stockyards Administration.
3. FOFSA: Federation of Oils, Seeds and Fats Associations Ltd.
4. AOCS: American Oil Chemists' Society.

RM in laboratories. Because the laboratory is informed of the value of the proficiency samples after the completion of the proficiency round, the samples have a known concentration. Care must be taken in using such materials. The uncertainty value of such standards can be difficult to ascertain, and their other characteristics (homogeneity, stability) will have to be established by the user. Measurements using such materials are not traceable back to source. Legal or contractual obligations may limit the use of PT materials for the described application and require disposal of remaining materials following the completion of the PT, and this should be considered before use in this manner.

Individual governments are producing limited amounts of some plasmid (DNA) RM for purely regulatory purposes in connection with their local requirements.

8.7 Reference material types

RM for GM analysis is often available in several forms. Ensuring the suitable RM choice, for example, type of material, matrices, concentration range, level of homogeneity, etc. is important in selecting the best RM for a specific testing need. CRMs for GM events in maize, soybean, potato, sugar beet, and cotton are available to testing and control laboratories, mainly from two sources (Table 8.1) that work in collaboration with the developers of GM crops. As of the time of writing, there are about 60 different CRMs of this type available. There are both advantages and disadvantages to the use of each type of RM. These are summarized in Table 8.2.

TABLE 8.1 Main sources of certified reference materials (CRM) for use in GM testing.

Source	Types of reference material	Address and web link
AOCS (American oil chemists' society)	• Seed (devitalized) • Biological powders from grains/seed • DNA isolated from plants	Urbana, IL, USA https://www.aocs.org/crm
Directorate F - health, consumers and reference materials (JRC)	• Biological powders from grains/seed • Plasmids	Geel, Belgium https://crm.jrc.ec.europa.eu/c/By-analyte-group/GMO-content/40481/also available via secondary sources such as Sigma–Aldrich

TABLE 8.2 Summary of the characteristics of reference material (RM) types.

RM type	• Pros	• Cons
Seed	• Readily available • Matrices are the same as many samples • Reasonable storage life and stability	• Issues around sampling variation and providing a true negative control. • Freedom of movement issues for viable seed as may require permits and phytosanitary measures • High IP risk
Grain/seed powder	• Readily available • Low IP risk • Reduced sample size • Ease of shipping • Reduced need for phytosanitary permits for shipping	• Must confirm on a crop by crop basis that powder is a representative and stable RM • Dependent on good grinding and homogenization of the samples
Plant DNA	• Suitable for many uses (identity, calibrant), if it is validated and stable • Low IP risk • Reduced/no permits for shipment	• Extraction chemistry and matrix may not be identical to the test substance • Could be unstable
Plasmid DNA	• Easy to produce, suitable as calibrant if validated and stable • Can be used as "spike" if validated for specific matrix • Low IP risk • Reduced/no permits for shipment • May offer advantages if other RM not available	• Does not fit the ideal of matrices for plant RM • May behave differently than the test substance (need correction factor) • Could be unstable • Cross-Contamination is more likely

8.7.1 Seed

The RM that is the most widely applicable is seed. Seed lots have been extensively tested for quality characteristics including germination, vigor, varietal and event purity, and adventitious presence for other transgenic materials. Seed lots used in grain production are homogeneous or heterozygous for the GM event(s) depending on the crop. The advantage of seed is that it can be used to characterize methods that require germination, and DNA and protein can be extracted for characterization of methods based on DNA or protein measurements.

Commercial seed lots are used to produce grain lots. Grain may also be substituted as an RM in certain cases, although it generally lacks the extensive quality and purity characteristics and it can also be segregating for the GM traits, making it less than ideal as an RM. While the parent seed or the grain progeny may look much the same, the quality criteria to propagate and check for seed and grain are generally quite different. Seed lots are grown under close supervision and inspected for purity. This type of rigorous quality stewardship of seed lots results in lower uncertainty when used as RM. In comparison, grain lots, whose size can be very large, commonly have less quality oversight, resulting in the purity standard for grain having higher uncertainty as RM.

The primary disadvantage of seed RM is that the material is drawn from a seed lot. The properties of the seed lot can only be characterized by mathematical probability to a particular certainty as these properties are determined by examining a sample of the lot, not every seed in the lot. Thus the certified values for seed lots are typically more uncertain than those for some other materials. Establishing a truly negative (conventional) seed sample is difficult in seed lots for this reason. Additional resources like the use of contained environments (greenhouses and growth chambers) for the production of pure conventional RM seed and the need for additional quality oversight and inspections are costly. Use of viable seed RM may also be inappropriate due to Biosafety constraints in moving viable seed between countries (or even regions of countries) and considerations around the intellectual property that is represented by the seed and stewardship needed for materials that are not allowed in commerce. Moreover, seed is rather bulky and costly to ship, and most countries require a certificate verifying the absence of pathogens, so the shipping process takes additional time and resources to complete for viable RM seed.

8.7.2 Nonviable seed

Methods for the production of nonviable (devitalized) seed that can be used as RM have been published (Schafer et al., 2008), and nonviable seed is being routinely used as a means to provide seed where there may be stewardship or intellectual property issues or concerns.

In addition, nonviable seed has the advantage that it cannot be germinated and used to breed new varieties and will not establish a population of plants that are not approved for release into the environment. The disadvantage of such nonviable seed is that in general it may be more suitable for use with DNA-based methods, as the devitalizing process can denature the proteins in the seed.

8.7.3 Grain/seed powder

Seed or grain can be used to produce ground materials (powder). These RM powders are easier to ship and subsample than seeds because of their reduced volumes and packaging and limited phytosanitary requirements. In addition, the powder can be characterized in a (relatively) large batch and small samples supplied to users. Seed powders offer the benefit of being nonviable and therefore cannot be used to generate plants. The powder can be sampled easily for the establishment of the characteristics of the batch reference material, and providing that the lot is sufficiently homogenous and of a fine particle size compared to the sample, the measurement of characteristics is relatively easy. Evidence for the absence of impurities (other events) in the lot will be limited only by the size of the samples examined, and the limit of detection of the analytical methods used for the impurity.

Powders of this type should be well ground, as a fine particle size helps ensure homogeneity. Additionally, the samples that are withdrawn for characterization must be representative of the material. This is particularly important for powders that are supplied as calibrated samples (such as 5% or 1% GM). The producer must confirm on a crop-by-crop basis that such powders are stable, including during shipment. It is also important that the particles in the powder not separate due to handling or are well-homogenized before being dispensed or used, as the particles may originate from different parts of the seed, which may have different specific gravity. Powders are commonly used with DNA-based assays and are the most commonly available CRM. They have been used in some cases to show identity in protein-based methods if the protein is sufficiently stable in the powder.

8.7.4 Powder produced from nonseed/grain plant parts

Powders (and DNA) may be produced from leaves, partial tubers, or stalks in addition to seed. This is of necessity in the case of species that do not produce seed or are not propagated via seed. Provision of whole tubers or other plant parts as reference materials is in almost all cases not feasible due to issues of grinding large tuber samples to accomplish homogeneity and storage requirements for large powder RM samples. However, the use of nonseed plant parts for producing an RM may be preferable due to the risk of the release of seed materials with traits (particularly for discontinued events) or due to the oil content of the seed that may prevent grinding of the seed into a uniform and stable powder. This nonseed source material is likely to be genetically identical, and sampling of such material before use to make the RM powder can be relatively easily performed. Each plant that contributes a leaf or stalk can be individually tested, or a portion of each tuber can be tested. Thus this RM powder has the advantage that purity (and impurity) can be established to a very high certainty in the source material.

The same issues around homogeneity of the powder, particle sizes, stability, etc., that are applicable to seed powders, also apply to this kind of RM.

8.7.5 Plant DNA

Plant DNA can be produced relatively easily in quantity from seeds, grain, leaf, or other plant parts. It is relatively easy to produce and ship DNA RM, and due to the fact that it is produced as a solution, it does not suffer from homogeneity issues. DNA can be produced with a high certainty of purity and lack of impurity if the source materials are tested as described for nonseed reference materials. It is a convenient material for a DNA testing laboratory as it does not require extraction. Like powder RM, DNA RM offers the benefit of not being able to generate plants.

As for any RM, the stability and suitability of the material must be established. When DNA RM is supplied as an alternate matrix to that being tested, it must be verified that the extraction method behaves sufficiently identically to that used for samples. A particular challenge for DNA RM is that there is no certified standard for plant DNA concentration or agreed international measurement method, and thus a laboratory may not agree on the concentration or amount of the DNA in the sample should they use a different method for measuring the DNA concentration. DNA RM is often supplied as a dry powder to facilitate shipment. DNA RM can be used as a calibrant and to validate the analytical step of a method (e.g., PCR), but is not suitable for calibration or validation of the extraction step.

8.7.6 Plasmid DNA

Plasmid DNA RM is easy and cheap to produce in large quantities, easy to ship, and is suitable as a calibrant if validated. Similar to plant DNA RM, plasmid DNA RM does not suffer from homogeneity issues and can be produced with a high certainty of purity and lack of impurity. Plasmid DNA can be used to spike a variety of matrices needed for testing validation controls such as those found in the food industry. DNA testing laboratories find plasmid DNA RMs convenient to use as they do not require extraction. From a stewardship point of view, plasmid DNA RM has an advantage in that it is not viable material and thus not able to escape into the environment (Caprioara-Buda et al., 2012).

The use of plasmid DNA RM as a calibrant has a number of challenges. Plasmid (nonplant) DNA has a different DNA conformation as compared with plant genomic DNA. Plasmids are super-coiled circular molecules (which can be linearized), and the DNA is not of nuclear origin. Thus plasmid DNA RM has been seen to behave differently from plant DNA in PCR reactions, and the question arises as to whether it is sufficiently similar or analogous to (commutable with) plant material CRM (Debode et al., 2010). A number of studies (Ballari et al., 2013; Burns et al., 2006; Caprioara-Buda et al., 2012; D'Andrea et al., 2009; Debode et al., 2010; Guan et al., 2011) have shown that users of plasmid RM must apply a correction factor to normalize the values to those obtained using genomic DNA derived from plant materials.

A further challenge of using plasmid DNA RM occurs if the material is produced in the same facility where it is to be used as an RM, or if multiple plasmid DNA RM is produced. The presence of replicable plasmids in a facility poses challenges to ensure that there is no cross-contamination or lateral gene transfer that could lead to false positives.

8.7.7 Proteins and plant extracts

Protein RMs are not generally available commercially but are used in analytical laboratories where quantitation of proteins is required. As the level of a recombinant protein in a plant is usually low, they are usually produced in other systems such as microorganisms (both eukaryotic and prokaryotic) using systems that have proven successful for the pharmaceutical industry (Gupta et al., 2017) and are produced for internal use. Even in these systems, they are challenging to produce in any quantity at the required purity and in the required conformation. Additionally, maintenance of protein RM stability requires a high level of expertise for most proteins. There is no certified protein RM available for transgenic crops. Such protein RMs are required for quantification of the amount of protein expressed in plants (such as in regulatory studies) and for the validation of protein detection methods. In certain cases, where identity confirmation of the protein is the intended outcome, powders can be used as RM (such as for lateral flow strips). However, the CRMs available from the sources described later are not certified for use as protein RM.

8.7.8 Intellectual property issues

In situations where there is some risk of release of viable GM seed into the environment, or use of material outside of those allowed by the intellectual property owner, forms of RM other than viable seed are shared to reduce risk. RMs such as powdered seed, devitalized/nonviable seed RM, DNA RM, and plasmid DNA RM are lower-risk RM materials (see Table 8.3).

8.8 Specific considerations for using reference materials for quantification

8.8.1 True negative or 100% positive reference materials

Truly negative RM can be difficult and expensive to obtain. If the RM originates from seed, then extreme caution must be exercised in producing the seed. For example, every plant that is used to produce the seed should be tested. Due to the fact that a high percentage of the commercial crop in certain countries is transgenic, the seed must be produced in a location where there is no possibility of infiltration of pollen from other plants that may be in the area.

Even so, grain or seed produced for use as an RM can only be tested to a certain level of confidence. The use of destructive seed testing methods renders testing every single seed impossible without depleting the seed lot. Thus

TABLE 8.3 Uses of reference materials in analytical methods and their associated risk to intellectual property.

Reference material type	Identity	Quantification	Validate/ develop[a]	Proficiency test materials	Intellectual property/ stewardship risk
Whole (live) seed RM	Y	Y	Y	Y	High
Whole (devitalized) seed RM	Y	Y	Y/N[a]	Y	Low
Ground/Powder seed RM	Y	Y	Y/N[a]	Y	Low
DNA RM from nonseed plant parts	Y	Y	Y/N[a]	Y	Low
Plant DNA RM	Y	Y	Y/N[a]	Y	Low
Plasmid DNA RM	Y	Y	Y/N[a]	Y	Low

[a]*Methods would need to be validated for each RM matrix if they differ from the test sample matrices.*

negative RMs are typically certified to an upper threshold, such as <0.1% of GM events at 95% certainty. One technology that may offer the option to test every seed for RM applications is seed chipping. Using a small sharpened blade, a portion of the seed is chipped/removed from the seed for testing, but the remaining seed remains viable (Amery et al., 2016). This technique can be applied to large seed where the orientation of the embryo can be easily defined (e.g., maize, soy) allowing cutting of nonessential material, but is difficult to apply to all seeds.

The same purity limitations arise for the positive RM. A typical seed lot cannot generally be proven to be 100% pure, and so the uncertainty around a nominal 100% RM may typically be a few percent. However, a higher level of certainty can be achieved by producing the RM seed, for example, in a contained greenhouse compartment, and testing each parental plant before flowering. In this way, the high purity of the produced RM seeds can be maintained, but the number of seeds that can be produced in this way is limited by practical considerations. In addition, these RMs are typically tested for the absence of other events that might be present but only to a practical level.

One approach used to produce a negative RM is to produce plant DNA from leaf tissue. Because of the small number of plants that are needed to produce this type of RM, the plants from which the DNA is derived can be isolated in a greenhouse and can be individually tested to show that there are no other (identifiable) GM materials in the DNA. The uncertainty of the nominal zero or 100% value is lower for this type of material than for RM produced from seed or grain.

Generally, larger quantities of negative RM are needed for the validation of methods intended to detect small quantities of transgenic material. The negative RM is useful for dilution of the positive RM when determining the validation detection limits of a detection method. Additionally, both negative RM and positive RM are required to be made available for method validations and controls in several regulatory jurisdictions (European Commission (EC) 2011).

8.8.2 Zygosity of reference materials

If the seed that is being used to produce the RM is a hybrid monocot seed then there may be special issues that should be taken into account, particularly when quantification is desired (Trifa and Zhang, 2004). It should be noted that different tissue types/matrices contain different DNA content and computing the relation between mass and DNA content is a complex undertaking. With differing DNA contents, RM derived from these different tissues/matrices can change and impact final analysis results. In maize, for example, the proportion of the DNA in the embryo originating from each parent is the same, but in the endosperm, about 2/3 of the DNA originates from the female parent (Zhang et al., 2008). Other parts of the seed such as the pericarp originate completely from the female parent. This can lead to proportions of DNA in a whole-seed powder derived from a heterozygous seed that is not equivalent to the number of seeds that contain the target DNA (Zhang et al., 2008). The proportions can also be impacted by the relative size of the embryo in the seed, which may vary with growing conditions. Thus the use of hybrid seed poses specific challenges that must be overcome. This is less of an issue with dicot species, where the embryo constitutes almost all the seed and the DNA values are not impacted by the direction/origins of the hybrid cross. Ideally, a homozygous seed should be used as RM, but this is not feasible in all cases (for example when one of the parents is a male sterile).

Other biological factors may also affect the RM. As well as zygosity of seeds, tissue ploidy, and parental origin of the GM plant are important factors that can have an impact on quantification and final analysis results (Holst-Jensen et al., 2006; Zhang et al., 2008; Liu et al., 2010).

8.8.3 Application of reference materials to stacked events

A breeding stack contains two or more independently created events brought together in the same plant, seed, or grain by conventional plant breeding. An example is the stacking of an event expressing herbicide resistance and an event expressing insect resistance. Stacked products are produced intentionally through conventional breeding or adventitiously by crossing of plants containing different two or more events during cultivation. "It is estimated that a total of 80.5 million ha were planted to biotech stacks in 2018. This accounts for more than 42 percent of the 191.7 million ha of biotech crops planted worldwide" (ISAAA, 2018). Stacked events produced by breeding are regulated separately from the individual events in some jurisdictions and not in others. Thus, the issue has come up whether an RM should be available for breeding stacks.

RMs for commercialized events are available for single events. Given that a stack is made up of two or more events for which RM are available, it is not necessary to have an RM available for each possible combination to validate DNA-based detection methods. As stacked seeds and plants consist of combinations of two or more individual events, which are in any case detected by distinct methods, the single-event RMs are suitable for calibration or validation of methods used to analyze the stacked events. There are no methods available or foreseeable (ENGL, 2014) that would enable the unambiguous detection of stacks in a mixture of grain or seed; it can only be established via analysis of individual seeds, or by statistical inference (Akiyama, 2005; Mano et al., 2011; ISTA Seedcalc). Despite this, certain jurisdictions have requested that control samples of stacked materials be submitted as a condition of registration.

8.8.4 Uncertainty considerations for quantitative analysis when constructing standard curves

RMs are often used to determine the concentration of the target DNA in the DNA sample in relation to the concentration of the species (taxon) DNA present. For a seed or powder, this involves extracting the DNA from the RM.

The quantitative value is expressed in relative copy number by comparison to a calibration curve produced from either a number of calibrated RMs or a serial dilution series, typically using an RM with a nominal mass fraction of 100%.

There are thus several different approaches to constructing the calibration curve.

- Use of a series of calibrated RMs (usually powder) that are certified for the % mass fraction of the target DNA: That is, if the mass fraction is achieved by grinding each powder (conventional and trait), producing the same particle size and homogeneity involves considerable effort. It is difficult to achieve a homogeneous mixture as seed lots are not identical. Homogeneity testing is typically carried out on the powders using a single DNA extraction method, and equal extractability for each component making up the mixture for every conceivable DNA extraction protocol is not guaranteed. Different laboratories do not all use the same DNA extraction methods, nor do they employ the same method as the RM producer employs, thus leading to the potential for additional uncertainty in the actual value for each RM.
- Dilution of a 100% mass fraction sample: Testing laboratories make serial dilutions starting with a nominal 100% DNA sample extracted from powders, or plant DNA derived from material with a 100% mass fraction. This allows the operator to obtain Ct (threshold cycle = the relative measure of the concentration of target DNA) values for both endogenous and trait targets from the same diluted DNA sample. The benefits of this approach are the elimination of (1) uncertainty due to variation in DNA extraction efficiencies between materials and (2) uncertainty of the DNA concentration measurements. Uncertainty comes from the dilution step and, if done properly, dilution introduces only a small level of uncertainty. Therefore, the total uncertainty using this approach can be less than when using a premixed calibrated series of RM.
- Calibration against plasmid DNA: A number of governments and RM manufacturers are using plasmid DNA as the calibrator, as it is convenient to produce. In certain cases, the producers (e.g., plasmid DNA RM) may calibrate the standard back to the RM. The plasmid may contain multiple targets, including the taxon DNA PCR target. To construct a calibration curve, it is diluted in a similar fashion as a 100% plant DNA. The actual DNA concentration in the sample will be very low compared with plant DNA. In addition, it has been found necessary in many cases to apply a "calibration (correction) factor" to the results to make them comparable with those obtained using plant DNA or powder RM (Debode et al., 2010).

8.9 International coordination and standardization

RMs provided by technology providers today meet the requirements of ISO for the production, provision, and use of reference materials as set out in ISO

17034 and ISO Guides 30, 31, 33, and 35. Most RMs are available from a single source, and thus any analyses that use these RMs should be comparable. However, individual governments and laboratories are also producing RM for limited or internal use, in some cases, for regulatory purposes. Cross-calibration of these RMs would be desirable to avoid conflicting results. In other cases, where multiple standards for the same measurand are offered, the Bureau International des Poids et Measures (BIPM) has participated in comparisons of the different sources. However, at present, the BIPM and National Measurement Institutes are not working toward standardization and harmonization for Generic Network Model standards at national and international levels.

Analysts that use the available commercial CRM available from AOCS and JRC can be confident that their values are comparable across laboratories and countries.

It should be noted that several countries have discussed developing their own RM production capabilities for commercialized plant biotechnology events to make these materials more accessible and less costly in their geographies. This practice is not in keeping with the originally defined need for traceability of RM back to a single source, nor does it adhere to the legal ownership of the IP associated with the trait materials. In addition, some commercial supply companies have placed RM on sale, also without proper regard to intellectual property interests. The use of multiple-origin RM could result in variable RM qualities and traceability, as well as a breach in legal IP rights. If, however, this desire for additional sources of RM comes from the need for better distribution of RM in some geographies, then this is a topic of discussion on the RM distribution model currently used by RM suppliers and trait providers.

8.10 Future of reference material and the drivers

Testing approaches are still evolving. With the proliferation of transgenic events in the market, and those being produced in universities and research institutes around the world, there is a tendency for regulatory agencies to move to a screening approach for GM testing. The EU and India, for example, have instituted systems where 96-well plates are standardized to contain a predetermined set of primers for genetic elements such that the process can be in some degree automated. Results from the analyses are compared with a database that lists which elements are present in which events. In this scheme, samples that show a pattern that cannot be reconciled with the presence of one or more unapproved events are assumed to be unapproved events. This approach is fraught with the chance of misidentification, especially when complex mixtures and matrices (such as food) are involved. RM can play a role in validating the tools, at least for well-characterized known commercial events.

Standard RMs for specific genetic elements have not been a focus, as laboratories will generally pick a CRM that contains that element and use it to calibrate their methods. Another approach that is increasingly used is to employ plasmids, especially in support of "national methods." As different countries or agencies produce plasmid DNA RM, or other RM, the possibility increasingly arises of multiple RM being available for the same event or genetic element. So far, this has not to our knowledge caused an issue of disagreement regarding results, but the possibility is increasing as more of these materials are produced. As described in Section 9, cross-calibration of RM or the reliance on a single RM source would be desirable to avoid conflicting results.

A further issue is the proliferation of entrants into the biotechnology market. Until recently, commercial events have arisen from a limited number of large agricultural biotechnology companies, and these companies have made RM available through AOCS and JRC. As the number of technology developers increases, the source for obtaining RM for their products is not clear. In many cases, these new entrants (that include governments, startups, and universities) may only pursue regional approvals of their products in countries where the provision of an RM is not a requirement for commercialization. These situations have the potential to cause difficulties in proving the absence of these events in supply chains.

The final question, for now, is whether RM will be necessary for analysis that is carried out using newer technology and approaches? Digital PCR has the potential to give a quantitative result without having to calibrate the assay using an RM (Burns et al., 2010; Corbisier et al., 2010; Milavec et al., 2014). If such methods eventually become the standard testing approach, will there still be a need for RM of the type we see today, at least for quantitation? In addition, digital PCR can be used to characterize reference materials (Bhat and Emslie, 2016) and is being used by the EU routinely for this purpose. However, should there no longer be a need for calibrated materials for quantification, an RM for event identification will still be a necessary part of the assay validation process.

Acknowledgments

We thank Benoit Maes (BASF) and Elizabeth Bates (Bayer) for advice and for suggestions during the preparation of this chapter.

References

Akiyama, H., Watanabe, T., Wakabayashi, K., Nakade, S., Yasui, S., Sakata, K., Chiba, R., Spiegelhalter, F., Hino, A., Maitani, T., 2005. Quantitative detection system for maize sample containing combined-trait genetically modified maize. Analytical Chemistry 77, 7421–7428.

Amery, M.A.L., Fukushigea, Serson, W., Hildebrand, D., 2016. Nondestructive DNA extraction techniques for soybean (Glycine max) seeds. Journal of Crop Improvement 30, 165–175.

Ballari, R., Martin, A., Gowda, L., 2013. A calibrator plasmid for quantitative analysis of insect resistant maize (Yieldgard Mon 810). Food Chemistry 140, 382–389.

Bhat, S., Emslie, K.R., 2016. Digital polymerase chain reaction for characterisation of DNA reference materials. Biomolecular Detection and Quantifcation 10 (2016), 47–49. https://doi.org/10.1016/j.bdq.2016.04.001.

Burns, M., Corbisier, P., Wiseman, G., Valdivia, H., McDonald, P., Bowler, P., Ohara, K., Schimmel, H., Charels, D., Damant, A., Harris, N., 2006. Comparison of plasmid and genomic DNA calibrants for the quantification of genetically modified ingredients. European Food Research and Technology 224, 249–258.

Burns, M., Burrell, A., Foy, C., 2010. The applicability of digital PCR for the assessment of detection limits in GMO analysis. European Food Research and Technology 231, 353–362.

Caprioara-Buda, M., Meyer, W., Jenyov, B., Corbisier, P., Trapmann, S., Emons, H., 2012. Evaluation of plasmid and genomic DNA calibrants used for the quantification of genetically modified organisms. Analytical and Bioanalytical Chemistry 404, 29–42.

Corbisier, P., Bhat, S., Partis, L., Rui Dan Xie, V., Emslie, K., 2010. Absolute quantification of genetically modified Mon810 maize (Zea mays L.) by digital polymerase chain reaction. Analytical and Bioanalytical Chemistry 396, 2143–2150.

D'Andrea, M., Coisson, J., Travaglia, F., Garino, C., Arlorio, M., 2009. Development and validation of a SYBR-Green I real-time PCR protocol to detect hazelnut (*Corylus avellana* L.) in foods through calibration via plasmid reference standard. Journal of Agricultural and Food Chemistry 57, 11201–11208.

Debode, F., Marien, A., Janssen, E., Berben, G., 2010. Design of multiplex calibrant plasmids, their use in GMO detection and the limit of their applicability for quantitative purposes owing to competition effects. Analytical and Bioanalytical Chemistry 396, 2151–2164.

ENGL, 2014. European Network of GMO Laboratories - Working Group for Identification of Stacked GM Events (WG-IGSE) Working Group Report. Joint Research Centre, Ispra (VA), Italy. https://doi.org/10.2788/87226.

European Commission, 2011. Regulation (EC) No L166/2011 Article 3 of the European Parliament and of the Council of June 24, 2011 concerning the laying down the methods of sampling and analysis for the official control of feed as regards presence of genetically modified material for which an authorisation procedure is pending or the authorisation of which has expired, Certified Reference Materials. Official Journal of the European Union L166, 9–15.

Guan, Q., Wang, X., Teng, D., Yang, Y., Tian, F., Yin, Q., Wang, J., 2011. Construction of a standard reference plasmid for detecting GM cottonseed meal. Applied Biochemistry and Biotechnology 165, 24–34.

Gupta, V., Sengupta, M., Prakash, J., Tripathy, B.C., 2017. Production of recombinant pharmaceutical proteins. In: Basic and Applied Aspects of Biotechnology. Springer, Singapore. https://doi.org/10.1007/978-981-10-0875-7_4.

Holst-Jensen, A., De Loose, M., Van den Eede, G., 2006. Coherence between legal requirements and approaches for detection of genetically modified organisms (GMOs) and their derived products. Journal of Agricultural and Food Chemistry 54, 2799–2809.

ISO Guide 30, 2015. Reference Materials — Selected Terms and Definitions.

ISO Guide 31, 2015. Reference Materials — Contents of Certificates, Labels and Accompanying Documentation.

ISAAA, 2018. Global Status of Commercialized Biotech/GM Crops: 2018. ISAAA Brief No. 54.

ISO Guide 33, 2015. Reference Materials — Good Practice in Using Reference Materials.

ISO Guide 35, 2017. Reference Materials — Guidance for Characterization and Assessment of Homogeneity and Stability ISO 17034:2016 General Requirements for the Competence of Reference Material Producers.

ISO 17043, 2010. Conformity Assessment- General Requirements for Proficiency Testing.

ISO Guide 80, 2014. Guidance for the In-House Preparation of Quality Control Materials (QCMs).

Linsinger, T., Pauwels, J., van der Veen, A., Schimmel, H., Lamberty, A., 2001. Homogeneity and stability of reference materials. Accreditation and Quality Assurance 6, 20–25.

Liu, D., Shen, J., Yang, J., Zhang, D., 2010. Evaluation of the impacts of different nuclear DNA content in the hull, endosperm, and embryo of rice seeds on GM rice quantification. Journal of Agricultural and Food Chemistry 58, 4582–4587.

Mano, J., Yanaka, Y., Ikezu, Y., Onishi, M., Futo, S., Minegishi, Y., Ninomiya, K., Yotsuyanagi, Y., Spiegelhalter, F., Akiyama, H., Teshima, R., Hino, A., Naito, S., Koiwa, T., Takabatake, R., Furui, S., Kitta, K., 2011. Practicable group testing method to evaluate weight/weight GMO content. Maize grains J agric. Food Chemistry 59, 6856–6863. https://doi.org/10.1021/jf200212v. Epub May 23.

Milavec, M., Dobnik, D., Yang, L., Zhang, D., Gruden, K., Žel, J., 2014. GMO quantification: valuable experience and insights for the future. Analytical and Bioanalytical Chemistry 406, 6485–6497.

Schafer, B.W., Cal, C.Q., Embrey, S.K., Herman, R.A., Song, P, 2008. Devitalization of transgenic seed that preserves DNA and protein integrity. *Journal of biomolecular techniques: JBT, 19*(5), 348–352. Journal of Biomolecular Techniques 19 (5), 348–352.

Trifa, Y., Zhang, D., 2004. DNA content in embryo and endosperm of maize kernel (Zea mays L.): impact on GMO quantification. Journal of Agricultural and Food Chemistry 52, 1044–1048.

Zhang, D., Corlet, A., Fouilloux, S., 2008. Impact of genetic structures on haploid genome-based quantification of genetically modified DNA: theoretical considerations, experimental data in Mon 810 maize kernels (Zea mays L.) and some practical applications. Transgenic Research 17, 393.

Further reading

Bogožalec Košir, A., Demšar, T., Štebih, D., Žel, J., Milavec, M., 2019. Digital PCR as an effective tool for GMO quantification in complex matrices. Food Chemistry 294, 73–78. https://doi.org/10.1016/j.foodchem.2019.05.029.

Cankar, K., Stebih, D., Dreo, T., Zel, J., Gruden, K., 2006. Critical points of DNA quantification by real-time PCR—effects of DNA extraction method and sample matrix on quantification of genetically modified organisms. BMC Biotechnology 6, 1–15.

Charels, D., Broeders, S., Corbisier, P., Trapmann, S., Schimmel, H., Linsinger, T., Emons, H., 2007. Toward metrological traceability for DNA fragment ratios in GM quantification. 2. Systematic study of parameters influencing the quantitative determination of Mon 810 corn by real-time PCR. Journal of Agricultural and Food Chemistry 55, 3258–3267.

Koeber, R., Linsinger, T., Emons, H., 2010. An approach for more precise statements of metrological traceability on reference material certificates. Accreditation and Quality Assurance 15, 255–262.

Seedcalc8: Kirk Remund, Robert Simpson, Jean-Louis Laffont, Deanne Wright, and Sylvain Gregoire. http://www.seedtest.org/en/statistical-tools-for-seed-testing-_content—-1—1143—279.html Accessed March 24, 2016.

Taverniers, I., Van Bockstaele, E., De Loose, M., 2001a. Use of cloned DNA fragments as reference materials for event specific quantification of genetically modified organisms (GMOs). Mededelingen Rijksuniversiteit te Gent 66, 469–472.

Taverniers, I., Windels, P., Van Bockstaele, E., De Loose, M., 2001b. Use of cloned DNA fragments for event-specific quantification of genetically modified organisms in pure and mixed food products. European Food Research and Technology 213, 417–424.

Taverniers, I., Van Bockstaele, E., De Loose, M., 2004. Cloned plasmid DNA fragments as calibrators for controlling GMOs: different real-time duplex quantitative PCR methods. Analytical and Bioanalytical Chemistry 378, 1198–1207.

Wang, X., Teng, D., Yang, Y., Tian, F., Guan, Q., Wang, J., 2011. Construction of a reference plasmid molecule containing eight targets for the detection of genetically modified crops. Applied Microbiology and Biotechnology 90, 721–731.

Yang, L., Guo, J., Pan, A., Zhang, H., Zhang, K., Wang, Z., Zhang, D., 2007. Event-specific quantitative detection of nine genetically modified maizes using one novel standard reference molecule. Journal of Agricultural and Food Chemistry 55, 15–24.

Chapter 9

Seed and grain sampling

Larry Freese
United States Department of Agriculture Grain Inspection, Packers and Stockyards Administration, Washington, DC, United States

9.1 Introduction

Bulk commodities in commerce are typically traded with a contract. The contract will likely specify the characteristics of the lot that the buyer wants to receive and the seller agrees to deliver. One of those characteristics may be the percent of the lot that is biotechnology-derived seeds (also known as GE or GMO seeds). Most often, the percentage of biotechnology-derived seeds is desired to be small. Sometimes, this small percentage of biotechnology-derived seeds in the lot is referred to as adventitious presence (AP) or low-level presence (LLP).

Both buyers and sellers will want to know that the lot meets the lot specifications. Lots may be as small as a producer's truck or as large as an ocean-going vessel. Examination of the entire lot is one way to determine the lot contents but is prohibitively expensive and time consuming for bulk commodity lots. When the tests require the destruction of the test material, trying to test the entire lot would have some obvious disadvantages.

An alternative to examining the entire lot is to examine some small fraction of the lot. This small fraction is called a sample. Testing a sample from the lot is practical, economical, and time efficient. The disadvantage of testing a sample is that the sample will rarely have the same characteristics as the lot. The sample is not likely to contain the same percentage of GE seeds as the lot. The deviations of the sample content from the true lot content are called variation, variability, or uncertainty.

When samples are taken using appropriate sampling methods, probability theory will describe the distribution of likely sample deviations from the true lot content. The distribution of the sample deviations from the true lot content is influenced by the sample size. As the sample size increases, the deviations from the true lot content are expected to be smaller. Sample size can be used to help manage sample variation.

Application of Sampling and Detection Methods in Agricultural Plant Biotechnology
https://doi.org/10.1016/B978-0-323-99293-0.00014-3

Sellers of lots will often want to sample and test a lot to have some reasonable confidence that the lot being delivered will meet the terms of the contract. Likewise, buyers will often want to sample and test to have reasonable confidence that the lot being received will meet the terms of the contract. Since the buyer and seller of a lot will have different samples, the estimates of lot content will likely be different. Knowledge of sample variation will assist the buyer and seller in choosing appropriate sampling plans. Ideally, the sampling plans that the buyer and seller will use are part of the contract process.

Sampling is not the only source of variation associated with measurements on a sample. The processing of the sample and the analytical method can also contribute to the uncertainty of the lot estimate, although uncertainty contributed by the analytical method is often considerably lower than that associated with sampling. Some analytical methods may have more uncertainty associated with them than others. The measurement uncertainty associated with an analytical method will also have to be taken into account when sampling and testing a lot.

9.2 Sampling procedures

Samples taken from lots will likely produce estimates that deviate from the true lot concentration. Probability theory can describe the distribution of these deviations when random sampling techniques are employed. A simple random sample is a sample taken in a process in which every possible sample has an equal chance of being selected. Random sampling can be employed with lots where the individual elements of the lot can be identified and random numbers from a uniform random number generator can be assigned to the elements. The identity of elements in some high-value lots may exist in a computer database and random sampling can be employed. However, bulk commodity lots usually consist of many elements, such as seeds, and identifying the individual elements is not practical or is prohibitively expensive. Practical sampling procedures have been developed for sampling bulk commodity lots. These procedures produce samples that reasonably approximate random samples.

Practical sampling procedures have been developed for sampling both static and flowing lots. A static lot is a lot contained in a rail car, barge, or truck. For purposes of sampling, a static lot is to be sampled while the bulk commodity is idly sitting in the container. A flowing lot may also be identified as the contents of a railcar, barge, or silo. For purposes of sampling, a flowing lot is a lot that is being sampled while the lot is being moved from one container to another. For example, a flowing lot may be the contents of a silo but the sample is taken from the flowing stream as the lot is moved from one silo to another silo.

Practical sampling procedures, whether for static or flowing lots, have common characteristics. Small quantities of commodities are taken systematically throughout the lot. Each small quantity taken from any of the numerous places in the lot is called an increment or increment sample. The increment samples are combined to create the bulk sample to represent the lot.

To sample static lots, the surface of the lot must be accessible to the person taking the sample. The increment samples are usually taken with a hand probe or a mechanical probe. A hand probe is a long tube that has portals running the length of the tube. The probe is inserted into the surface of the lot and must reach the entire depth of the lot. A handle on the top of the probe is turned to open the portals on the side of the probe. This allows a small quantity of the commodity to enter the probe. The handle is then turned to close the portals and the probe removed from the lot. The contents of the probe comprise one increment sample.

Mechanical probes are used similarly to the hand probe. Several different designs of mechanical probes are available. The probe is mechanically inserted into the entire depth of the lot. The commodity that enters the probe is usually automatically removed to a collection box using a pneumatic process.

Multiple increment samples are taken from the lot according to a predetermined pattern for the particular lot configuration. Fig. 9.1 gives an example of a probe pattern for a hopper bottom truck trailer. Different probe patterns would be used with different truck trailer configurations and with various barge and railcar configurations. The increment samples are combined to form the bulk sample for the lot.

Large commercial facilities will usually sample large lots as flowing grain streams. An automatic sampler will usually be installed in a grain spout to systematically take increment samples from the flowing commodity stream. A typical automatic sampler is called a diverter sampler or a cross-cut sampler. This type of sampler has a device that periodically traverses the flowing commodity stream. As the device crosses the commodity stream, a small cross-section of the stream is taken. The small cross-section, or increment sample, is diverted to a collection box. A timer is used to determine how often the diverter sampler takes an increment sample from the flowing lot. The diverter sampler is started before the lot starts flowing. The diverter will continue taking increment samples until the entire lot has passed through the sampler.

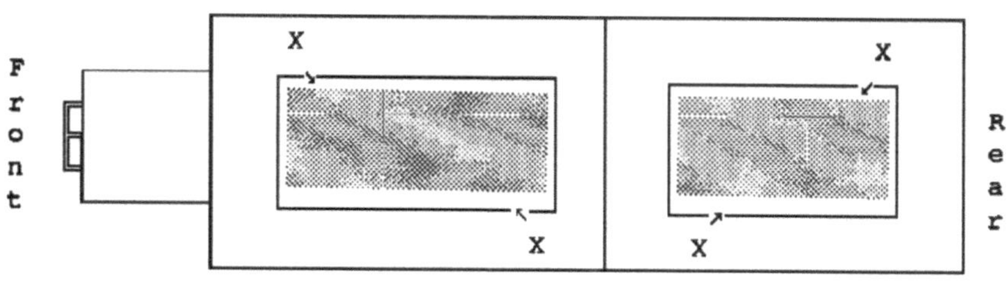

FIGURE 9.1 Example of a probe pattern for sampling a hopper bottom truck trailer.

Manual methods of sampling a flowing commodity stream can also be employed. A flowing commodity stream can be manually sampled at any point the flowing stream can safely be accessed, such as at the end of a spout or tailgate of a farm truck. The concept for manually sampling a flowing commodity stream is the same as the automatic sampling of a stream. The stream should be periodically traversed with a suitable collection device to take a cross-section of the stream. Each pass of the stream collects an increment sample. The composite of the increment samples constitutes the bulk sample for the lot.

Bulk samples from these practical sampling procedures are not random samples but are generally treated as random samples. One situation where practical samples do not perform similar to random samples is where the lot is highly heterogeneous. Heterogeneous means that the lot has areas that contain significantly higher levels of the characteristic of interest and areas that contain significantly lower levels of the characteristic of interest. Bulk samples from this type of lot will, on average, estimate the lot content, just as a random sample will. Potentially, estimates from bulk samples from highly heterogeneous lots can deviate more from the true lot content than random samples. This can happen if the lot has many more high and low concentration areas than increment samples, resulting in greater uncertainty of estimates. If enough increment samples are taken such that each of the high and low areas is sampled, then an increase in uncertainty is not expected. This is one reason for taking increment samples from throughout the lot.

Additional information on the equipment and practical sampling procedures for grain and seed can be found in ISO 24333, ISTA, 2015, USDA (2020a), USDA (2020b), and USDA (2020c).

9.3 Probabilities of sampling

Probabilities can be computed for many possible results from random samples. Suppose a lot contains 2% GE seeds and a random sample is taken to estimate the percentage of GE seeds in the lot. As the lot contains 2% GE seeds, a 100-seed sample would be expected to have two GE seeds. However, the actual sample can have anywhere from 0 to 100 GE seeds. Probabilities can be computed for each of the possible outcomes. Fig. 9.2 gives the probabilities for the most likely outcomes. The sample is most likely to contain less than 8 GE seeds.

As the sample size for probabilities in Fig. 9.2 is 100 seeds, converting the number of seeds to percent is easy. The two most likely outcomes are samples with one or two GE seeds (an estimate of 1% or 2%, respectively). Each of the two outcomes has about a 0.27 probability (probability is usually expressed as a fraction from 0 to 1 but can also be expressed as a percent from 0 to 100). The probability of getting no GE seeds in the sample is about 0.13, or about one out of every eight samples will have no GE seeds. The probability of

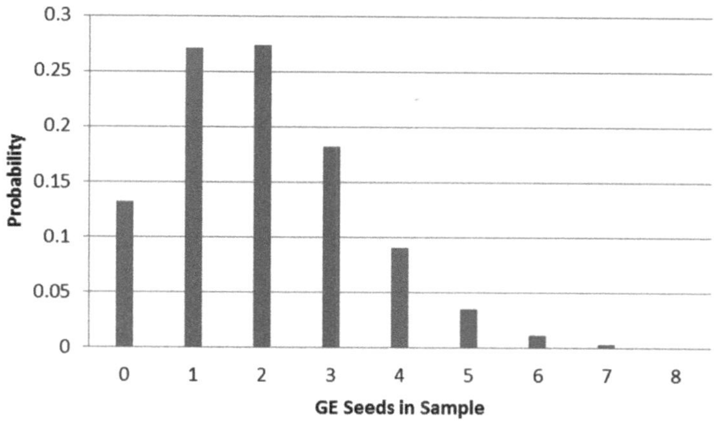

FIGURE 9.2 Probabilities for outcomes for a 100-seed sample from a lot with 2% GE seeds.

getting five or more seeds (an estimate of 5% or more) in the sample is about 0.05.

These probabilities help describe the likely deviations from the true lot concentration for a 100 seed sample.

Fig. 9.3 gives probabilities of likely results from a sample of 200 seeds from a lot with 2% GE seeds.

The most likely outcome is a sample with either three or four seeds in the sample (an estimate of 1.5% or 2%, respectively). Each outcome has a probability of just under 0.2. The probability of getting no seeds in the sample is about 0.02. The probability is about 0.05 of getting eight or more seeds (an estimate of 4% or more) in the sample.

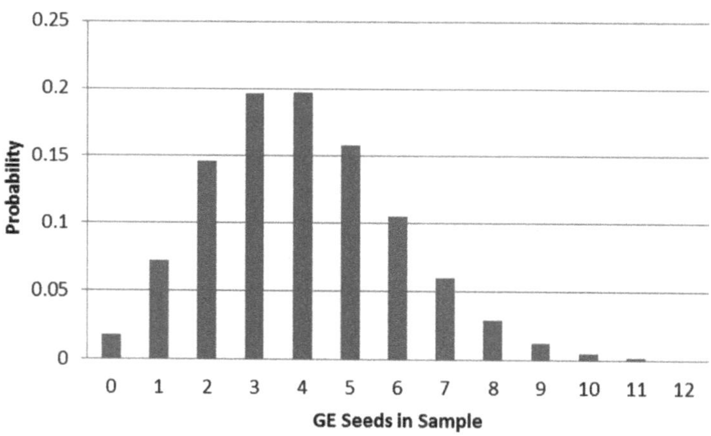

FIGURE 9.3 Probabilities for outcomes for a 200-seed sample from a lot with 2% GE seeds.

These two examples start to demonstrate the effects of sample size. One observation is that the probability of getting no GE seeds in the 200 seed sample is somewhat less than getting no seeds in the 100-seed sample. Larger samples are more likely to detect GE seeds when GE seeds are present in the lot. Another observation is that about 0.05 probability exists of getting a 100-seed sample with 5% or more GE seeds. The 200-seed sample has about 0.05 probability of getting 4% or more GE seeds in the sample. The percent GE seeds in the 200-seed sample is not as likely to have as many low estimates or as many high estimates as the 100-seed sample. The 200-seed sample has a narrower range of likely estimates than the 100-seed sample.

This observation can be extrapolated to larger sample sizes. As the sample size increases, the range of likely estimates from the sample will decrease.

9.4 Testing of the laboratory sample

The discussion of probabilities gives the probabilities associated with a sample size and lot concentration. This discussion may imply that only one sample is taken from the lot. In practice, only one bulk sample is likely taken from the lot. However, many different measurements are usually required for commercial lots. Some of the tests are destructive in that the sample can only be used for that single test. More than one sample is likely needed to perform all needed tests on a commercial lot.

The sample that a test is conducted on is the test or working sample. When more than one test is to be conducted on a lot, more than one test sample will likely be needed. The sample size that is determined with probability theory is the size of the test sample. The sample that the laboratory receives, called the laboratory or submitted sample, must be large enough such that it can be divided into all the required test samples. In addition, many laboratories maintain for a limited time a reserve sample (sometimes called a file sample) for future testing needs. The laboratory sample must then be large enough to obtain all the required test samples and reserve samples.

For practical reasons, laboratories usually do not want to receive laboratory samples that greatly exceed the needs of the laboratory. Laboratories usually do not have sufficient storage to hold much excess commodity on their premises. Excess commodity will result in additional costs for handling and disposal. Part of the planning process is to deliver a laboratory sample of adequate size without too much excess commodity.

As described in the section on sampling procedures, a bulk sample is obtained by combining the increment samples taken systematically from the lot. The laboratory sample is obtained from the bulk sample. The sampling procedure must include enough increment samples such that the bulk sample is at least as large as the required laboratory sample. Collecting a bulk sample that is exactly the size of the required laboratory sample is difficult. The usual practice is to collect a bulk sample that exceeds the required size of the laboratory sample. After thoroughly mixing, the bulk sample is reduced to the laboratory sample using an appropriate device (see ISO 24333; ISTA, 2015; USDA, 2020 a,b) at the sampling site and the excess commodity can then be immediately returned to the lot.

When testing lots for low levels of GE seed, two types of tests are used. One type of test is called qualitative testing. This type of test produces a positive (i.e., detected) result when GE seed is detected in the test sample. A negative (i.e., not detected) result is obtained when no GE seed is detected in the sample. A sample with one or more GE seeds should produce a positive, or detected, result from testing.

The second type of test is called quantitative testing. This type of test produces an estimate of the quantity of GE seed in the lot (e.g., % GE in the sample).

9.5 Acceptance sampling with qualitative testing

Acceptance sampling with qualitative testing simply means that a lot is acceptable for a commercial transaction if a sample tests negative for GE seed using a qualitative test. A positive test result would result in the lot being rejected for transaction purposes. A sample should test negative if no GE seed is in the sample. The probability of accepting a lot is the probability of selecting a test sample that contains no GE seed.

The probability of accepting a lot will depend on the size of the test sample and the percent of the lot that is GE seed.

If lots are being tested with a test sample of 100 seeds, the probabilities of accepting various lot concentrations can be computed. If a lot contains 0.1% GE seed, the probability of accepting the lot is about 0.90. Another way of stating the probability is that 90% of the 100 seed samples selected from the lot will contain no GE seeds. If the lot contains 1.0% GE seed, the probability of accepting the lot is 0.37. The probability of accepting other lot concentrations can be computed, and Table 9.1 gives probabilities of accepting more lot concentrations.

Probabilities can also be presented as a graph. Fig. 9.4 gives the probabilities as a curve. This type of curve allows the assessment of the probabilities of a wide range of lot concentrations.

Placing the curves for multiple sample sizes on the same graph is useful to see how the probability change as the sample size increases. Fig. 9.5 gives the probabilities for sample sizes of 100, 150, and 300 seeds.

From Fig. 9.5, the probability of accepting a lot with 1% GE seed is about 0.05 with the 300-seed sample, about 0.22 with a 150-seed sample, and about 0.37 with a 100-seed sample. If lots with 1% GE seeds are desired to be rejected 95% of the time, a sample size of 300 seeds may be an appropriate sample size. Fig. 9.5 also shows that lot concentrations of 2% have about 0.05 probability of being accepted with the 150-seed sample and that lot

TABLE 9.1 Probability of accepting selected lot concentrations with 100-seed sample.

% in lot	0.5	1	1.5	2	2.5	3	3.5	4	4.5
Probability of acceptance	0.61	0.37	0.22	0.13	0.08	0.05	0.03	0.02	0.01

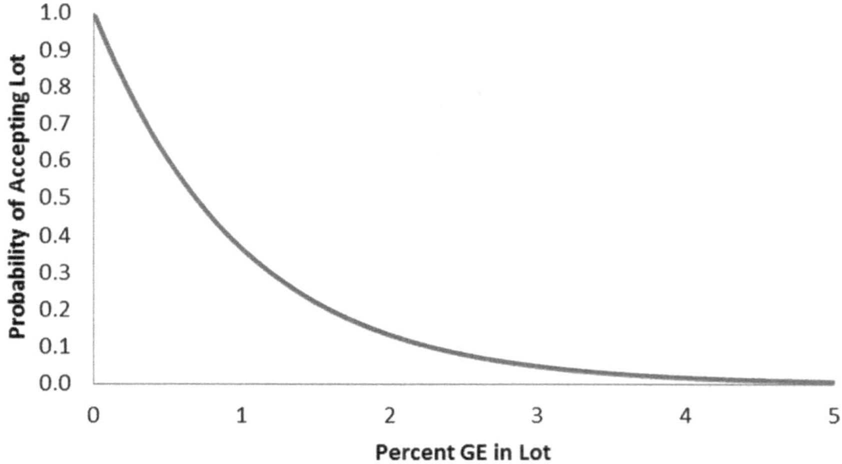

FIGURE 9.4 Probabilities of accepting various lot concentrations with a 100-seed sample.

concentrations of 3% have about 0.05 probability of being accepted with 100 seeds. If 3% lot concentration is desired to be rejected most of the time, a 100-seed sample may be an appropriate sample size. If 2% lot concentration is desired to be rejected most of the time, a 150-seed sample may be an appropriate sample size.

Sample size determines the risk of accepting any lot concentration. Conversely, a sample size can be computed for any chosen probability of

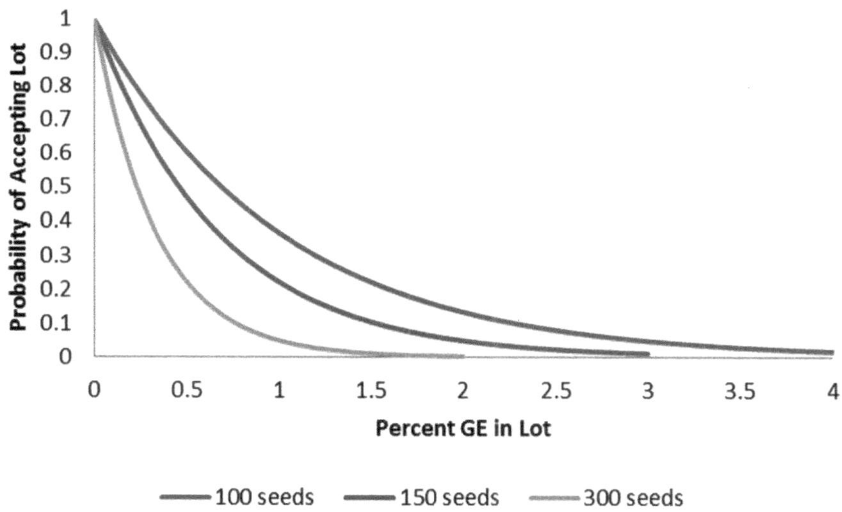

FIGURE 9.5 Probabilities of accepting various lot concentrations with 100-, 150-, and 300-seed samples.

accepting a particular lot concentration. A probability of 0.05 is often chosen as the risk of accepting lot concentrations that are to be accepted infrequently but any small probability, such as 0.01, can be chosen. If a lot concentration of 0.1% is to be accepted with a probability of 0.05, a sample size of 2995 seeds is needed. The sample size is based purely on the selected probability of accepting a particular lot concentration. However, some test methods specify a maximum limit on the number of seeds in the sample. This limit is based on the limit of detection (LOD) for the test method. For example, if a test method has a specified maximum sample size of 800 seeds, the test method is expected to reliably detect one GE seed in an 800-seed sample. The LOD for the method is expected to be around 0.125% (one in 800 seeds). The test sample size based on the desired risk can be greater than the maximum sample size for the test method. The test sample can still be tested with the test method. The test sample will have to be divided into multiple seed pools. Each seed pool will have to contain fewer seeds than the maximum for the test method. Each pool of seeds is then tested with the test method and all pools must test negative to be acceptable. If the test sample contains 2995 seeds and the test method has a maximum sample size of 800 seeds, the test sample can be divided into four pools of about 749 seeds. All four pools are tested, and all tests must be negative to be accepted. Similarly, if the LOD of the detection method is 1% (one in 100 seeds), the test sample can be divided into 30 pools of about 100 seeds (a total of 3000 seeds). All 30 pools are tested, and all tests must be negative to be accepted.

9.6 Acceptance sampling with quantitative testing

Quantitative testing provides an estimate of the lot content based on the content of the test sample from the lot. Quantitative testing is useful when the buyer and seller in commercial transaction agree that some level of GE concentration in the lot is acceptable. The advantage of quantitative testing over qualitative testing is that acceptance sampling plans can be devised that increases the chances of accepting low-level concentrations that are acceptable. Acceptance sampling plans for quantitative testing have two components: a specified test sample size and an acceptance limit (AL). A test sample of the specified size is obtained from the lot and a measurement is made on the sample. If the measurement on the sample is less than the AL, the lot is accepted. The lot is rejected if the measurement is greater than the AL.

Acceptance sampling with qualitative testing is actually a special case of acceptance sampling with quantitative testing. The AL is always zero with qualitative testing.

Uncertainty from sampling is always a consideration when measurements are made on samples. Both qualitative and quantitative measurements must take sampling uncertainty into consideration. Quantitative testing has an additional source of uncertainty that must be taken into consideration. This additional source is measurement uncertainty associated with the test method. The various test methods may have different measurement uncertainties. Measurement uncertainty for a test method is typically estimated experimentally. For this reason, taking measurement uncertainty into account when computing acceptance probabilities is more difficult. Consulting an expert in statistics is advised when developing quantitative sampling plans.

Once a sample size and AL have been selected, the probabilities of acceptance can be computed for various lot concentrations. When many probabilities for a range of lot concentrations have been computed, these probabilities can be plotted against the lot concentrations as was done with qualitative testing. This plot is called an operating characteristic (OC) curve. Fig. 9.6 gives the typical shape of the OC curve and the relationship of the AL to the OC curve. Lots with concentrations at the AL will have a probability of approximately 0.5, regardless of the sample size. Intuitively, a lot with a concentration at the AL should have about the same chance of a sample below the AL as above the AL. The probability will deviate slightly from 0.5 if the probability distribution is not what is called a symmetric distribution.

Sometimes the buyers and sellers in a transaction agree on a lot concentration that should be accepted most of the time, say 95% of the time. The concentration that is to be accepted most of the time is called the acceptable quality level (AQL). If an AQL is selected, the AL cannot be set at the AQL because the AL will have a probability of acceptance of approximately 0.5. The AL must be selected higher than the AQL at a point that produces an OC curve with the appropriate probability at the AQL. Fig. 9.7 shows the relationship of the AQL to the AL.

Sometimes the buyers and sellers in a transaction agree on a lot concentration that should be accepted infrequently, say 10% of the time. The concentration that is to be accepted infrequently is called the lower quality level

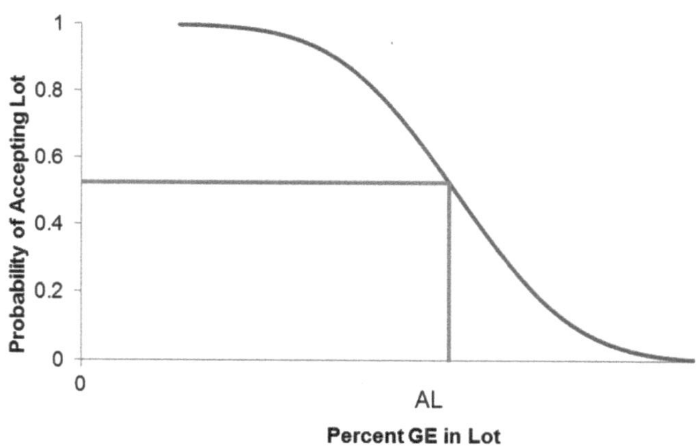

FIGURE 9.6 An example of an OC curve and the relationship to the AL.

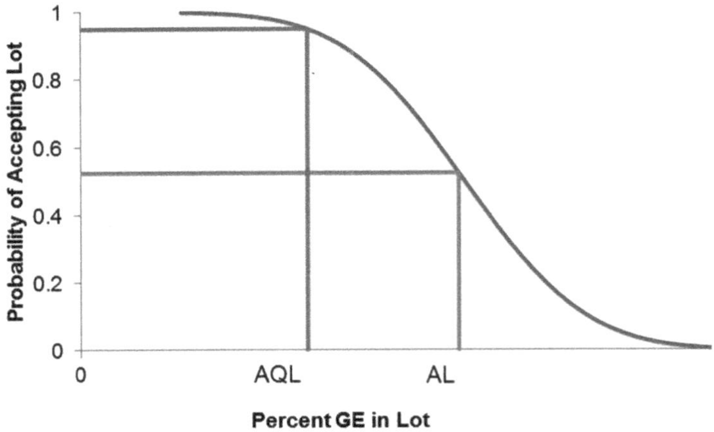

FIGURE 9.7 An example of the relationship between the AL and the AQL.

(LQL). If an LQL is selected, the AL has to be selected lower than the LQL at a point that produces an OC curve with the appropriate probability at the LQL. Fig. 9.8 shows the relationship of the LQL to the AL.

To reduce the probability of erroneously rejecting good lots or of erroneously accepting bad lots, the AL is desired to be as close to the AQL or LQL as possible. To do this, the uncertainty must be reduced. The uncertainty of sampling can be reduced by increasing the sample size. Practical considerations may limit the size of the sample or increased costs may result from increasing the sample size. The measurement of uncertainty can be reduced by making multiple measurements and averaging the resulting measurements. Increasing the number of measurements will almost certainly increase the cost of testing.

As previously mentioned, consulting an expert in statistics is advisable.

9.7 Acceptance sampling with qualitative testing on multiple subsamples

As discussed in Section 9.5, qualitative testing can be used to limit the risks of accepting lots with high concentrations of GE seeds. For example, a sample of 300 seeds has about a 0.05 probability of accepting lots with 1% concentration of GE seeds. Lots must have concentrations very near zero to have a high probability of being accepted. When some lot concentration above zero is acceptable, qualitative testing is not very good at limiting the erroneous rejection of good lots. Quantitative testing can reduce the erroneous rejection of good lots.

Situations may occur where some nonzero lots are acceptable but quantitative testing is not available, not available to give timely results, or too

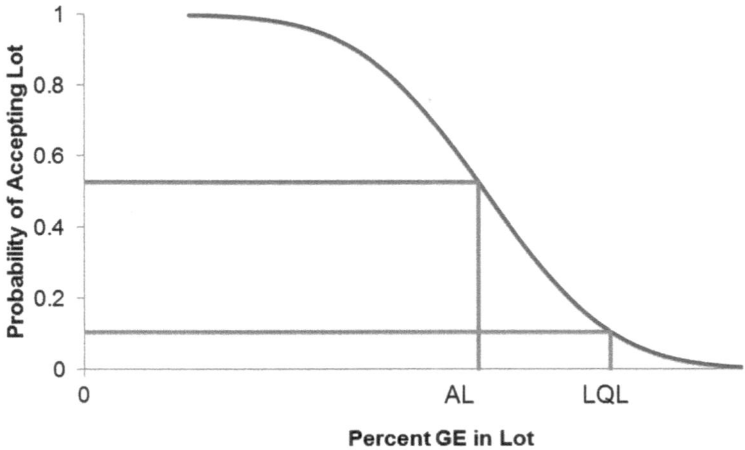

FIGURE 9.8 An example of the relationship between the AL and the LQL.

expensive. A low-cost qualitative test may be available. When a qualitative testing method is the only practical alternative, sampling plans can be devised that use qualitative testing but result in probabilities similar to quantitative testing. In effect, qualitative testing is used to obtain quantitative results. These sampling plans involve testing multiple subsamples with qualitative testing and are known as group or pool testing (Remund et al., 2001).

An acceptance sampling plan with multiple subsamples has three parameters. These parameters are the number of subsamples to be tested, the size of each subsample, and the acceptable number of positive results. The number of subsamples specifies the number of subsamples to be divided from the laboratory or working sample. Each subsample will be of the same size as specified in the plan. The lot will be accepted if the number of subsamples testing positive is less than or equal to the number of positives specified in the sampling plan.

An example of a multiple subsample plan is to divide 39 subsamples from the laboratory or working sample. Each subsample contains 100 seeds. The lot is accepted if 19 or fewer subsamples test positive. Fig. 9.9 gives the OC curve for this sampling plan. For comparison, a single sample plan with 300 seeds is also given on the graph. If the objective is to detect lots with 1% GE seeds, either plan represented in Fig. 9.9 would be adequate. And, if the objective is to detect lots with 1% GE seeds, testing a single sample of 300 seeds is certainly the quickest and most cost-effective test. However, if some low level of GE seeds is acceptable in the lot, the single sample plan will reject significantly more lots at some of the low lot concentrations.

For example, suppose lots with up to 0.5% GE seed concentration are acceptable. A single sample plan of 300 seeds has a 0.22 probability of accepting lots with 0.5% concentration. The multiple subsample plan has a

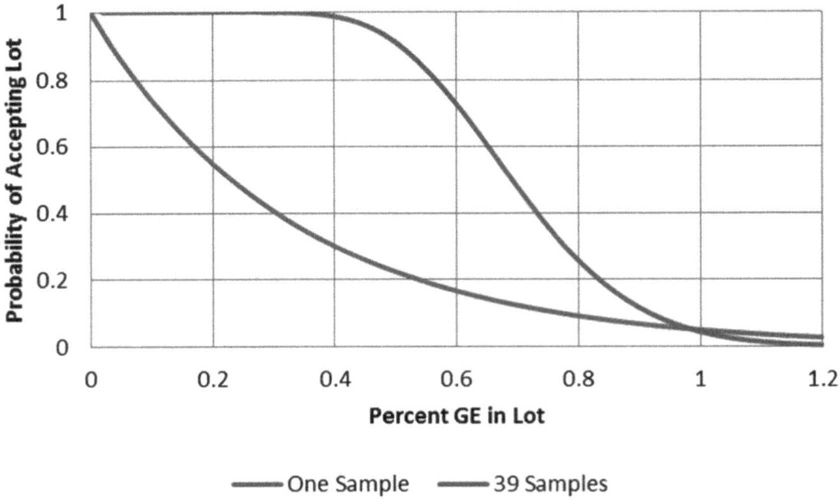

FIGURE 9.9 Probability of accepting lots with a single sample plan and a multiple subsample plan.

0.91 probability of accepting the lot. The multiple subsample plan has a significantly higher probability of accepting a lot at 0.5% concentration. Fewer good lots will be rejected with the multiple subsample plan.

The shape of the OC curve for multiple subsample plans is influenced by the number of subsamples, the size of each subsample, and the number of acceptable positive results. Many combinations of these three parameters can be made. A plan can be designed to achieve most goals of acceptance sampling. However, because so many combinations exist, finding a plan that achieves the desired goals can be complicated. Consulting an expert in statistics is highly recommended. Also, the International Seed Testing Association has developed a useful program (Seedcalc, 2007) to assist in designing sampling plans.

9.8 Conclusion

Sampling and testing are the only practical means of estimating the characteristics of bulk commodity lots. Unfortunately, samples taken from a lot will usually not have the same characteristics as the lot. The sample content will likely deviate from the lot content. Variability is unavoidable when measurements are made on samples from a lot. Probability theory can describe the distributions of these sample deviations when the samples are taken with appropriate methods. Risks can be managed by choosing appropriate acceptance sampling plans. Acceptance sampling plans for GM seeds and grains typically consist of a sample size and an acceptance limit (Freese et al., 2015).

When the acceptance limit is zero, qualitative testing is appropriate. Quantitative testing is needed when the acceptance limit is greater than zero. Multiple subsample plans with qualitative testing can also be used when the acceptance limits are greater than zero.

Selecting an appropriate sampling plan can be complicated, and consulting an expert in statistics is advisable.

References

Freese, L., Chen, J.W., Shillito, R.D., 2015. Sampling of grain and seed to estimate the adventitious presence of biotechnology-derived seeds in a lot. Cereal Foods World 60 (1), 9–15. https://doi.org/10.1094/CFW-60-1-0009.

ISO 24333, 2009. Cereals and Cereal Products – Sampling. International Organization for Standards. ISO 24333:2009.

ISTA, 2015. International Seed Testing Association Rules for Seed Testing—Chapter 2-Sampling vol. 2015.

Remund, K.M., Dixon, D.A., Wright, D.L., Holden, L.R., 2001. Statistical considerations in seed purity testing for transgenic traits. Seed Science Research 11, 101–119.

USDA, 2020a. Grain Inspection Handbook, Book 1, Grain Sampling. United States Department of Agriculture, Agricultural Marketing Service. Available on line at. https://www.ams.usda.gov/sites/default/files/media/Book1.pdf.

USDA, 2020b. Equipment Handbook. United States Department of Agriculture, Agricultural Marketing Service. Available on line at. https://www.ams.usda.gov/sites/default/files/media/EquipmentHB.pdf.

USDA, 2020c. Mechanical Sampling Systems Handbook. United States Department of Agriculture, Agricultural Marketing Service. Available on line at. https://www.ams.usda.gov/sites/default/files/media/MSSHB.pdf.

Seedcalc, 2007. Statistical Tests for Seed Mixtures. International Seed Testing Association. Available online at: https://www.seedtest.org/en/services-header/tools/statistics-committee.html?msclkid=2ea21d96cf2811ec99fc07b1dd7819ec.

Chapter 10

Plant and field sampling

Doug Miller
Illinois Crop Improvement Association, Champaign, IL, United States

10.1 Introduction

This chapter will focus on tissue sampling techniques used by the seed and crop production industries. Determining the genetic makeup of a growing plant has applications throughout the value-chain starting with plant breeding. By definition, tissue sampling is the collection of cells from a specific part of an actively growing plant. Leaf punches, leaf discs, entire leaves, or leaflets are the most common tissue sample types used in the industry. Unlike testing bulk seed, bulk grains, and related commodities, fresh tissue has actively growing, metabolizing, and replicating cells. This has a direct impact on the techniques, handling requirements, and application of the methods. Sampling for plant health, such as in nutrient analysis and plant pathogen identification, is not addressed in this book. The techniques described in this chapter lend themselves to both high-volume and low-volume sampling. Tissue samples are used by plant breeders to accelerate their work and ensure a clean hand-off of varieties and lines to parent seed producers. Crop advisors use tissue samples to verify that a crop is tolerant to or has a trait to resist a nonselective herbicide, before application. Tissue samples are also used to verify the identity and purity of crops and outside contaminants in seed and specialty production. The following will cover breeding applications as well as seed and commercial crop testing.

10.2 Plant breeding

Plant breeders can generate hundreds to hundreds of thousands of tissue samples as part of a project or breeding program. Plants are labeled by various means to maintain the identity of each plant, row, or group. A tissue sample is then carefully collected for testing, typically for laboratory analysis. Data from the lab are returned to the breeder prepollination if plants will be selected before mating. The best way to facilitate the removal of undesired plants, especially those plants classified as contaminants, from a nursery or small seed increase is also to receive and act on test results prepollination. Removing unwanted plants before flowering eliminates the chance of unintentional cross-

Application of Sampling and Detection Methods in Agricultural Plant Biotechnology
https://doi.org/10.1016/B978-0-323-99293-0.00011-8

pollination that would perpetuate unwanted genetics. Even in self-pollinated species where all means of preventing unwanted cross-pollination are used, preflowering removal of plants is part of most stewardship programs.

Genetic markers for specific traits or entire genetic profiles can be run on the tissue samples collected. The use of molecular markers applies to any heritable characteristic. Breeders have embraced molecular markers for their ability to accelerate breeding work, as well as a means of addressing contamination. Historically, plant breeders would fight the odds of limited information about the genotype by increasing the number of plants in the field as well as increasing the number of pollinations or crosses they made. The ability of molecular markers to eliminate the need to preform blind selections or blind backcrossing is a significant driver in modern plant breeding.

A blind backcross derives its name from using a segregating population of individuals for crossing. Disease resistance genes are a good example of heritable qualities that are not readily apparent by phenotype in a segregating population. Artificially inducing disease can be problematic or impossible to use as a consistent selection tool. Therefore, the chance of selecting an individual with the desired trait is unknown or blind.

With informative genetic markers, the chance of using desired plants in the breeding process is dramatically increased. Only the natural segregation in the gametes of desirable plants selected from a breeding population will affect the genetic makeup of the next generation of seeds. Breeders also accelerate the rate of genetic gain through the use of complete genetic profiles. A complete genetic profile can be used to rapidly identify backcross progeny that matches the desired recurrent parent thereby eliminating another source of uncertainty and inefficiency in the selection process.

10.3 Seed and commercial crop testing

Inspectors that are charged with determining crop identity and purity may also collect or test tissue samples. Tissue testing can be used to confirm the identity and check the purity of seed production and grain crops during the growing season. Inspectors may randomly test plants or test variants and off-types identified as potential contaminants in a seed crop. When permissible, nearby crops can be tested to determine if they will be an undesirable source of pollen for the seed crop. Tissue testing can also be used to meet requirements for regulated field trials. Postseason checks are required to determine if regulated materials persist at the trial site or if off-site movement of materials has occurred. Where crops with plant incorporated protectants are prevalent, inspectors can also verify the crop trait(s) and its refuge. Integrated refuge products, such as "refuge in the bag" or "integrated refuge" products, can also be sampled and tested in the field to identify the typical 5%–20% refuge component of the blend. Crop advisors, those tasked with crop protection and production services, can use tissue samples to manage several issues. By confirming the tolerance trait before herbicide application, crop damage or crop destruction can be avoided. Crop advisors may also test samples for nutrient analysis, crop health, pathogen detection, and phytosanitary purposes that will not be addressed here.

10.4 Direct testing in the field

The first step in sampling tissue is to understand the requirements of the assay or laboratory analytical method. Immunoassay test kits that use lateral flow test sticks or strips are commonly used in the field for rapid trait checks and pathogen detection. Lateral flow devices or strips come in ready-to-use kits that include the reagents and materials necessary to conduct the test in the field or an on-site work area. Small test tubes (e.g., Eppendorf tubes) are common in many of these test kits. Such test tubes are typically small, 1.5–2.0 mL in volume, plastic tubes with hinged caps (Fig. 10.2A). It is a single-use disposable container that is suited for direct use in the field with lateral flow devices but can also be labeled and shipped to a laboratory for analysis. The hinged cap makes these test tubes a sampling tool and container all in one. The lip of the cap fits into the tube and when closed over the leaf tissue to be sampled, a leaf disc is created. For some protocols, the leaf is folded over to achieve a double leaf disc. Analysis can be conducted in the field by using a disposable mini-pestle designed to grind the tissue for extraction and analysis. The tissue can be processed fresh and can stay with the plant during field analysis. Eppendorf tubes can also be submitted to a laboratory for analysis. Laboratory submission requires coordination with the lab regarding the method of submission. The labeling of plants, tubes, and racks are all considerations for laboratory submission. The test-tube sampling method is not typically used for high volume sampling due to the time it takes to label individual tubes and collect the sample. For direct testing in the field, be sure to use a kit that is specific to leaf and other fresh tissues. Grain and seed kits should be avoided as they may not have been developed or validated for fresh tissue. When large amounts of chlorophyll are present, test lines may turn green resulting in a false positive. Antibodies may be rendered ineffective resulting in a false negative. Once analysis is complete, the results can be recorded and the plant can be removed if undesirable or labeled as desirable, etc. depending on the result.

10.5 Sampling for laboratory testing

The first step in sampling is to understand the requirements of the assay. The lab conducting the processing and testing should be your primary resource for sampling and submission requirements. Test results may be affected and resampling may be required to achieve useable results if sampling requirements and protocols are not followed. Good laboratories will have

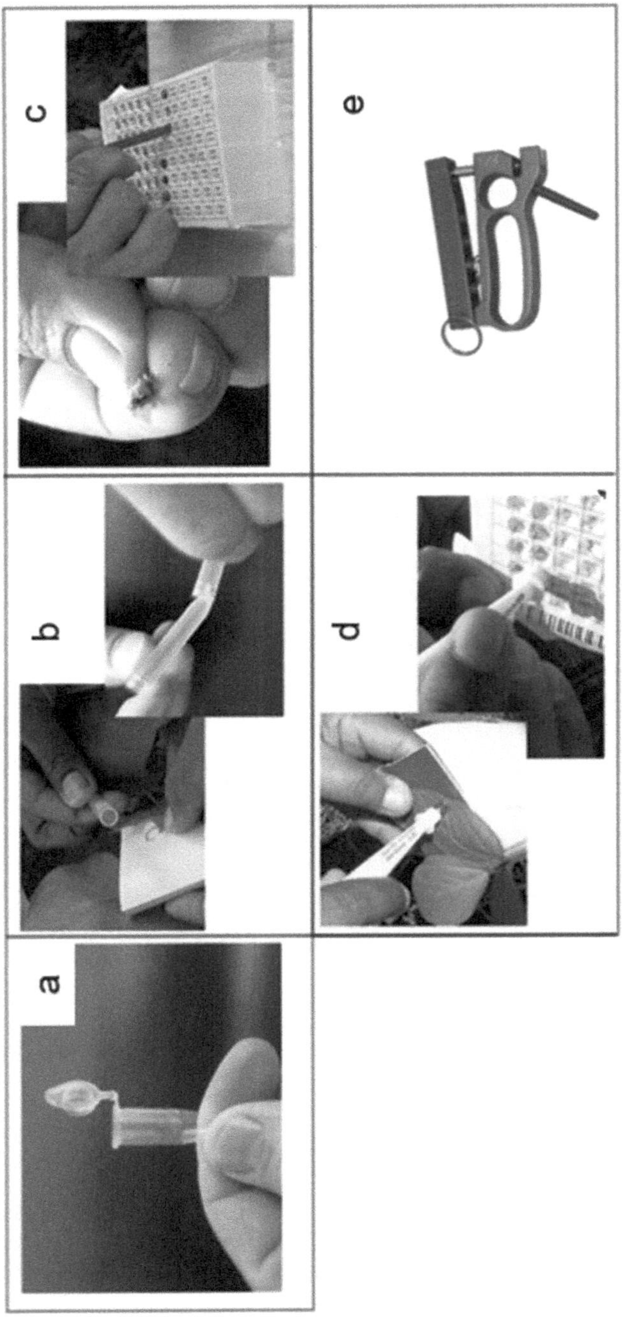

FIGURE 10.2 Equipment and techniques used for tissue sampling.

predefined sampling requirements that cover sampling technique, containers, handling, and shipping methods. In some cases, laboratories will supply specific containers to use. Always consult and confirm the appropriate process with the laboratory to ensure success in sampling and testing. A discussion of containers or "plastics" should be part of the presample planning.

The term "plastics" has been used by laboratories and field personnel as a general term that applies to any tube, honeycomb box, or similar container used to house samples. Plastic zip-top bags are an obvious choice for relatively few samples or whenever entire leaves or leaflets are collected. For high volume testing the subject of "plastics" can be brand and catalog-number specific to ensure that robotic handling equipment performs properly in the laboratory. Plates, blocks, or racks typically have an 8 × 12 pattern of wells (holes) or individual tubes for a total of 96 wells per block or 96 tubes per rack. There are also 4 × 6 (48-well) and 16 × 24 (384-well) plastics that occupy a similar footprint as a 96-well block. The wells or holes of a 48-well block are significantly larger and more suited to individual seed testing. The 384-well block is significantly smaller and designed primarily for liquids and high-throughput laboratories. However, the 8 × 12 (96-well) system is the middle-ground and is almost universally used to collect fresh leaf tissue in the form of a leaf punch or leaf disc. The terms plate, block, and tube-rack in the 8 × 12 patterns are also used interchangeably. For simplicity, the term "block" will be used throughout the text for any of the 8 × 12 pattern plastics.

10.6 Marking and labeling

Data are rendered useless when results cannot be traced directly back to plants from which samples were collected. This applies to fields, rows, individual plants, and in some cases, specific tissues of an individual plant. Therefore, the next step to successful tissue sampling is the proper labeling and identification of the material being sampled.

Labeling for low volume sampling can be as simple as manually writing on the sample container with an indelible marker. As the volume of samples increases, the procedures need to change to maintain accuracy within the time constraints of the work. Printed tags, labels, and blocks that include barcodes for the laboratory's Laboratory Information Management Systems are common in high-volume sampling programs. The 8 × 12 blocks and racks are typically labeled on the south or west side with Row A serving as north (Fig. 10.1). The laboratory should clearly define the sampling procedure as well as the handling and shipping requirements. To be clear, the laboratory requirements must be followed to avoid delays, imperfect data, or resampling requests.

Plants can be labeled individually, and breeding organizations generally have high-speed printers that use rolls of plastic slip-lock plant labels. Stakes of various sizes can be used to mark individual rows or a series of plants in a

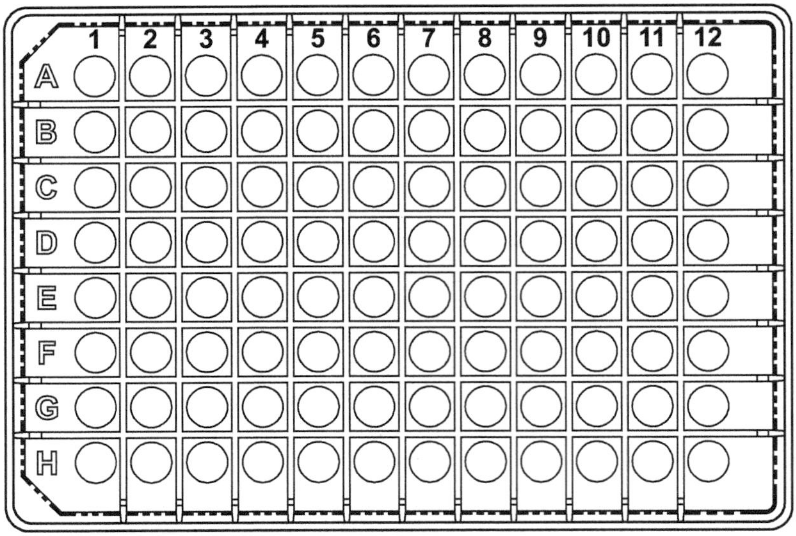

FIGURE 10.1 Standard 96-well layout.

row for bulk sampling. The proper labeling of plants and sample containers is key to preserving data integrity and preventing mix-ups. Samplers must ensure the right data comes back attributed to the correct plant, the correct row, and the correct field. The loading template or pattern must, therefore, also be established with the laboratory when using 8 × 12 blocks.

The 8 × 12 pattern plastics use an alphanumeric system of labeling that is typically molded into the block; rows A to H and columns 1 to 12 (Fig. 10.1). As indicated previously, the depth, style, and manufacturer of the blocks are often specific to the laboratory, and substitutions should be avoided or be approved by the laboratory. Samples can be numbered left to right, west to east, A1, A2, A3 … H10, H11, H12 or top to bottom, north to south, A1, B1, C1, D1 … F12, G12, H12. In addition to the loading template orientation, specific check wells may need to be left empty for the laboratory. Each lab varies with the number and placement of its check wells. Paper sampling templates are one of the most common ways of properly labeling and subsequently loading blocks from well A1 to H12 while in the field. An 8 × 12 paper template is created, printed, and affixed to the block. The paper template is punctured as the sample tube is placed in the proper well. The sampler moves plant to plant in the field puncturing the template to place each new tissue sample according to the predesigned plan.

10.7 Collecting the sample

Young leaves are far and away from the most common tissue type that is sampled for protein and nucleic acid (DNA, RNA) analysis. Good quality tissue ensures rapid and reliable results. Tissue should be actively growing and turgid (not wilted) when sampled. Sampling stressed tissue should be avoided due to the low quality and quantity of both proteins and nucleic acids that are targeted by most assays. Excessively damp or wet leaves can be blotted dry before collection. When using a lyophilizer to freeze-dry tissue, avoid wet or unevenly wet leaves as they can produce uneven products for the lab to test or store. For tissue samples that are shipped fresh, degradation should be a concern. Samples that are heavily colonized with actively growing saprophytes should be avoided during sampling. Diseased tissue, infected with live plant pathogens, should also be avoided. Infected tissues can possess foreign proteins, DNA, and metabolites that may affect testing. The presence of plant pathogens may also prevent the shipment of samples due to plant protection and quarantine regulations. Shipping prohibited plant materials, live pests, and soil may violate the plant protection and quarantine regulations set by the destination's designated authority. This is especially true when shipping between states, countries, and regions that have established phytosanitary requirements.

Cleanliness is also important for successful sampling. Sampling tools should be kept clean of cell exudates that can create cross-contamination. Cleaning of the sampling tool can be as simple as punching a clean piece of paper or wiping the cutting surfaces. For some applications, a solvent such as alcohol may be required. A simple rubbing alcohol, such as 70% EtOH (Ethanol), is effective but results may vary across tissue types and plant species. The use of 70% percent of alcohol is recommended over stronger concentrations. Pure alcohol may cause proteins to coagulate, effectively sealing single-celled organisms such as bacteria or stray plant cells to the surface you are trying to clean. The higher osmotic pressure of 70% ethanol also allows the alcohol to cross cell membranes and destroy bacterial cells by denaturation. A 2% solution of NaClO is another cleaning solvent that can be found in the literature. Regardless of the method, it is important to validate the cleaning process. This can be achieved by subjecting equipment to known positive plant tissues followed by known negative tissues. Testing the resulting samples will reveal the persistence or carry-over with and without cleaning.

Along with clean sampling, it is also important to protect samples from sun, heat, and other elements that can cause sample degradation. Collect samples as close as possible to the time you will be shipping or lyophilizing them. All types of tissue samples should be free of moisture, soil, and insects. Blocks should also be protected from foreign material, moisture, and physical damage. Coolers and cold packs are often used to keep blocks cool until they can be brought into a controlled environment. However, even if a cooler is used, the time between collection and the next step in the process (test, ship, or lyophilize) must be minimized.

10.8 Safety considerations

The safety of sampling personnel is an important consideration during fieldwork. Basic exposure risks include sun, heat, and other elements inherent to working outside. There are some additional issues specific to handling tissue that should also be addressed to protect workers. Pesticides are commonly used in agricultural fields and in greenhouses, and restricted entry intervals must be respected or addressed through proper personal protective equipment (PPE). PPE includes, but is not limited to, long sleeves/pants, socks, close-toed shoes, a hat, and sunscreen as well as a consideration for protective eyewear. Eyewear should be considered for UV protection as well as physical protection when working with taller plants/crops (e.g., maize). As all tissue sampling involves intimate contact with plants and soil, pesticide residues should be considered. Pesticide residues are especially a concern where hand sampling is employed. Depending on the height of the crop to be sampled the ergonomics of sampling can be an issue. Constant stooping or bending can be a concern along with any number of repetitive actions. Working with a low-growing crop can be facilitated by working in teams of two. The sampler sits or kneels and moves along the row handing the sample or sampling tool to the loader. Two-person teams can also be of benefit for all crops by allowing better focus on the plant by plant sampling and loading samples into containers. Periodically, the two should also switch roles to avoid mental and physical fatigue. When collecting samples from taller crops, PPE such as protective eyewear and long sleeve shirts should be considered to shield eyes and skin from abrasions. Sampling tools can also pinch or cut depending on the type of tool and its design. Some tissue sampling tools used for sampling plants were originally designed for mammalian histology, dissection, and biopsy applications and are very sharp. Examples include the Harris and Whatman Unicore samplers. Worker protection is an important consideration when planning any type of tissue sampling program or project.

10.9 Specific procedures

Sampling can be as simple as placing entire leaves or leaflets in plastic bags. It can be done entirely by hand or with more sophisticated leaf punch tools that create 4–6 mm leaf discs, not unlike a paper punch. High volume sampling is most often used by breeders interested in sampling a large number of plants. The following sections describe some, but by no means all, of the common tissue sampling procedures.

10.9.1 Tube sampling

Tube sampling uses the 8×12 pattern plastics that hold individual removable tubes in a rack system. Each individual tube is removed from the rack to collect a leaf disc cut from the plant with the open end of the tube. To perform the collection, the open end of the tube is pressed into a leaf that has been placed against a backer to cut the leaf disc. Simple card stock or small note pads can be used as a backer. The leaf disc typically remains in the mouth of the tube when the tube is returned to its rack. The next tube selected from the rack can be used to push the previous leaf disc down into its tube (Fig. 10.2B). The process is then repeated until the rack is filled. The mouth of the tube is not the ideal cutting tool, and care should be taken to avoid cross-contamination with leaf exudates. Advantages of this system include a uniform sample size with no equipment other than the backer and the tubes. Also, this system is capable of collecting a large number of samples rapidly. One of the disadvantages of tube sampling is the crushing of tissue on the lip of the tube. This can result in plant extracts that may cross-contaminate other samples. Another disadvantage is the use of loose tubes in a rack. While most tube systems use a cover or lid to lock tubes in place when not in use, loose tubes can be misplaced in the rack or the entire rack of tubes can be spilled.

10.9.2 Hand sampling

Sampling by hand is not necessarily a low volume approach and can be used for high throughput automated laboratories. However, hand sampling requires samples of equal size to avoid clogging or overwhelming the liquid handling activities of an automated lab. To sample by hand, a small amount of leaf tissue is pinched off and rolled between the fingers before placing it in the sample container (Fig. 10.2C). The rapid nature of the technique lends itself to high volume sampling that uses 8 × 12 blocks. A preprinted paper template should be used as described in the Marking and Labeling section of this chapter. The paper template that is affixed to the 8 × 12 block is punctured before placing the tissue sample in the proper well in the block. A mini pestle or similar rod is then used to push the tissue sample down into the well. The advantage of this method is the rapid sample collection without special tools. As a disadvantage, the technique can be subject to sample size variation that creates processing issues in the laboratory. Too much or too little sample can impede or obstruct the DNA extractions and subsequent reactions performed. A high degree of dexterity and practice is required for hand sampling, and exposures to pesticide residues should be considered.

10.9.3 Disposable test-tube cap method

The disposable test-tube cap method of sampling uses a specific tube that is common in laboratories and test kits. The test tube has a hinged cap making it a sampling tool and container all in one. The lip of the cap fits into the tube, and when closed over a leaf, a leaf punch or disc is created. For some protocols, the leaf is folded over to achieve double-leaf discs. Analysis can be conducted in the field by macerating the leaf disc(s) using a mini pestle. Depending on the kit protocol, a buffer or similar solvent is added, and the extract can be tested with a lateral flow device directly in the tube. The tube can either stay with the plant during field analysis or can be placed in a rack to be tested elsewhere. Once analysis is complete, the results can be recorded, and the plant can be labeled or removed if undesirable within minutes of sample collection. Test tubes can also be submitted to a laboratory for analysis. Laboratory submission requires coordination with the lab regarding the method of submission. The labeling of plants, tubes, and racks are all considerations for laboratory submission. Advantages of the test-tube sampling technique are the lack of tools required to clean the cutting tool, the ability to obtain uniform sample sizes, immediate sealing of the sample containers, and that the tubes can stay with the plant being analyzed in the field. The tubes are often included in test kits simplifying purchasing. One disadvantage is that cap may not cleanly cut the leaf tissue during the closure process, making for an untidy sampling process. The method requires good dexterity and is not typically used for high volume sampling due to the time it takes to label individual tubes and collect the sample. When testing leaf tissue, be sure to use a kit that is specific to leaf tissue testing. Grain kits should be avoided as they may not have been developed or validated for fresh tissue. As mentioned previously, leaf tissue contains large amounts of chlorophyll; therefore, test material may turn green resulting in a false positive, or antibodies may be rendered ineffective by the matrix resulting in a false negative.

10.9.4 Core samplers

The Harris Uni-core and similar sampling tools can be used for any volume of sampling where a high degree of precision in leaf disc size is desired. The open end of the tool is pressed against a backer and twisted to cut the leaf disc. Simple card stock or small note pads may be used as a backer; however, a specialized self-healing cutting mat is recommended due to the razor-sharp cutting edge of the Harris Uni-core tool (Fig. 10.2D). This technique can be used for high-volume sampling that uses 8×12 blocks. A preprinted paper template is used as described in the Marking and Labeling section of this chapter. The paper template is punctured before placing the tissue sample in the proper well on the block. The Harris Uni-Core design includes a plunger that pushes or ejects the leaf disc out of the cutting tool. This tool can also be used to rapidly collect multiple leaf punches, 7–10 depending on leaf thickness, which can be tested in bulk as a more economical and higher throughput screening process. With bulk sampling, positive results may condemn the entire subset of plants or a second round of sampling may be applied after the initial results are returned. The tool should be kept clean of cell exudates that can create cross-contamination. As mentioned in the Collecting the Sample section of this chapter, a simple wiping of the cutting surfaces by twisting the tool on a paper towel may be sufficient. However, for some applications, a solvent such as alcohol or 2% NaClO may also be required. Advantages of the Harris Uni-core are uniformity and speed. Disadvantages range from cleaning to safety. The tool is razor sharp for the collection of muscle tissue samples. A degree of dexterity and proficiency with a clear understanding of safety is required when using razor-sharp cutting tools. As the tool dulls, cleaning may need to be intensified and staff need to be aware of when to discard a dull tool. Additionally, there are tools similar to the Harris Uni-core tool, which can be sharpened with a specialized sharpener rather than discarding when the tool becomes dull.

10.9.5 Tissue punch

Tissue punches include any of a number of designs that mimic a single-hole hand-held paper punch. While a paper punch can be used as a sampling tool, most are not suited to cutting soft, moist materials cleanly or uniformly into a complete disc. Midco Global, a supply company, offers a punch designed specifically for leaf tissue sampling (Fig. 10.2E). This device can hold a tube for the leaf disc to fall into upon punching. There is also a rod extending from the bottom of the tool that can be used to push leaf discs into the sample tube before returning it to the sample rack. This tool can also be used to rapidly collect multiple leaf punches that can be tested in bulk as a more economical screening process. With bulk analysis, positive results may condemn the entire subset of plants or a second round of sampling may be applied after the initial results are returned; therefore, the tool should be kept clean of cell exudates that can create cross-contamination. Cleaning may be achieved by punching clean paper or wiping down the cutting parts of the tool, but for some applications, a solvent such as alcohol may also be required. Most designs allow for the tool to be disassembled for regular cleaning. Simple rubbing alcohol such as 70% EtOH is effective but results may vary across tissue types and plant species. Pure alcohol may cause the protein to coagulate effectively sealing single-celled organisms such as bacteria or stray plant cells to the surface you are trying to clean; thus, the use of 70% of alcohol is recommended over stronger concentrations. The higher osmotic pressure of 70% ethanol also allows the alcohol to cross cell membranes and destroy bacterial cells by denaturation. A 2% NaClO solution is another option that can be found in the literature. The advantages of using a tissue punch include the rapidity of sample collection and the precise size of the leaf discs generated. Disadvantages range from cleaning to possible limitations in sample containers. This type of tool is primarily designed to hold single tubes or vials (see the tube sampling process detailed in this chapter). With some high throughput labs requiring solid blocks for samples, the purchase of a tissue punch should be considered only after determining what plastics will be required.

10.10 Shipping samples

Before shipping, blocks are typically sealed with a plastic film, gas-permeable membrane, sealing mat, or cap system. Some types of sealing mats may be designed for easy piercing with a pipette tip or the shaft of a syringe. Solid sealing mats are designed for removal by the lab before processing the sample block. Again, the type of sealing mat or film may also be prescribed by the laboratory. Laboratories may also have specific packing requirements for shipping. For example, fresh leaf samples may need to be double bagged and shipped with no heavy objects such as ice packs that may crush tissue or puncture bags during shipping. The method for sealing blocks should be specified by the lab before sampling, and the materials must be on-hand for immediate shipment. In certain instances, caps or sealing mats can separate from the block during shipment due to improper sealing or the pressure changes that occur during shipping. Letting blocks warm to room temperature after lyophilization helps ensure that caps are not displaced due to the expansion of warming air. The use of a large ink roller, cap press, or similar tool that helps apply the sealing mat or caps securely can ensure delivery without spillage. Samples should be sealed and prepared for shipping in a clean, preferably air-conditioned, environment. For lyophilized samples, static electricity may cause dried tissue samples to move unexpectedly and cling to surfaces. Consult with the laboratory if the issue of static electricity persists and ask if they have a recommended way to remedy it. The remedy may range from a household antistatic product to a benchtop air ionization unit.

Samples shipped internationally or between territories that have phytosanitary requirements will require additional preparation, inspection, and/or permits. Samples that are incorrectly shipped can be stopped by inspectors at the port of entry and refused entry or can even be destroyed. Laboratories with permits to receive such shipments may not be able to receive unsolicited samples. Improperly shipped samples that do happen to make it past a port of entry may be destroyed as a condition of the laboratory's permit. In all cases, communication between those shipping samples and those receiving them must be clear. Package samples in sturdy containers that will survive the type of mishandling that can occur in commercial shipping. Always use a reputable carrier capable of tracking packages.

10.11 Summary

This chapter has outlined several tissue sampling techniques, how they are used, and for what purpose. Unlike testing bulk seed and bulk commodities, plant and field analysis involves taking a sample from a growing plant. Testing the tissue sample establishes the identity of a crop or the identity of a specific plant within it (e.g., transgenic and conventional). Plant breeders have eliminated much of the guesswork in selecting parent materials and progeny through the use of tissue sampling and testing. The high volume of samples generated, along with the turnaround times necessary to make selections of actively growing plants, has necessitated the development of highly automated laboratories. These automated laboratories require specific sampling procedures and materials to perform the necessary tests properly. Coordination with the testing laboratory is key to achieving actionable data in a timely fashion. Sampling details need to be established well ahead of any sampling in the field. A wide range of techniques, containers, and processes exist within the broader subject of tissue sampling. While plant breeders are by far the biggest users of tissue sampling techniques, seed inspectors and crop advisors can also use tissue sampling and testing techniques. Inspectors and advisors are more likely to use test kits in the field to determine if a specific trait is present or absent. Lateral flow devices are available for many of the traits currently on the market allowing for rapid identification in the field. While purity testing can be done with these test kits, the industry focuses on testing tissue during the breeding and parent seed phases of crop development. The number of plants is relatively small, more manageable, and under the strict control of the breeder or breeding organization. Any errors or unwanted genetics that contaminate the breeder's work will be perpetuated during the commercial production process. With the number of plants involved in commercial production, testing bulk seed and grain becomes the most effective method of addressing undesired traits in crops. It is crucial to understand how different sampling techniques and testing plans can be implemented in the crop improvement and production process.

Sampling for the sake of sampling serves no real purpose and may return misleading results. Be sure to evaluate your objectives and determine if tissue sampling is even necessary. What is your role? Are you enforcing a regulation or performing a quality assurance service? Are you looking to detect low-level contamination of a crop or gross contamination and misidentification? The amount of resources expended to find a minor issue, such as adventitious presence, is typically used by breeders to ensure the purity of new varieties and hybrids. Census testing of plants in the nursery followed by bulk seed sample testing provides a critical control point in the development and commercialization of new seed products. Parent and commercial seed producers may use tissue sampling to a lesser degree. Looking for Genetically Modified Organism (GMO) contamination in a commercial field by leaf tissue analysis may be uninformative and economically prohibitive. Consumer, environmental, and economic considerations motivate sampling in many parts of the industry. It is important to determine what can be effectively achieved or resolved at the field level. Be sure to sample and test when and where it will be most effective for your goals or program.

Chapter 11

Testing laboratory design and management

Margit Ross[1] and Reetika Rawat[2]
[1]*Corteva Agriscience, Johnston, IA, United States;* [2]*Joyn Bio, Boston, MA, United States*

11.1 Introduction

The establishment and maintenance of a laboratory intended for the detection of genetically modified organisms (GMO) is a comprehensive process involving much more than just the spatial arrangement of the laboratory. Effective management to sustain data quality is reliant on the ability to identify the proper instrumentation suitable to meet the upcoming analyses demands, establish standards and structure regarding data, instrument, and personnel management, and develop robust assays, controls, and scoring capabilities. In this chapter, these areas are outlined, with examples provided for developing various approaches for comprehensive laboratory management.

11.2 Laboratory design

A primary consideration in the design of a laboratory is the layout of the laboratory. Forecasting, based on expected throughput, personnel, data management, allocated budget (operating and capital), and equipment is recommended to be completed before setting up the laboratory. This type of planning and information helps to maximize the use of dedicated space, minimize the potential for cross-contamination or carry-over contamination from reagents, samples, aerosols, etc., purchase equipment that will meet the laboratory demand, and prepare the laboratory for future growth.

Assays to detect GMOs are highly sensitive and can detect even low-level presence. Contamination from reagents, other samples, assay products, etc. are a risk to any analytical laboratory and can result in the production and/or release of false-positive and invalid results. The design decisions for the analytical laboratory represent a big factor in controlling this issue and should include efforts to minimize all sources of contamination risks.

Application of Sampling and Detection Methods in Agricultural Plant Biotechnology
https://doi.org/10.1016/B978-0-323-99293-0.00009-X

Within the expected workflow and dedicated workspace(s), consideration of the level of throughput, necessary turnaround times for data, and number of personnel can help to determine what type of equipment and instrumentation is needed. Careful thought needs to be put into the laboratory design based on the following questions.

i. How many samples will need to be analyzed in a given period of time? Are there peak periods of time for analyses? If sample numbers vary from limited numbers, at certain time-points, to high numbers at others, it is important to plan and design for handling the higher numbers, especially if timing of the released data is critical. Workload for analyses of high sample numbers in the laboratory, provided there is a long period of time to completion, can be spread across instrumentation and labor resources to allow for the balance of the work. Where timing is critical, for example, completion of data for decision making on which plants to pollinate, etc., planning and design of the laboratory to accommodate the demand is imperative. Instrumentation is discussed later.

ii. Are the samples to be analyzed internal or external of the facility? Samples submitted for analyses may require different handling procedures within the laboratory if derived from an external source versus an internal source. For instance, plant samples propagated in a controlled environment may be grown in a more aseptic environment than those harvested from a field setting. If samples are to be received from a field environment, managing dirt and other contaminants, such as mold, dust, and insects, carried in with the samples, should be expected and will need to be handled appropriately in the sample receiving area of the laboratory. Mitigation of potential sources of contamination derived from the receipt of samples, regardless of their source, is imperative to prevent issues with the analyses.

iii. What technology platform(s), equipment, and instrumentation will accommodate the workload? The processes for receipt of samples, the number of samples per a given time period, the level of data quality expected, and the time expectations for analyses provide guidelines as to what type(s) of technology platform(s) should be considered, how much capacity each instrument, or instruments, will need to provide as well as, potentially, what specific instruments may be needed. Some example scenarios are provided later.

a. If a high throughput laboratory is needed and analyses require significant pipetting, robotic instrumentation for liquid handling could be considered. Liquid handlers, in different forms, can help with extraction processes, reagent distribution onto plates, and template delivery.

b. If there are peak times for receipt of high sample numbers and required analyses, compare the number of personnel resources to the amount of work to be completed using manual pipetting as well as the amount of work that could be completed with the inclusion of liquid handlers to determine which approach would allow the work to be completed on time.

c. Investment in and use of instrumentation that includes the capability for multiplexing is beneficial to cost control and can lead to faster raw data production. Included with this calculation should be the time necessary to develop and validate the required assays.
d. If data quality is a factor, determine if robotics will improve consistency and minimize errors or if there are other issues like pipetting inconsistencies that need to be addressed.
e. GMO detection technologies are rapidly evolving and advancing, special thought should be given to accessibility to different technologies. If a specific technology is available and is identified to meet the needs of the laboratory, adequate technical service support and provision of raw materials required by the instrument both need to be available to maintain the technology.

iv. What are the stop points, or potential bottlenecks, within the laboratory that may slow the workflow? Points in the process that can be identified as stop points are areas where the workflow may become time consuming or confusing. In designing a laboratory, example areas where this could happen are listed below with some solutions.

a. Inability to receive, or manage, large groups of samples at one time. Consider receiving the samples on a multi-well plate rather than as individual tubes. This can help with sample management as well as help to organize multiple groups of samples together.
b. Inconsistent sample size, requiring special handling to normalize for analysis work. Inconsistent sample sizes may require additional instrumentation for quantification and dilutions. Consider requiring a specific size per tissue type or sample type to enable the workflow process.

v. What other responsibilities will lab personnel have outside of general laboratory work? Lab personnel with outside responsibilities will impact turn-around time on the release of analysis data. The design should take this aspect into consideration, especially where it impacts workload, expected timelines for data, and types of instrumentation to select.

An ideal laboratory design is one that minimizes waste and maximizes resource usage while satisfying the requirements to minimize contamination. Using this type of approach in the design is helpful, especially regarding consumables, capital, and the overall budget. Purchasing equipment and instrumentation that serves multiple purposes can help with this. For example, if the laboratory is to conduct real-time PCR, an instrument that has filter sets to allow for the conduct of multiplex assays allows the user to run one or multiple assays in a reaction well. Liquid handling robots that have fixed, washable tips, and/or allow for decontamination of the pipette tips maximize the use of the pipette tip through multiple uses and minimize the consumable cost and plastic waste. Similarly, software supporting the selected instrumentation that has user programmable methods allow for in-house process improvement designs and greater flexibility in usage.

The budget required to design, set up, and maintain a laboratory will always have limits. Careful analysis of spending on equipment and reviewing what purpose(s) the equipment will serve in the laboratory can help to justify the capital cost. Equipment that is multipurposed is ideal, as described earlier. From an operating cost, before purchasing equipment, determine if the equipment usage will save money in the long run regarding consumables.

A good laboratory design for GMO detection must be sustainable over time and must make sense to the personnel. Steps to provide guidance and clarity can be provided through both process and equipment Standard Operating Procedures (SOPs) as well as supporting worksheets for documentation. Training for each SOP and process is critical for safety, data quality, and maintenance of the instrumentation. In writing an SOP, criteria to consider for inclusion are the following:

a. Purpose and scope
b. Definitions
c. Safety
d. Data security requirements, if software is involved
e. Cleaning/inspection
f. Operation/calibration
g. Routine and nonroutine maintenance
h. References
i. Responsible person

Another factor to keep in mind is if the laboratory must be certified for quality systems like ISO 17025 or ISO 9000 series. Most laboratories use quality systems as a method of assuring consistency of service to a defined set of standards or customer expectations. There are many quality systems developed in different countries and some of these are accepted internationally. Planning which quality system would be appropriate and understanding the quality management requirements for that system would be very important. Understanding and incorporating these requirements at the time of the laboratory design and planning stage would make the certification process easier and more seamless.

Laboratories also need to have a process in place for the labeling of regulated material and to make sure they are compliant with the governing agencies. All regulated material should be tracked, stored, and disposed of appropriately after testing. Any reference material that will be used for method development and as controls should be stored and inventoried appropriately to avoid the unintended release of any regulated material in the environment.

11.3 Laboratory workflow

It is important to know the proposed laboratory workflow to get a good understanding of the space requirements before designing and managing a GMO testing laboratory. Some of the factors include sample types to be tested, contamination risks, and direction of the workflow. DNA-based testing including PCR is more sensitive to contamination when compared with protein-based testing, hence understanding the contamination risks and minimizing these risks is imperative for the success of any testing lab. Another important consideration is the work process flow of people working in the laboratory and the resulting traffic patterns based on their responsibilities.

i. **Sample type**: Sample types can generally vary from food, fiber, grain, oilseed, or leaf tissue. Preprocessing and sample extraction protocols used in the laboratory will largely depend on different crop and tissue types. Sample type will also determine the equipment requirement and the infrastructural needs for sample containment. For example, processing leaf tissue may produce less dust when compared with ground seed flour, hence measures for minimizing and controlling dust in a seed testing lab are crucial.
ii. **Contamination risks and minimization**: One of the benefits of PCR-based testing methods over other DNA- or protein-based testing is high sensitivity and detection of even trace levels of a target sequence. However, a major drawback of this highly sensitive technique is that even minimal amounts of contamination can result in false positives, and special care needs to be taken to minimize these risks (Rys and Persing, 1993). Some of the sources for cross-contamination in PCR can be amplicons/plasmids, aerosols, contaminated reagents, proximity to other labs, and even personnel. These topics are discussed below in detail:

  a) Amplicon/plasmid: Amplicon generated after PCR, and plasmids used as controls could be a source of contamination for any new PCR experiments being set up in the same area and as feedback contamination. Hence, the opening of the used PCR tubes/plates for further analysis or any other reason should be discouraged. One option is to eliminate the traditional gel-based post PCR analysis, and instead use florescent-based PCR detection methods like TaqMan, or SYBR green. A second option is to practice the "Forward flow" that suggests separating the areas of activities of post-PCR from pre-PCR (Mifflin, 2007).

  b) Aerosols: If a testing lab is repeatedly testing the same target sequence, it can lead to the accumulation of PCR amplicons in the laboratory environment. In each PCR reaction, about 10^{12} copies of the target sequence are generated, and even the smallest amount of aerosol, that is, 10^{-6}μL would contain at least 10^5 copies of the target sequence (Persing, 1991). If uncontrolled, this could be a huge problem leading to the contamination of reagents, equipment, and even ventilation system (Persing, 1991). To prevent aerosol contamination, physical as well as chemical methods can be applied. Use of positive displacement or filter tips when pipetting and the use of a laminar hood to prepare PCR samples and reagents are some physical isolation methods. PCR plates may also be permanently sealed to prevent contamination from post-PCR amplicons. Uracil-DNA-glycolase enzyme (UDG or UNG) is very effective at reducing contamination in PCR and can be used in the master mix (Longo et al., 1990).

c) Contaminated reagents: Assays to detect GMOs are highly sensitive. Contamination from reagents, other samples, assay products, etc. are a risk to any laboratory and can result in the production and/or release of false-positive and invalid results that can lead to excessive repeat testing. Autoclaving the laboratory-made extraction buffers or filtering the reagents through two 0.45-μm nitrocellulose filters where possible is highly advisable (Mifflin, 2003). It is advised to aliquot sterile reagents bought from vendors into smaller aliquots, ideally for one-time use, to prevent large batches from becoming contaminated.

d) Proximity to other laboratories: The specific location of the GMO testing laboratory and proximity of any other laboratory must be considered, especially if similar samples are processed in the two laboratories or even if they use similar assays. For example, the proximity of a laboratory processing samples for protein assays to a DNA laboratory can easily contaminate the DNA samples leading to false positives, if proper containment measures are not in place. The air supply to the GMO testing laboratory should be kept separate from any other laboratories nearby if possible. The movement of people, equipment, and reagents should be highly discouraged between laboratories.

e) Personnel: Scientists should not only be provided with training on the protocols used in the lab but should also be encouraged to follow good lab practices to minimize contamination risks. Personal protection (lab coats and surgical gloves) should be worn and changed frequently, especially between specific processes. Hands should be washed frequently. The laboratory should be consistently maintained as clean and free of dust. Any tools, benches, or equipment that may come in contact with samples should be cleaned before, during, and after each use with a mild detergent, 10% sodium hypochlorite solution, as applicable. Dust in the laboratory (floor, shelves, drawers, etc.) can harbor contamination and negatively impact sample quality. Filtered pipette tips should be used to prevent aerosol contamination, and they should be changed when necessary. Labware should also be suitably decontaminated, with verification of decontamination before reuse. Avoiding unnecessary movement of supplies to and from outside the laboratory is a good practice.

iii. **Direction of workflow**: Process workflow is also a decisive factor when designing the GMO testing laboratories. Some of the design elements include separate rooms for different steps in the process, positive and negative pressure of the airflow, and how samples move from one room to the other (NATA, 2006; Nolan, 2013; U.S. EPA, 2004).

Dedicated space(s) for specific processes: Ideally different steps of testing should be separated by space and should be performed in different rooms with their own reagents and equipment (NATA, 2006). However, if that is not possible, areas in a laboratory should be assigned and physically separated for key steps with dedicated resources. It is also advisable to perform some high contamination sensitive steps, like PCR assay setup and reagent dispensing, in a laminar flow hood.

General processes in any GMO lab, and some considerations for designing the designated areas of those processes, are:

Sample receipt: An area in the lab should be assigned to receive, track, and store samples until they are processed. As soon as samples are received, they should be scanned/recorded into the database along with the date received so that they can be tracked and processed according to priority. It is important to have proper conditions and enough space for the storage of the samples before they get processed. Some sample types like leaves are perishable and should be stored in refrigerators or freezers. Generally, these kinds of samples must be processed as soon as possible; otherwise, the quality of extracts might be compromised. There should also be a dedicated dust-free area for storing excess nonperishable samples like seed or flour in case a retest is required.

Sample homogenization: It is highly advisable to perform sample homogenization in either a separate room or an isolated area in the lab. Any spills during the process of homogenization should be documented in case the samples are compromised, and the area/equipment should be cleaned immediately with mild detergent, 10% sodium hypochlorite solution, as applicable. Measures should be in place to contain any spill to avoid cross-contamination of the samples. There should not be a trace of dust in the area or the equipment.

Sample extraction: For most of the protein and DNA based GMO testing, sample extraction is required. Extraction buffers might require special storage, hence enough space for refrigerators and freezers should be planned when designing the lab.

Reaction set-up: This area/room is used for preparing master mix or buffers for setting up the testing reaction. Controls and reference standards should also be stored in this area to avoid unnecessary walking between labs. While designing this area of the lab, storage and usage requirements for refrigerators, freezers, and centrifuges should be planned. The DNA or protein samples after extraction can be stored here until the analyses are complete; however, it is important to separate the control storage from the extracted samples. The need for a laminar floor hood should be assessed, and space assigned if required.

Raw Result Production: The production area of the laboratory should have enough space for equipment like PCR platforms, gel running apparatus, and plate readers. All the waste generated in this room should be properly stored until disposal to avoid spills and the contamination of new samples. There should be no movement of plates and reagents from this area to any other room. Any spill should be cleaned up immediately. This room should have dedicated lab coats, and gloves should be changed frequently.

Data analyses and release: Once the raw results are generated, they can be further analyzed outside the laboratory before release to the customer.

Positive pressure to negative pressure: Within the laboratory, airflow should move from positive (higher) pressure at sample handling and extraction, to negative (lower) pressure at analysis to minimize the aerosol carry over and contamination from the analysis lab back into the sample handling, and extraction labs. The source of air supply between the preanalysis and analysis lab should be kept separate when possible (Mifflin, 2007).

Unidirectional transfer: The laboratory layout for GMO detection should have a primary emphasis on minimizing any potential for cross-contamination. Dedicated space(s) and workflow direction should have samples moving in one direction, for example, samples move from receipt into the lab, to extraction, to assay set up, and to analyses (NATA, 2006). The samples from the analyses room should never move to set-up or extraction laboratories. The equipment should be marked clearly to indicate in which area they belong.

iv. Work Process Flow: In the design of the laboratory, workflow patterns become established based on where equipment and instrumentation may be located as well as how personnel use them. To maximize productivity, minimize the potential of contamination, and allow for work efficiencies, it is recommended to examine the proposed lab layout of benches and equipment and have the working area, or group, in the laboratory draw their workflow patterns on the map, defined by color, as shown in the following examples (Fig. 11.1).

Visible in Fig. 11.1 are the following:

a) Five working group processes within the laboratory, each with cross movement
b) Single point of entry and exit
c) Linear lab layout of benches

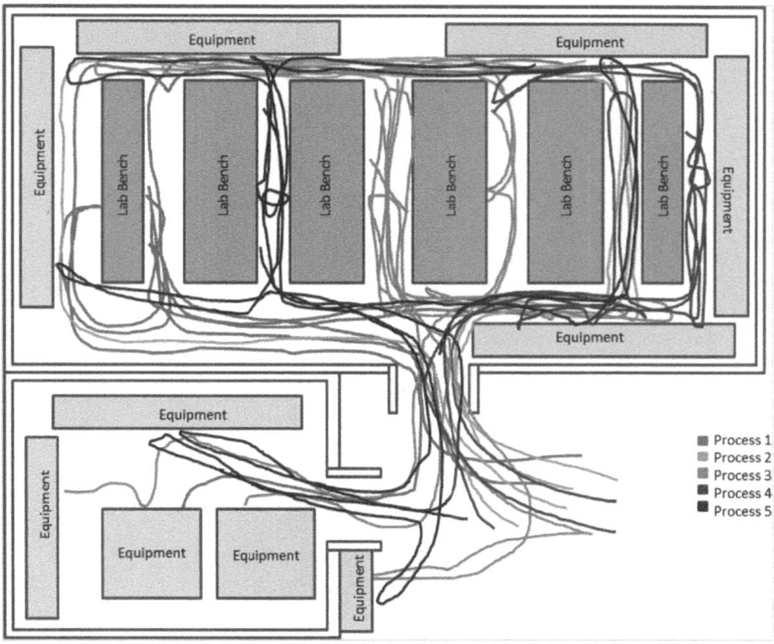

FIGURE 11.1 Workflow and traffic patterns of a "standard" laboratory.

d) Common use items are spread across lab and located on outside walls
e) Work processes migrate toward corners and generate lots of motion
f) Some work processes use the entire length of the laboratory, based on workflow patterns

Following an assessment of workflow and traffic patterns, a revision or optimization of the design can occur to reconsider the issues listed earlier regarding equipment/instrument placement, cross movement patterns, and bench layout, as in the following example (Fig. 11.2).

Visible in Fig. 11.2 are the following:

a) Decreased cross movement through creation of common use equipment areas
b) Recognition of the need for an additional entry between rooms
c) Creating main door as an exit only to decrease traffic and potential for contamination
d) Nonlinear bench layout determined and set based on work process
e) Designated areas for specific lab processes

Consideration of the workflow patterns in lab design can create greater efficiencies in the laboratory. More importantly, a good design can maximize personnel satisfaction with their work environment, yield higher productivity, reduce contamination, and improve data quality.

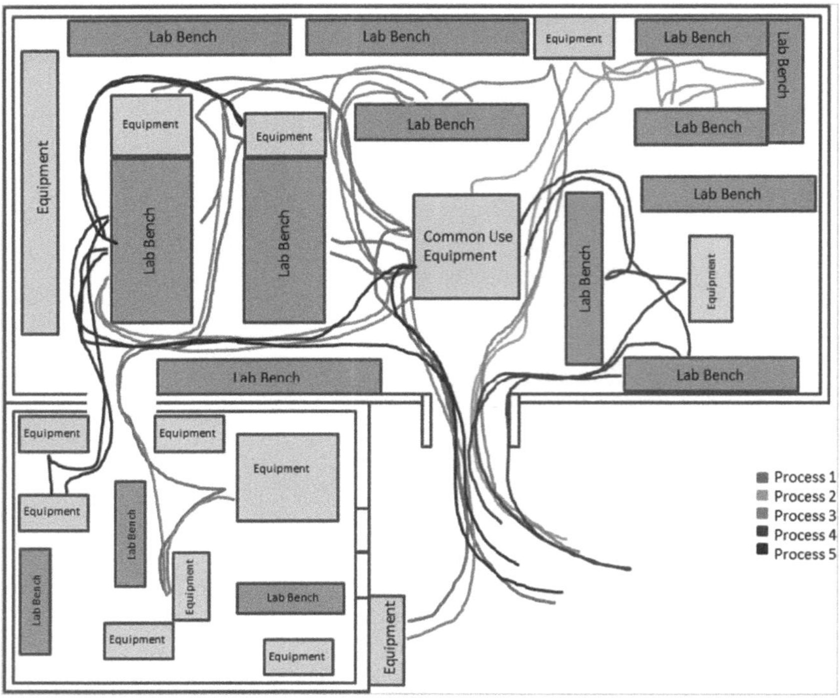

FIGURE 11.2 Revised workflow and traffic patterns to minimize cross-movement.

V. **Data management**: One of the requirements of the GMO testing laboratory is to be able to record, analyze, and retrieve data associated with all the steps right from sample receipt up to data release. Data management is important for detailed tracking and troubleshooting to increase the overall quality of the results. Many times, retrieval of raw data is required to answer customer questions. Links of all information to the sample should be through a unique ID for efficient retrieval of data. Some of the information that is important to track and record is as follows:

- Sample information that includes unique sample ID, location/customer information, sample type, date of collection, shipping date, storage, condition, etc.
- Tests/assays to be performed.
- DNA and protein extraction methods, including lot numbers and expiration dates of critical reagents.
- Operator information for each step
- Equipment maintenance data like dates of major service or calibration and results.
- Raw score and information on tracking assay plates.

- Time and date when samples were processed in the lab, and when data analysis and interpretation were completed.

Data can either be recorded manually and stored in paper records using an indexing system or can be recorded electronically. The electronic systems can be maintained as simple databases and spreadsheets or can use a software specific for laboratory information management (LIMS). The decision on which data management process to use depends on the laboratory requirements such as data throughput and complexities in the testing step. Some testing laboratories use paper records or databases and spreadsheets for recording the validation steps but use LIMS to track production samples and analyses of the raw data to generate test reports.

11.4 Physical laboratory processes

i. **Documentation Practices**: Before starting physical work within the laboratory, documentation practices for sample receiving, storage, handling, and management should be in place. Bar-coding of the samples can streamline the process and eliminate any human errors when handling a large number of samples. Inspection of all samples should be done at the time of receiving to check for sample integrity during shipping. Sample deterioration or any loss of sample containment, for example, broken tubes or opened tubes or tube covers resulting in loss of sample and cross-contamination, should be documented. It is highly recommended to have a process in place to handle compromised samples and requesting of resamples if required.
ii. **Sample Management**: For any type of analysis chosen, it is important to have the right sample size and sampling plan to have the required sensitivity, and to avoid any issues due to nonhomogeneity of the samples (Gilbert, 1999). As statistical significance achieved with a small sample size is often weak, sampling plan should ensure statistical representation, and sample size should have the required sensitivity (Gilbert, 1999). Optimum sampling strategy is a balance of sensitivity, confidence, and cost (Ahmed, 2002). Sample management will also include correct assignment of testing to each sample and division of samples that require protein and/or DNA testing. Sometimes samples would require sharing of extracted protein or DNA between laboratories for multiple testing, and care should be taken to separate the tissue or make an aliquot before performing any testing to avoid cross-contamination.

Samples should be inventoried and stored under recommended conditions as soon as they are received and checked for integrity. Any delay or improper storage especially of perishable samples like leaf tissue could result in sample deterioration and negatively affect both protein and DNA quality. Whenever possible, a portion of the original sample should be stored for any reanalysis if necessary. Depending upon the sample type and requirements of the testing, arrangements should be made for short/long term storage and tracking of original samples or partially processed tissues

iii. Sample extraction: Methods used for sample extraction will depend upon the type of testing required (DNA/protein) as well as the sample type, that is, whether raw materials (e.g., leaf/seed) or processed products (e.g., food) are analyzed. Soybean and canola seed are high in oil content, whereas corn seed is high in starch content. Each crop may potentially require a different extraction method. These methods may not only be crop specific but also may be tissue specific; for example, corn starch could be a challenge during extraction from seed but not from corn leaf. Food processing like canning at higher temperature and additional liquids like brine, oil, vinegar, or tomatoes leads to degradation of DNA and should also be factored in when deciding on an extraction method (Bauer et al., 2003), and data suggest that a minimum DNA size of 400 bp is required for successful PCR (Meyer, 1999). The extraction methods should be evaluated and validated for the yield, purity, degradation, downstream performance, and the sensitivity of the testing (Elsanhoty et al., 2011).

iv. Cross-contamination: Care should be taken to prevent any cross-contamination during protein or DNA isolation. Estimation of the quality and quantity of DNA or protein should be done before conducting any testing. Depending upon the number of samples tested, the use of liquid handlers and automation of the extraction steps can not only increase the efficiency of the laboratory but can also decrease human error. Controls for the PCR or ELISA-based testing can both be extracted before and validated before use or in real time along with the samples; having controls for the extraction process also is highly recommended. Tracking of the samples and reagents used in the extraction process is imperative for troubleshooting and quality control. Also having a checklist of the various steps in the process is highly recommended. *Type(s) of testing:* Raw plant materials or derived food can be tested for the presence of the introduced DNA or for the presence of expressed transgenic protein. The decision for the type of testing needed will depend upon the nature of sample, traits tested, sensitivity required, the methods available for quantification and qualitative (presence/absence) analysis, and also cost. Both protein-based and DNA-based testing has advantages and disadvantages and all should be given weight when deciding upon the type of testing (Table 11.1). More specific information is provided for both DNA-and protein-based testing in Chapters 4 and 6.

a. *Lateral flow strip:* is a very quick test and setup for this test is fairly easy and requires dipping of the test strip in the crude plant lysate for 5–10 min, followed by reading of the control and test line. This test is very rapid, requires no laboratory equipment, and can be done on site or in the fields directly.

TABLE 11.1 Advantages and disadvantages of different GMO testing platforms.

Type of testing	Advantages	Disadvantages
Protein: Lateral flow strip test ELISA	It is a quick test and can be done at the site/field with almost no lab equipment required. Mostly qualitative. Both qualitative and quantitative assays available.	Protein of interest might not be expressed in all plant tissue types, the levels of protein expressed might vary, not ideal for processed food, sensitivity of the assay might not be ideal, availability of the antibodies, and time required for developing the method is usually more than DNA based tests. Cost of testing can be high depending upon number of traits tested.
DNA: PCR	Both quantitative and qualitative assays available. High level of sensitivity and some assays can even detect up to 0.01% of the DNA of interest. Assays can be designed for almost any gene or any DNA fragment. Assays can be designed for the genetic elements common to multiple transgenic events/traits and require a single test to detect multiple traits. Cost of developing an assay and testing is much lower.	Due to high sensitivity, even small amounts of contamination can give false-positive results. Testing is done in a laboratory and not on the field.

b. *ELISA:* This test can either be done in a microtiter plate (Brown, 2011) or in a coated microtube format. Set up for both require the addition of the diluted samples to the wells/tubes coated with protein-specific antibodies. Among the various ELISA formats, sandwich ELISA is most commonly used for the detection of transgenic proteins as shown in Fig. 11.3. a) Plates precoated with monoclonal capture antibodies specific to the protein of interest are incubated with extracted samples. b) The capture antibody will bind to the antigen (peptides from the protein of interest) and the unbound components of the sample are removed by washing the wells multiple times. c) After washing, a secondary antibody (detection antibody), which binds to the second antigenic site on the transgenic protein and is linked to the enzyme conjugate is added to the wells, incubated, and washed to remove excess antibody (d). e) After washing, the detection substrate is

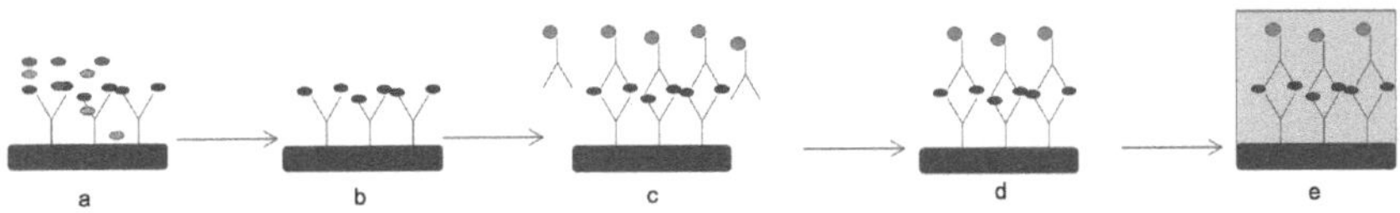

FIGURE 11.3 Sandwich ELISA process.

added, and plates are read on a plate reader. The color signal is related to the concentration of antigen and can be used for quantitative analysis.

All the steps in the ELISA can be automated to increase the efficiency and also to reduce human error. Prevalidated methods and appropriate controls should be used in the testing. A standard dilution curve can be used for the quantification of the targeted protein.

c. *PCR:* Although several DNA-based testing methods including Southern blot are available, PCR is the most common and popular DNA method used for detecting foreign DNA. Set up includes the addition of the PCR mix which includes PCR buffer, forward plus reverse primers, and Taq DNA polymerase in each plate well/microtube as shown in Fig. 11.4. As PCR is a very sensitive assay, to avoid any cross-contamination, it is advisable to add the PCR mix to each well before adding any template DNA. After the PCR set-up is complete, the plates are sealed and kept for thermal cycling in either individual PCR cycling units or in water-bath-based PCR instruments. PCRs can either be quantitative or qualitative. For quantitative analysis, a DNA standard/internal control is amplified along with the target DNA (McPherson and MØller, 2000). The PCR can be analyzed by either gel electrophoresis or by measuring relative fluorescence when performing Taqman PCR.

11.5 Equipment management

A GMO testing laboratory uses a variety of equipment including robotics that should be routinely calibrated and cleaned to ensure correct performance. Standard operating procedures to manage each instrument should be documented and followed. Laboratory personnel should not only be trained to use the equipment but also on how to clean the equipment and execute some basic

FIGURE 11.4 The PCR process.

maintenance. It is important that equipment meets the laboratory's specification requirements and complies with relevant standards. Calibrations of equipment, such as liquid handlers, PCR instruments, pipettes, balances, and gel tanks, should be performed routinely by either a qualified service provider or by the onsite staff trained for this task. Intermediate internal checks of calibration should also be done from time to time by the laboratory personnel to ensure accurate operation of the equipment. Records of internal and external calibration checks, routine maintenance, and cleaning should be maintained. Spills on any equipment should be thoroughly cleaned before any use to prevent contamination of samples.

For GMO detection by PCR, there are many choices of technology platforms to consider. Choices will be made based on needs for:

- Detection sensitivity
- Precision
- Qualitative or quantitative analyses needs
- Throughput requirements
- Laboratory space availability
- Cost

There are four main technology platforms: gel-based, real-time PCR, array PCR, and digital PCR. Each technology platform has both advantages and disadvantages to consider for implementation in a laboratory, as discussed in Chapter 4.

In making determinations for what type of platform is suitable for the laboratory needs, also consider the ability to source the instrumentation with supplies to run and maintain the instrumentation and the availability of technical support to correct issues that arise. Real-time, array, and digital PCR are all precision technologies for which the data quality is high; however, the instrumentation has to be carefully maintained. Some technical support will train users how to conduct routine maintenance on instrumentation, for example, calibrations, diagnostics, etc., which can help both the user to have a deeper understanding of the technology and allow for quicker maintenance.

11.6 Reagents

Laboratories should have a procedure in place for purchasing, inventory management, and storage of reagents. Reagents are expensive, thus maintaining the correct inventory is critical for the smooth operation of a GMO testing lab. Any improper storage and/or batch contamination of reagents will lead to repeat testing, time loss, and also increase the cost of testing samples.

i) **Inventory management**: Database of all available reagents and stocks should be maintained and updated regularly. An inventory management system can help the lab avoid reordering something that already exists in the lab, avoid excess inventory, and also prevent delays associated with running out of reagents necessary to conduct sample testing. Based on the throughput of the laboratory, inventory management can be done either in Microsoft Excel and/or an Access database or use specialized software programs like LIMS. Besides the number of aliquots available, inventory management should also include information such as supplier, lot number, and date received, date put in use, verification date, and date of expiration. Knowing the expected sample numbers ahead of time will be important to effectively manage reagent inventory without wasting expensive reagents due to expiration.

ii) **Hydration and dilution**: Some reagents arrive as lyophilized dried powder and need to be hydrated and diluted before use. As some of these reagents can degrade or precipitate more easily when hydrated, it is advisable to store the dried powder for long-term storage and hydrate only when needed. Any special precautions in preparation or use of reagents, like stability of reagents to light, heat, and other chemicals, should be documented.

There is always a risk of contaminating the reagent; hence it is advisable to have pre-measured aliquots suitable for the type of testing routinely completed in the laboratory. Single-use aliquots will also prevent multiple freeze–thaw cycles for some sensitive reagents.

Quality checks for all critical reagents should be performed and documented before use in production. Verification procedures, criteria for acceptance, shelf lives, and special storage conditions should also be well documented.

iii. Labeling and storage: Standards and reagents must be labeled or bar coded with the name of the reagent, concentration, aliquot number, preparation date, name/initials of the preparer, and the expiration date. Ideally, this information would be linked to the reagent inventory to better manage the available stocks. Bar coding of the reagents also helps in tracking the aliquots used in each experiment for easy troubleshooting.

Proper storage of all reagents is imperative, and manufacturer's recommendations should be followed strictly to avoid any retesting and wasting of resources. Storage areas like freezers and refrigerators should be monitored for temperature fluctuations on a regular basis to prevent any problems.

Organizing the reagent storage area for easy access is also important. As a reagent is placed into the freezer, recording the location by shelf, container, and rack, for example, should be considered to allow easy retrieval.

11.7 Assay quality control

Detection of the presence of GMOs is reliant on the performance of the assays that are used. Whether the assay used is protein or DNA based, amplification and the quality and reliability of the amplification detected are of critical importance to provide valid data. From a laboratory management perspective, having established and validated procedures for addressing the raw data provides a critical management structure for yielding consistency in training analysts, for data interpretation, and for reporting.

Regardless of the assay type selected, having a quality control structure is of utmost importance. Methods to distinguish and control the release of data containing false-negative, false-positive, and nonspecific amplification are necessary to insure the release of reproducible datasets (Nolan, 2013; Brookman-Amissah et al., 2011–12; ISO Technical Committee, 2006). Any assay can be subject to interferences that can influence the resulting data, including, but not limited to, contamination, inhibitors such as polysaccharides, phenolics, chlorophyll, etc., instrument calibration, instrument software, normalization, and fluorescence variability (U.S. EPA et al., 2004).

As PCR is a predominant detection platform for GMO detection (Zel, 2012), examples of how to manage these situations within the laboratory are presented later as guidance to establish the determination of the accuracy of an amplicon as well as how to implement corrective actions should false-positive/false-negative/nonspecific amplification be detected.

For PCR, the areas to focus on within the laboratory are as follows:

- Genomic and no template controls
- Data analyses
- Sample performance
- Robotics
- Instrumentation

For genomic and no template controls, it is important to run controls in sets, where multiple technical replicates per set can be evaluated. Each control has an expected result and should produce consistent and expected amplification, whether in real-time PCR analysis (consistency, where expected, across all dye layers tested) or in gel-based analysis.

Failure to amplify or amplification where not expected in genomic controls can indicate one or more of the following (U.S. EPA et al., 2004):

- incorrect lot number used
- error in plating the controls, for example., pipetting or liquid handler error
- poor DNA quality
- genomic DNA contamination

The no template control (NTC) is used to verify that no contamination has been introduced during the assay setup process. An expected result is no amplification, meaning no linear amplification of any dye layer in real-time PCR and no detected band produced in gel-based PCR.

Presence of NTC linear amplification can be an indicator of genomic DNA contamination. At certain times, nonlinear amplification that is not DNA contamination can be detected. These instances can be an indicator of a software algorithm baseline setting where the default setting is allowing the amplification to cross the threshold. After evaluation of the quality of the amplification, the analyst can decide if it is appropriate to adjust the baseline.

In some circumstances, if adequate replicates of controls of all forms are used, it may be acceptable to exclude <25% of the specific control replicates. An instance where this may be acceptable is where there is proof that there may be have been a pipetting or liquid handling issue affecting the consistency of the plating of the assay components in a distinct well. A second instance is where the software algorithm is falsely scoring a negative control as positive. All other replicates should produce an expected score to allow for exclusions. Some pictorial examples of acceptable exclusions from real-time PCR (Figs. 11.5 and 11.6) are shown later.

Unacceptable circumstances, as listed below, can occur where the genomic controls or NTC controls do not produce an expected result. In the cases described later, it is not acceptable to exclude these technical replicates from the analysis. Rather, the analysis should be rerun once the cause for the unexpected results is understood and rectified.

a) A negative genomic control that scores positive for the gene of interest with demonstrated linear amplification or produces a band on a gel is an indication of DNA contamination.

b) An NTC control that produces linear amplification or band(s) on a gel for the endogenous gene, or for both the endogenous gene and the gene of interest is an indication of DNA contamination.

c) A positive genomic control that scores negative for the gene of interest is an indication that an incorrect control was selected and questions the integrity of the positive scoring sample data.

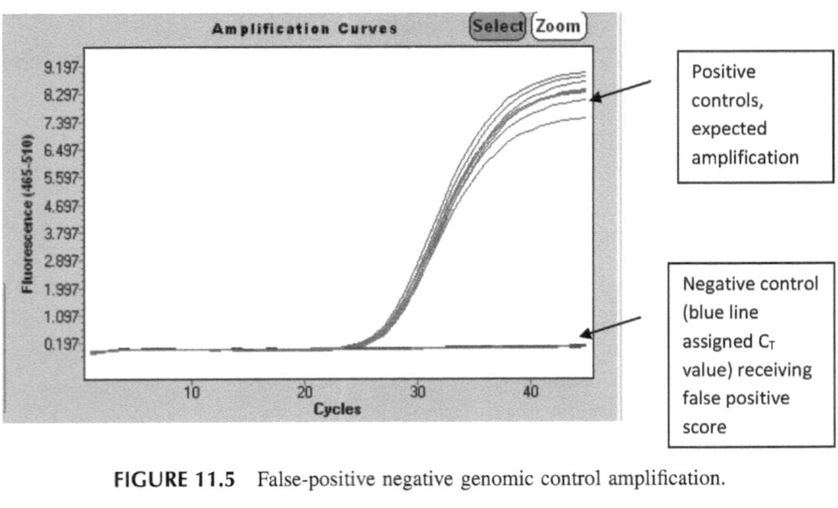

FIGURE 11.5 False-positive negative genomic control amplification.

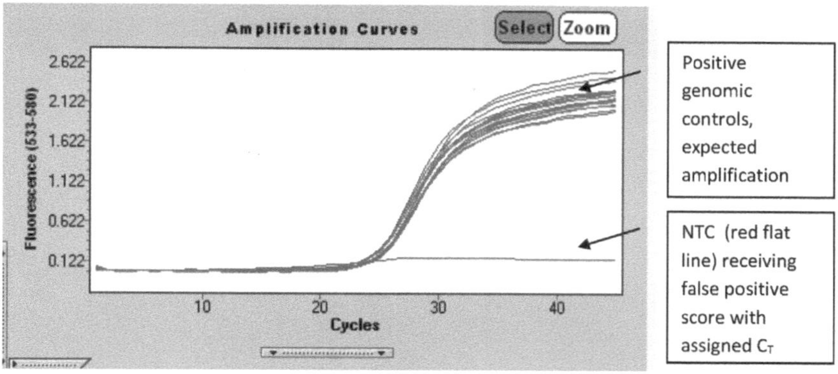

FIGURE 11.6 Software algorithm issues with non-linear amplification in NTC control.

Analyses of the raw data, be it real-time PCR data or observations from gel electrophoresis can cover several areas:

a) Master mix performance
b) Liquid handler performance
c) Sample performance
d) Instrument performance

For master mix, there should be consistency, for example, expected results produced for all controls, in all dye layers as well as consistency across the positive scoring samples, based on C_T values or band amplification.

Patterns of inconsistency should be checked regarding robotics/liquid handling instrumentation. Issues with liquid handlers present themselves in distinct patterns across multiple plates, for example, specific wells and/or rows across multiple plates may indicate the effect of low DNA concentrations or no amplification due to pipetting errors.

Consistency within a set of sample replicates should be evaluated using the relative standard deviation (RSD), or standard error, where the expected value should be $<2\%$. This value is calculated by taking the standard deviation of the C_T values of three or more technical replicates/mean C_T value *100. In general, an RSD value will be $>1.5\%$ if one or more of the technical replicates of a sample differs by 1 C_T value (U.S. EPA et al., 2004). Sample replicates also should present consistency in their amplification curves, as shown in Fig. 11.7.

Instrument performance consistency is a necessary component for producing reproducible data within the laboratory. Continual observations should be made regarding edge effects, increased background, and general anomalous amplification. Any of these observations may provide an indication of problems with the thermal cycler block (dirty wells due to spills or lint within the heating block) or to the calibration of the instrument and should be carefully

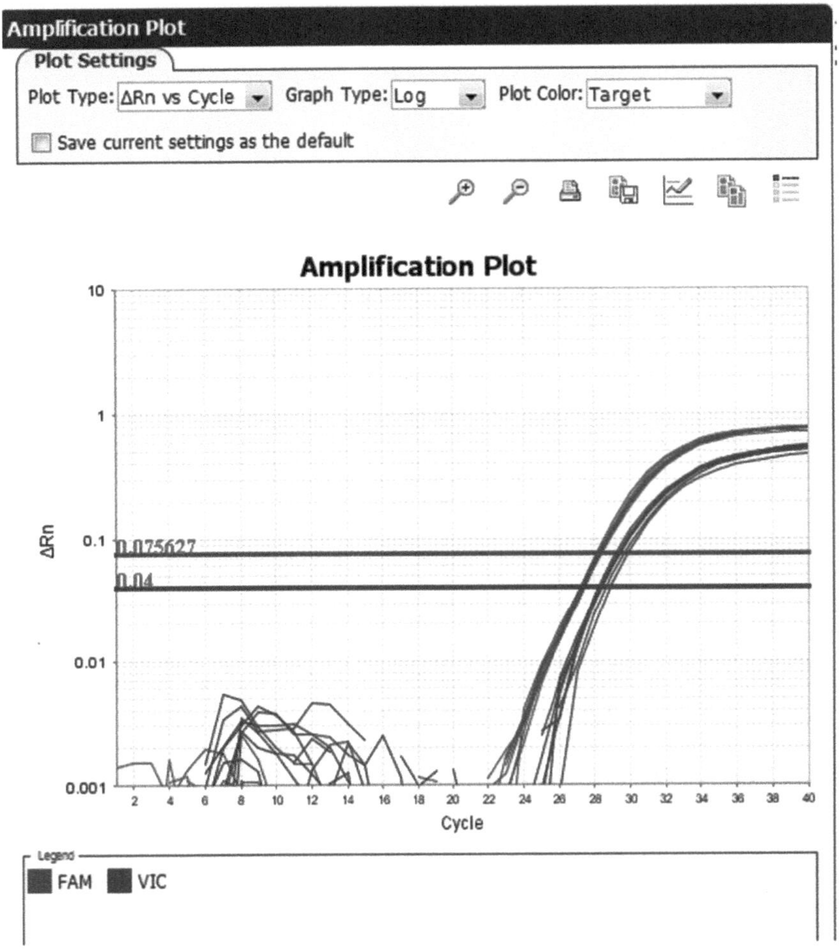

FIGURE 11.7 Consistency of sample replicate amplification of gene of interest and endogenous reference.

investigated. Some examples where each sample replicate can be looked at individually and corrected, due to anomalous amplification, are shown in Figs. 11.8–11.10.

In scoring the data, regardless of the PCR technology platform selected for the laboratory, one must remember that with PCR, there are many variables and anomalies that may occur during an experiment due to the complexity and sensitivity of PCR. Due to these factors, acceptability criteria need to be established and each experimental set of raw data needs to be carefully examined for any required adjustments in relation to the performance of control samples. Control performance for PCR is described in more detail in Chapter 5.

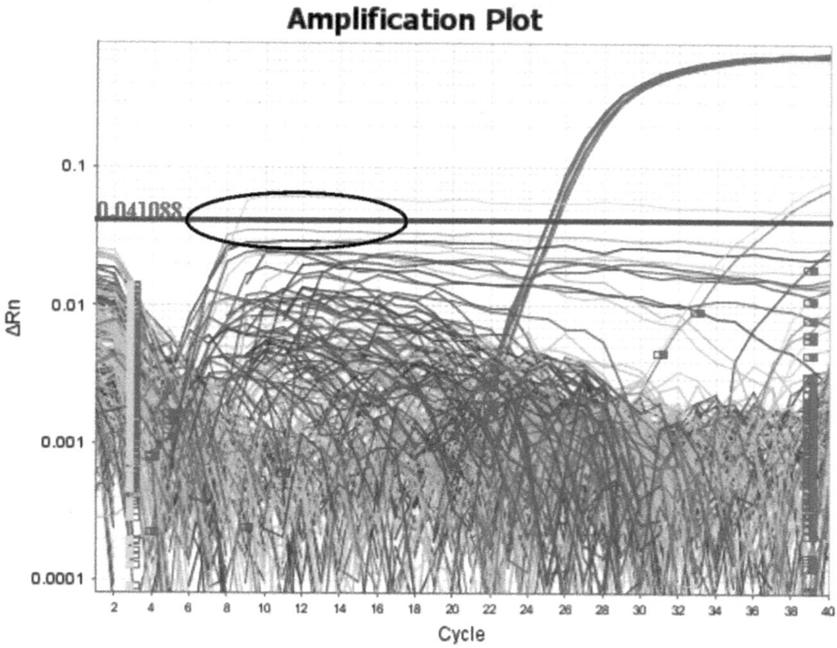

FIGURE 11.8 High background with C_T value assignment (circled) is corrected by a baseline change.

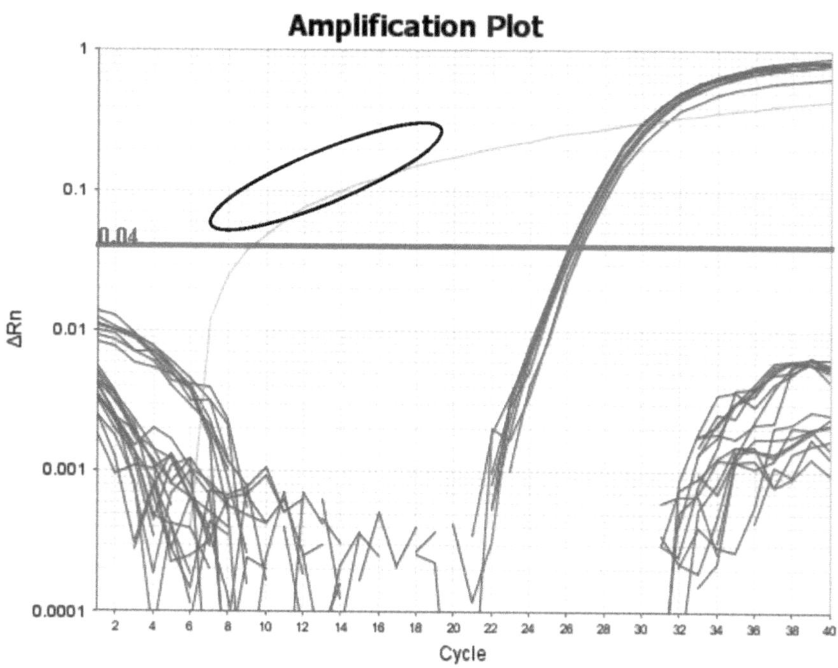

FIGURE 11.9 Anomalous amplification with C_T assignment (circled) is corrected by a baseline change.

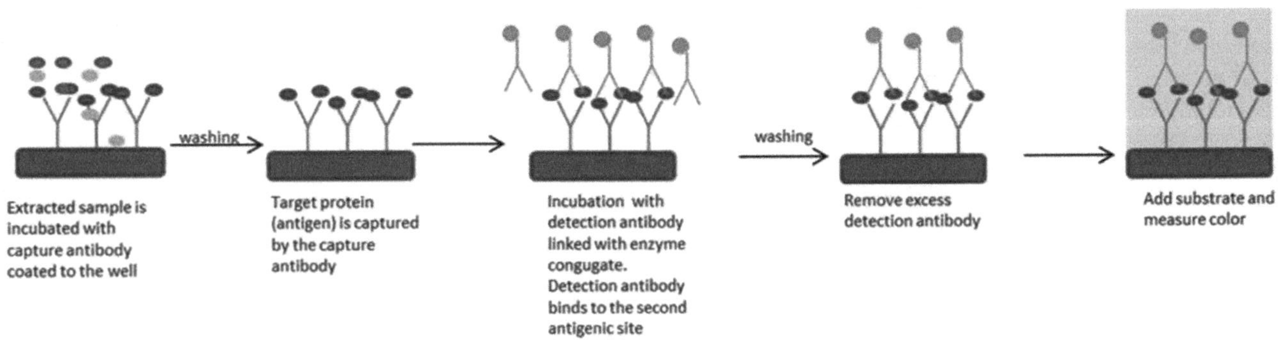

FIGURE 11.10 Nonlinear amplification with C_T assignment (circled) is corrected by eliminating replicate.

11.8 Personnel management

To effectively manage a laboratory for GMO detection, personnel are required to have an adequate scientific background, understanding, and suitable training to perform the required procedures and analyses. These factors are necessary to provide confidence that laboratory safety and data quality standards are understood and that the data that are released are of the highest quality and reproducibility (U.S. EPA et al., 2004).

The first aspect of personnel training should be on laboratory safety. The laboratory should be managed according to safety standards that are adhered to in all situations. Laboratory safety training should address chemical safety, personal safety, and fire safety. Every person who works within a laboratory needs to receive this type of training before performing any work in the laboratory.

Second, personnel should be trained on laboratory procedures. Personnel background and training are reliant on educational experiences as well as physical or hands-on experiences. The training should be guided by methods and procedures (SOPs) that provide a framework for each laboratory procedure to be completed. Along with the written information, each person should receive individual training by a recognized trainer to demonstrate competency both with and without supervision. This training should be completed in steps and should be documented both to help the analyst receiving the training to know what skills they are ready to manage on their own and to help the laboratory manager know how to distribute responsibilities.

To provide laboratory training, it is a good idea to use known samples, rather than unknowns, to demonstrate competency. From any assay standpoint, known samples are genomic and NTC controls for which performance in assays is established to allow the analyst to try to repeat expected datasets. These types of training provide accurate measurements of skill. Additionally, unexpected results provide a clear basis from which to assess levels of scientific understanding regarding troubleshooting.

Regardless of the educational background or experience of an analyst working in a laboratory, training is imperative. Each location may have different safety rules and each laboratory may follow different approaches and methods and use different instrumentation; all of which require proper training according to specific procedures. Participation in annual proficiency testing and laboratory auditing by either internal or external quality control groups is also highly recommended.

11.9 Summary

In summary, the establishment of a GMO testing laboratory is a comprehensive process that requires careful attention to multiple factors. Sustainability through management of the laboratory is an ongoing process that should be viewed as fluid, where process changes can and will happen to create ongoing improvements for data quality, technology, and personnel development. These changes should be welcomed and expected and will help to foster an environment of quality, productivity, and continuous improvement.

References

Ahmed, F.E., 2002. Detection of Genetically modified organisms in food. Trends Biotechnology 20, 215–223.

Bauer, T., Weller, P., Hammes, W.P., Hertel, C., 2003. The effect of processing parameters on DNA degradation in foods. European Food Research and Technology 217, 338–343.

Brookman-Amissah, N., Packer, H., Prediger, E., Sabel, J., 2011-2012. qPCR Application Guide, Experimental Overview, Protocol, Troubleshooting, third ed. Integrated DNA Technologies Inc www.IDTA.com.

Brown, M.C., 2011. Microtiter plate elisa. In: Shan, G. (Ed.), Immunoassays in Agricultural Biotechnology. John Wiley & Sons, Inc., Hoboken, NJ, USA, pp. 59–89. https://doi.org/10.1002/9780470909935.ch4.

Elsanhoty, R.M., Ramadan, M.F., Jany, K.D., 2011. DNA extraction methods for detecting genetically modified foods: a comparative study. Food Chemistry 126, 1883–1889.

Gilbert, J., 1999. Sampling of raw materials and processed foods for the presence of GMO's. Food Control 10, 363–365.

ISO Technical Committee, 2006. ISO 24276, Foodstuffs – Methods of Analysis for the Detection of Genetically Modified Organisms and Derived Products – General Requirements and Definitions, first ed.

Longo, M.C., Berninger, M.S., Hartley, J.L., 1990. Use of uracil DNA glycosylase to control carry over contamination in polymerase chain reactions. Gene 93, 125–128.

McPherson, M.J., Møller, S.G., 2000. PCR. BIOS Scientific Publishers, Oxford.

Meyer, R., 1999. Development and application of DNA analytical methods for the detection of GMOs in food. Food Control 10, 391–399.

Mifflin, T.E., 2003. Setting up a PCR laboratory. In: Dieffenbach, C.W., Dveksler, G.S. (Eds.), PCR Primer: A Laboratory Manual. Cold Spring Harbor, N.Y, pp. 5–14.

Mifflin, T.E., 2007. Setting up a PCR Laboratory. CSH Protocols 2007. https://doi.org/10.1101/pdb.top14. Available at: http://cshprotocols.cshlp.org/cgi/content/abstract/2007/14/pdb.top14.

National Association of Testing Authorities, Australia (NATA) March, 2006. Criteria for accreditation of laboratories testing genetically modified organisms. Guidelines for Assessment (Issue 2).

Nolan, T., Huggett, J., Sanchez, E., 2013. Good Practice Guide for the Application of Quantitative PCR (qPCR). LGC.

Persing, D.H., 1991. Polymerase chain reaction: trenches to benches. Journal of Clinical Microbiology 29, 1281–1285.

Rys, P.N., Persing, D.H., 1993. Preventing false positive: quantitative evaluation of three protocols for inactivation of polymerase chain reaction amplification products. Journal of Clinical Microbiology 31, 2356–2360.

U.S. EPA, Sen, K., Fout, S.G., Haugland, R., Moulton, C., Grimm, A., Giovanni, G.D., Feige, M.A., Birkenhauer, J., Lott, Scheller, J., Reilly, E., Connell, K., Marshall, M., 2004. Quality Assurance/Quality Control Guidance for Laboratories Performing PCR Analyses on Environmental Samples. Available online at: http://www.epa.gov/microbes/documents/qa_qc_pcr10_04.pdf.

Zel, J., et al., 2012. How to Reliably Test for GMOs. Springer Briefs in Food, Health and Nutrition, pp. 1–95.

Chapter 12

International standards and guidelines

Ray Shillito
BASF Corporation, Morrisville, NC, United States

12.1 Background and purpose of standards

The first question to be answered is "Why are Standards necessary?"

When goods and services are being traded, exchanged, or sold, they are done so to a set specification. Such specifications tend to be thresholds and may for example include the moisture or aflatoxin content of grain measured using an agreed standard method using a sample obtained in a standard way, or the maximum or minimum GM content of a grain or seed lot. Agreement on how to measure these specifications is critical to allow international trade to work effectively, as both sellers and buyers must have the same basis on which to agree specifications. The need for specifications dates from the beginning of trade and led to the development of measurement standards and money.

In the present day, a standard is a published specification that establishes a common language and contains a technical specification or other precise criteria. Standards cover a vast array of subjects and topics and include products and services of almost every description. Some libraries contain over half a million standards. ISO alone has more than 21,000 standards.

In addition to standards agreed upon in commerce, governments (e.g., Japan, China) may themselves set standards, in the form of official methods. These are based on methods that may or may not be subsequently submitted as international standards, or governments may adopt ISO standards for example as a national standard.

For this chapter, we will focus on standards relevant to the sampling and detection of GM materials in agriculture and food.

12.2 Standards setting process

The impetus to develop a standard can have several sources and is generally market driven. Standards developing organizations (SDOs) develop standards as part of their business model, as they sell copies of the standards. They will

Application of Sampling and Detection Methods in Agricultural Plant Biotechnology
https://doi.org/10.1016/B978-0-323-99293-0.00004-0

start a standards-setting process when a need is perceived. Another impetus may be the development of regulation that requires, for example, declaration of the presence or absence (below a threshold) of a food component. A recent example is the development of methods and standards for measuring gluten, when this became a popular label for food products, or for dietary fiber. In these cases, a kit manufacturer developed a test, and an SDO validated the method.

Standards may be developed for methods for measuring an analyte, or a process such as a quality management system for managing a laboratory. Our first focus will be on methods standards.

12.3 Standards organizations

Due to the extensive need for standards, a number of organizations have taken on the role of developing, distributing, and managing standards as SDOs. Standards are also set by governments and regional and intergovernmental agencies. Usually (and preferably), standards are set with input from all those that may be impacted, from suppliers, to manufacturers and traders to end-users. A number of different types of organizations are involved in standards setting.

Bodies that develop standards are termed SDOs. Many are science-based associations, and include, for example, AOAC, AACCI, AOCS, ICC, ISTA, IUPAC, and ISO (see box). AOAC, AACCI, and AOCS were founded in the late 1800s or early 1900s as trade became less local and a need for standards was appreciated. IUPAC was an independent association of chemists, and this was followed by the formation of ISTA for seed method standardization. ISO was founded in 1947 by 25 governments, and international cereal chemists came together in 1955. The key to these acronyms is included in the text box.

Organizations such as ISO and ISTA provide methods standards, and also develop management standards that concern procedures such as quality assurance.

SDOs are listed in order of their founding: (see references for URLs).

AOAC: Known since 1991 as AOAC International, the AOAC was founded in 1884 as Association of Official Agricultural Chemists—an organization that was formed by USDA to provide uniform chemical analysis methods for fertilizers, and sponsorship was changed in 1927 to the US FDA. https://www.aoac.org

AOCS: AOCS was formed in 1910 as Cotton Products Analysts, and changed its name to The American Oil Chemists' Society in 1927. https://www.aocs.org

Cereals & Grains Association: Known since 2005 as AACC International was founded in 1915 as the American Association of Cereal Chemists to standardize methods used in the grain industry. They are now known as the "Cereals and Grains Association". https://www.cerealsgrains.org

ISTA: The International Seed Testing Association was founded in 1924, with a focus on seed testing, rather than grain. https://www.seedtest.org

IUPAC: The International Union of Pure and Applied Chemistry was founded in 1919 to unite a fragmented, global chemistry community. Its focus is not on food-related products. https://iupac.org

ISO: The International Organization for Standardization was founded in 1947 by 25 countries interested in standardization and now comprises 161 member countries. Each country is represented by its national standards body. ISO has a very broad range of standards. https://www.iso.org

ICC: The International Association for Cereal Science and Technology was founded as an international organization for cooperation between cereal chemists in 1955. https://www.icc.or.at

12.3.1 Scientific association SDOs

AOAC, Cereals and Grains Association, AOAC, and ICC are scientific organizations that develop standards as part of their mission to their members. They carry out validation and standardization of methods and publish the methods as a fee-based service for use by analysts. They achieve method harmonization through internationally recognized procedures in which methods are compared globally across laboratories following a harmonized protocol (Horwitz, 1988) to make sure that they are robust and able to be replicated. Methods are then published as "Official," "Approved," or "Standard" methods. A limited number of methods covering the detection of GM products are available from these organizations.

12.3.2 ISO

ISO, established in 1947, is composed of the national standards bodies, one from each of the 161 member economies. ISO has a very broad range of standards, and together with the International Electrotechnical Commission and the International Telecommunication Union comprise the World Standards Cooperation alliance. ISO has published the most methods of any SDO concerning detection and quantification of GM plants and plant materials. Other ISO standards cover sampling of grain, general method performance, and operation of laboratories; none of these are specific to GMOs. Although nongovernmental ISO operates on a one-country-one-vote principle. Individual experts from a wide range of sources (e.g., industry, government, academia, social organizations) collaborate to inform each country's input and voting position. ISO does not cover standards for seeds as ISTA covers that area.

12.3.3 Government-led organizations

Two key intergovernmental bodies are the Codex Alimentarius (Codex) and International Seed Testing Association (ISTA). These bodies are made up of governmental/country delegates who decide standards based on their governments' positions, with the input of experts.

12.3.3.1 Codex Alimentarius

Codex Alimentarius Commission (CAC or Codex) is a government-led organization established in 1963 to protect the health of consumers and ensure fair food trade practices. It operates under the Joint FAO/WHO Food Standards Program. The Codex Alimentarius, or "Food Code" is a collection of standards, guidelines, and codes of practice (Codex Alimentarius Guidelines)

Codex has committees that cover both subject areas (e.g., Codex Committee on Methods of Analysis and Sampling, CCMAS) and commodities (e.g., Codex Committee on Cereals, Pulses, and Legumes). Codex does not generally develop methods standards itself, but adopts standards developed by SDOs. It does publish specifications which it calls standards. However, Codex does develop guidelines. Recent guidelines are criteria based, rather than prescribing specific methods. This approach was adopted due to the proliferation of specific methods, and the fact that the technologies are constantly changing.

12.3.3.2 ISTA

The International Seed Testing Association (ISTA) is the internationally recognized standards body for seed testing. Its voting body is its 75 member countries. ISTA publishes a set of rules for sampling and testing of seed, including a chapter on Trait testing. It has several technical expert committees and the GMO committee is also developing a handbook on GMO testing. In addition, ISTA has a very useful and unique proficiency testing program for GM events in seed, and a statistics committee that has developed a number of useful tools for seed testing (as referenced in chapter 9).

12.3.4 National and regional organizations

12.3.4.1 National organizations

Most countries have their own national standards bodies. These organizations are responsible for developing national standards at the request of their governments, academia, method or kit developer, affected industry, or other stakeholders. A country may have more than one standards institute, in which case, one of them is responsible for representing their country's position in developing standards at the international level through ISO. This is the case in the United States, where ANSI is the ISO representative, but NIST is also responsible for standard setting in some areas.

Standards developed by a national body may subsequently be submitted for consideration as an international standard, commonly through ISO, but may also be submitted to Codex, ISTA, or other standard-setting organizations. Alternatively, they may be considered for adoption as regional standards.

12.3.4.2 Regional organizations

There are regional standards bodies that operate in collaboration with the ISO system. Examples are CEN (European Committee for Standardization), PASC (Pacific Area Standards Congress), COPANT (Pan American Standards Commission), ARSO (African Organization for Standardization), and subregional bodies such as AMN (MERCOSUR Standardization Association), and CROSQ (CARICOM Regional Organisation for Standards and Quality). Some of these may have agreements with ISO regarding standards codevelopment such as the "Vienna Agreement" between CEN and ISO that prevents simultaneous development of competing standards.

12.4 Development of a standard

Standard development in all these organizations operates in similar ways. Generally, at the SDO level, those developing the standard will be a panel of experts knowledgeable in the area in which the standard is being set. They may be for example from the industry that is to use the standard, academics or from government. At the initial stages of the process, they will be expected to have expertise in the subject matter.

A standard starts as a request for a standardized protocol, or as a useful method that may or may not already be published. Each organization has its own committee and administrative structure, which involves the review of the proposed standard at various points by other experts in the field. Methods intended for standardization must be tested in multiple laboratories before being adopted. This process established the characteristics of a method—whether it can be applied by many laboratories to obtain the same result, and what the range in the variability of that result will be. Methods standardization bodies all follow an agreed harmonized protocol published by IUPAC. This validation process also tests the readability of the written method so that it can be understood by the analyst that has to use it. The relevant committee is also responsible for writing the method to the standards body's format and generating and reviewing the data.

SDOs such as AOAC, AACCI, ICC, and AOAC are scientific societies that have standards committees. These committees generally have subcommittees that concentrate on the different standards areas. For example, AACCI has 21 technical committees that develop standards, and their work is overseen by an approved methods committee. Membership of the committees is on a volunteer basis. The same kind of process occurs in other SDOs, for example, AOCS has a Uniform Methods Committee, and AOAC has Expert Review panels. In the case of AOAC, AACCI, and AOCS, an accompanying publication is prepared that describes the method validation and includes the data generated. Once a proposed method or standard is developed, it will be reviewed before publication by a voting process of experts (e.g., in scientific organizations) or by countries that are participants (such as in ISO, ISTA, and Codex). standards undergo periodic review to ensure that they remain useful and may be updated as appropriate. They may be adopted by other SDOs as there are agreements in place between SDOs to allow for this to occur. SDO standards may become ISO standards provided there is sufficient interest and agreement between the SDO and ISO, as both sell standards to support their activities. Development of standards in SDOs can typically take 2–3 years, and at ISO, there is a maximum timeline limitation in most cases of 3–4 years.

ISO develops standards through its more than 300 technical committees and their subcommittees. Membership in a committee requires the country involved to pay a membership fee, so committees only consist of a subset of countries and experts that are particularly interested in the outcome. A standard is proposed and then developed by expert groups within those committees, and then moves to a committee level where member countries suggest changes and vote on whether to move the standard forward in the process. Once a vote to publish the standard is taken, it is published. Some countries have a requirement that an ISO standard is adopted as a country standard. ISO also develops and publishes a less stringent level of documents, technical specifications, which may not be required to be adopted by certain member countries, as well as technical reports that are information documents. Other ISO documents include guidelines, meant to help with the implementation of standards, and these may eventually be developed into standards in their own right.

Codex guidelines are developed by expert groups and then adopted by consensus of governments, meeting every 1 or 2 years. Because of the consensus nature of guideline development, and the politics involved in a government-led process, they can take years to complete (e.g., CCMAS guideline CAC/GL 74-2010 took 10 years). Methods developed by SDOs can be adopted by Codex as the defining method for a guideline.

ISTA develops and approves methods for testing of seed, through a government and expert-led process. Like the Scientific SDOs, ISTA has subject–matter expert committees, and these committees develop and refine the rules for seed testing. The rules are amended each year at the annual meeting, where country delegates vote on whether to adopt changes in the rules proposed by the committees.

12.5 Standards relevant to GM detection

ISO has published the most methods standards for GM detection. Subcommittee 16 of Technical Committee 34 (ISO/TC 34/SC 16: Horizontal methods for molecular biomarker analysis https://www.iso.org/committee/560239.html) has developed general guideline standards for the performance of the methods accompanied by a series of specific method standards. ISO/TC 34/SC 16 was established in 2008 and has 24 countries participating in developing standards, and 17 more countries acting as nonvoting observers. At the time of writing this committee has 20 published standards and 16 more under development. These methods cover PCR and protein-based methods, and standards for sampling and for isothermal DNA amplification are among those under development. Other committees with relevant activities are ISO/TC 34/SC 9 (microbiology), ISO/TC 69/SC 6 (statistical methods), and ISO/TC 276 (Biotechnology).

Codex published a guideline for the detection of specific DNA and proteins in 2010 (CAC/GL 74–2010). This guideline was developed through the Codex Committee on Methods and Sampling. It provides guidance on the use of methods developed to detect specific DNA sequences or specific proteins in foods (as Codex only has jurisdiction over foods) and is thus not restricted to GM detection methods. The Codex guidelines provide a good basis for judging the applicability of methods.

The following is a list of internationally relevant standards for GM detection. Readers looking for local and/or regional standards should refer to their local standards setting organizations. These standards are updated at regular intervals so the operator should obtain the most recent version. The year of the latest version at the time of writing is given.

12.5.1 Sampling

- ISO 21294: 2017 — Oilseeds—Manual or automatic discontinuous sampling
- ISO 24333:2009 — Cereals and cereal products—sampling
- ISO 6497:2002 — Animal feeding stuffs—sampling
- ISO 22753: 2021 — Molecular biomarker analysis—Method for the statistical evaluation of analytical results obtained in testing sub-sampled groups of genetically modified seeds and grains—General requirements

12.5.2 Guidelines and standards regarding analytical methods

- CAC/GL 74-2010 — Guidelines on performance criteria and validation of methods for detection, identification and quantification of specific DNA sequences and specific proteins in foods.

• ISO 16577:2016	Foodstuffs—molecular biomarker analysis—terms and definitions
• ISO 16578:2013	Foodstuffs—molecular biomarker analysis—general definitions and requirements for microarray detection of specific nucleic acid sequences
• ISO 21569:2005/ Amendment 1:2013	Foodstuffs—methods of analysis for the detection of genetically modified organisms and derived products—qualitative nucleic acid-based methods
• ISO/TS 21569 parts	The additional parts of this standard provide as either technical specifications, or reports, detailing methods for the detection of specific DNA sequences
• ISO 21570:2005/ Amendment 1:2013	Foodstuffs—methods of analysis for the detection of genetically modified organisms and derived products—quantitative nucleic acid-based methods
• ISO 21571:2005/ Amendment 1:2013	Foodstuffs—methods of analysis for the detection of genetically modified organisms and derived products—nucleic acid extraction
• ISO 21572:2019	Foodstuffs—molecular biomarker analysis—immunochemical methods for the detection and quantification of proteins
• ISO 24276:2006/ Amendment 1:2013	Foodstuffs—methods of analysis for the detection of genetically modified organisms and derived products—general requirements and definitions
• ISO 22942-1: 2022	Foodstuffs—molecular biomarker analysis—isothermal nucleic acid amplification methods—part 1: general requirements

12.5.3 General analytical methods standards

• ISO/TS 16393:2019	Molecular biomarker analysis—determination of the performance characteristics of qualitative measurement methods and validation of methods
• ISO 5725:1994 parts 1–4, 6	Accuracy (trueness and precision) of measurement methods and results—parts 1–4 and 6.

12.5.4 Sources of methods

Not all methods that are used for GM detection have been converted into standards. Methods can be found in a number of places other than in standards. Availability of a method online does not guarantee that there is a license to use the method as many of the event-specific methods are the intellectual property of the method developers.

Methods for GM events commercialized by the major technology providers are available with a restricted license via the CropLife database at http://www.detection-methods.com/.

The Cartagena Protocol on Biosafety under the Conventional on Biodiversity established a network of laboratories with the goal of establishing or identifying easy-to-use and reliable technical tools available for the detection of unauthorized LMOs (living modified organisms). The network has yet to develop guidelines. Specific methods can be referenced through the Biosafety Clearing house https://bch.cbd.int/database/organisms/which will refer you to other databases for more information.

The EU has invested in a certification process for EU-based laboratories (the EU network of GMO laboratories (ENGL)) through their Joint Research Center (EU-RL GMFF). The JRC publishes guidelines (http://gmo-crl.jrc.ec.europa.eu/guidancedocs.htm) for use by the ENGL network; these documents essentially agree with Codex and ISO guidelines and standards. The methods that have been submitted to the EU are available at http://gmo-crl.jrc.ec.europa.eu/gmomethods/. In addition, the EU has published a database of genetic-element-specific methods at http://www.euginius.eu/euginius/pages/home.jsf.

12.5.5 Laboratory accreditation standards

Accreditation is the procedure by which an authoritative body gives formal recognition that a body or person is competent to carry out specific tasks. Laboratories may be accredited to international or regional standards. Many countries have national standards.

Accreditation demonstrates that a laboratory operates under a quality management system to a globally or locally recognized standard that it has passed a rigorous examination of its processes and methods as well as staff training. Its reporting of results must also meet the requirements of the standards.

The most common standard applied to laboratories is ISO 17025, General Requirements for the Competence of Testing and Calibration Laboratories. Laboratories are accredited by accreditation bodies that are in the case of ISO17025 themselves accredited to ISO 17011, Conformity Assessment-Requirements for Accreditation Bodies Accrediting Conformity Assessment

Bodies. To maintain equivalence accreditation bodies around the world perform accreditation of each other. This is intended to normalize the interpretation of standards. However, it has been observed that there are differences in interpretation, particularly in terms of flexible accreditation of methods. For example, some bodies will require that each and every PCR method has to be inspected separately, whereas others accept that demonstration of the ability to perform the method in a crop matrix for one analyte infers proficiency for another similar method in the same crop and matrix (a Blanket or Flexible Accreditation Scope). These differences have not been resolved as of today.

ISTA also operates an accreditation process for seed sampling and for testing. The ISTA accreditation is based on ISO17025 adapted for use in seed operations. ISTA accredited laboratories can issue analytical certificates that are globally recognized by governments.

Laboratories that are accredited to produce reference materials (Chapter 8) should meet the requirements of ISO 17034, General Requirements for the Competence of Reference Material Producers as set out in ISO guides 30–34, and those providing proficiency samples should meet ISO 17043, Conformity Assessment—General Requirements for Proficiency Testing.

12.6 Conclusions

Technological advancements in diagnostics and testing present an evolving set of challenges to the testing community. A large number of standards and guidelines available play an essential role in establishing and maintaining international norms that allow trade. These standards provide the best consensus ways of carrying out accurate and repeatable analyses and provide critical guidance to laboratories, and their operators. These standards and resulting best practices can only exist with the active participation and consensus of the experts and organizations that develop them.

Acknowledgment

The author thanks Nancy Brooker for the review of and suggestions for improving this chapter.

References

Horwitz, W., 1988. Protocol for the design, conduct and interpretation of collaborative studies. Pure and Applied Chemistry 60 (6), 855–864.

Codex Alimentarius, 2010. http://www.fao.org/fao-who-codexalimentarius/codex-texts/guidelines/en/. (Accessed 6 May 2022).

Further reading

AOAC International: https://www.aoac.org/

AOCS: https://www.aocs.org/

Cereals & Grains Association: https://www.cerealsgrains.org/

Codex Alimentarius guidelines: http://www.fao.org/fao-who-codexalimentarius/codex-texts/guidelines/en/

ICC: https://www.icc.or.at/

ISO: https://www.iso.org/standards.html

ISTA: https://www.seedtest.org/

IUPAC: https://iupac.org/

Chapter 13

Analytical strategy and interpretation of results

Yongcheng Wang[1], Lucy Liu[1], Helen Mero[1] and Tim Perez[2]
[1]*Bayer Crop Science, Chesterfield, MO, United States;* [2]*Conagra Brands, Omaha, NE, United States*

13.1 Introduction

After more than 25 years (1996–2022) of the commercialization of biotech crops, the global area planted with biotech crops has increased more than 110-fold from 1.7 million hectares in 1996 to 190.4 million hectares in 2019 per the International Service for the Acquisition of Agri-biotech Applications (ISAAA). Such an increased adoption rate speaks about the multiple benefits delivered to farmers, both small and large, and consumers globally (ISAAA, 2019).

Commercialization of a new biotech crop happens after clearing different regulatory processes depending on the country where the crop will be planted as well as the key export countries where the applicable crop commodities will be traded. Due to differences in regulatory timelines for reviews and approvals in different countries, asynchronous approvals happen regularly. Hence, this asynchronous approval system can result in unapproved biotech products being present at low levels in commodity shipments going to a country where the regulatory process has not been finalized. This can result in trade disruptions with a variable severity based on the limits of tolerance established and can be more severe when zero tolerances policies are employed (Wilson and Dahl, 2016).

As the development and cultivation adoption of biotech crops expands globally, so does the need to confirm the presence of the intended trait(s) in the biotech crop product as well as confirming the absence of unintended traits in seed or grain commingled in seed or grain lots that are nationally or internationally commercialized. Therefore, this chapter discusses such analytical strategies.

13.2 Confirming the presence of intended trait

Trait purity in biotech crop testing is very important during seed production and quality control functions of technology companies, the expected result of such testing is the quantification of a high percentage of presence of the

Application of Sampling and Detection Methods in Agricultural Plant Biotechnology
https://doi.org/10.1016/B978-0-323-99293-0.00015-5

specified trait. Seed standards exist to ensure the availability of genetically pure seeds with limited impurities and contaminants. Organizations like the Association of Official Seed Certifying Agencies working with seed producers and organization for Economic Cooperation and Development (OECD), publishing the OECD seed schemes that establish rules for seed certification for the global seed movement, provide some of the existing guidance on seed certification that assures genetic purity as well as measures to maintain the identity of certified seed (Roberts et al., 2015).

Although genetically modified organism (GMO) trait purity testing can be performed on either individual seeds, bulked seeds, or grouped seeds, a representative sample of individual seeds is preferred and most commonly used. This is especially true while evaluating the purity of the combined traits, and pooled samples will complicate the interpretation of the testing results. Statistical consideration must be taken into consideration to make sure the samples are representatives of the lot bulk characteristics to minimize the sampling uncertainty per different testing schemes (Remund et al., 2001).

13.2.1 Herbicide bioassay

13.2.1.1 Introduction

Herbicide bioassays have been conducted for a number of years before the release of genetics targeted to make a cultivar tolerant to a specified herbicide. One of the first soybean strains identified to be resistant to what is now known as sulfonylurea resistant soybeans was in 1988 (Pomeranke and Nickell, 1988). Sometime thereafter, seed laboratories were being asked to conduct testing to determine what percent of a seed lot was susceptible to a herbicide. The company supplying seed to the customer must know the risk associated with a customer spraying a field with an herbicide, resulting in an unacceptable proportion of the population dying or hindering yield potential.

13.2.1.2 Methodologies

Herbicide bioassays can be divided into two types: preemergent and postemergent bioassays. Greenhouse bioassay methodologies had long been a means of screening for soil contamination and herbicide screening development. From there it was a logical jump to adapt to a postemergent bioassay. Test duration (Goggi and Stahr, 1997) and space constraints play into the selection of the suitability of this method. Goggi and Stahr (1997) abandoned the methodology due to needing 7 days to grow the seedling and the period of time required to identify the symptoms. Disease in a laboratory postemergence test may be difficult to control, especially under low light conditions (unpublished personal observation). Some laboratories such as Illinois Crop Improvement Association with greenhouse facilities continue to offer a postemergent herbicide bioassay in a sterile sand media, or where pre-emergent methods are not appropriate.

The critical hint as to the potential of a preemergent bioassay is likely to have stemmed from the development of sulfonyl-urea tolerant soybeans in 1987 (Sebastion and Chaleff, 1987). The mutagenesis of tolerant plants involved a direct soak of soybeans. This method was used with the introduction of the Roundup Ready soybean (Goggi and Stahr, 1997) and corn (Sinning, 1995). Seed analysts in production laboratory settings find this to be a less than desirable option, due to the 24-hour imbibition in a dilute solution of herbicide as well as difficulty planting swollen wet seed and the potential to damage imbibed seeds while planting. Evaluation is typically in the same time frame as a germination test (bib_AOSA_Rules_for_Testing_Seeds_2014aAOSA Rules for Testing Seeds, 2014a,b edition).

Substrate imbibition was the next logical step. Cotton seeds were planted on germination toweling wetted with a dilution of herbicide (Savoy et al., 2001) and evaluated to determine susceptible seedlings. Gutormson (1998) published a procedure to plant Roundup Ready corn on creped cellulose paper wetted with a solution of Roundup.

13.2.1.3 Seedling morphology of susceptible seedlings

The key to a successful herbicide bioassay is based on the dilution of chemicals to enable seedling morphological distinction between tolerant and susceptible seedlings (personal unpublished observation). The Non-Tolerant Seedling Study Guide (2016) (https://analyzeseeds.com/wp-content/uploads/2016/06/non-tolerant-seedling-study-guide.pdf) provides descriptions of some of the seedling morphology descriptions for nontolerant seedlings. The Society of Commercial Seed Technologists (SCST) Herbicide Bioassay Working Group has provided images. Key morphological traits of a non-tolerant seedling may vary with herbicide concentration. Thus, it is imperative for a lab to standardize its operating instructions.

13.2.1.4 Training resources

The SCST has developed a certification system to ensure thresholds of knowledge. The title Registered Genetic Technologist denotes mastery of three or more areas of genetic technology. The title Certified Genetic Technologist denotes mastery of one or two areas of genetic technology. The SCST Genetic Technology Committee has several working committees that provide workshops and other training aids for a variety of methodologies. SCSTSCST manual (2016), Chapter 14 provides an overview of the genetic technologies associated with testing seeds in a laboratory.

13.2.2 Immunoassay

Immunoassas areideal for the detection of target trait protein(s) in very complicated matrices (Brett et al., 1999, Chapter 6). It can provide quantitative, semiquantitative, and qualitative data for trait purity testing at the protein level. Immunoassay is amenable to automation that enables high-through data generation for large numbers of samples and minimizes human errors during sample handling and testing. All data management including data recording, analysis and interpretation, and archival and retrieval can be easily handled by the commercially available Laboratory Information Management System (LIMS).

For trait purity testing, the assay sensitivity should be defined in terms of known concentrations of standards. If positive and negative samples need to be called, then a detection limit should be defined based on the historical data or statistical analysis. Consideration should be taken to address potential cross-reactive antigens/proteins present in the seed sample extracts by choosing the proper antibodies and optimizing the assay parameters. Once the key reagents and assay parameters are chosen, validation work (e.g., addressing matrix effects from nontarget proteins, evaluating dilutional agreement of diluted seed samples, etc.) should be executed to ensure the assay is accurate, consistent, and reliable. Also, practical consideration of user skill, the environment, and assay duration should be considered to ensure the assay is robust in either a laboratory or in a field.

The immunoassays are either commercially available in a kit format or can be developed in-house to assure product purity, seed production, and quality assurance. A typical immunoassay workflow for GMO trait purity testing starts with a clear testing objective (either quantitative or qualitative) and is followed by proper preparation of working samples (e.g., bulked seeds for qualitative; bulked, grouped, or individual seeds for quantitative). After sample grinding and target trait analysis, the immunoassay data are collected and interpreted. A statistical analysis is sometimes performed to facilitate the completion of a final report if needed.

Immunoassays have a great capacity to detect target traits at the protein level using a microwell plate. The plate can be 96-well, 384-well, or even 1536-well formats. For 384-well and 1536-well assay formats, the automatic liquid handler is necessary to minimize the difficulties in manual handling. The current immunoassays for trait purity testing take on two major forms, enzyme-linked immunosorbent assay (ELISA) and lateral flow strips (LFS). With more and more combined traits in commercial crop production, other immunoassays with multiplexing capacity start to emerge as alternative assay platforms including Meso Scale Discovery (MSD) and bead-based Luminex assays.

13.2.2.1 Enzyme-linked immunosorbent assay

Depending upon the property of the target analyte (trait protein) and the availability of proper antibodies against it, ELISA for trait purity testing can be in competitive or sandwich formats. Although the competitive assay format is applicable to analytes of any molecular weight with different numbers of epitopes, the establishment of such a competitive assay is not as straightforward as the sandwich format. The competitive assay is most commonly established for target analytes with a molecular weight of 10 kDa or less, for the analytes' lack of proper antibody pair for the sandwich format, or for other analytes with specific attributes that make it impossible to develop a sandwich assay.

Competitive assay has two different orientations: (1) antibody coating and (2) antigen coating. In the first orientation, microwell plate is coated with antibody of trait of interest. Target protein (trait of interest) is prelabeled with enzymes or other chemical probes (e.g., biotin). Target analyte is removed from seed samples and transferred to the plates, either before or coincident with the labeled protein. The analyte extracted from the sample competes with labeled protein for binding to the coated antibody, resulting in a decrease of the signal compared to the maximal signal from the labeled protein alone. In the second orientation, the target protein is coated onto the microwell plate. The target analyte extracted from seed samples is then added to the well either preincubated or together with the labeled antibody. The coated protein and the target analyte from seed extract compete for binding to the labeled antibody, the resulting signal is related to the labeled antibody bound to the coated protein.

Sandwich ELISA is the most commonly used format for target protein quantification in complicated matrices in genetically modified crops (Stave, 2002). Sandwich ELISA should be considered when the size of the target trait protein is greater than 10 kDa. Sometimes, such a format is applicable when the target protein is close to 5 kDa, when the configuration of protein, the binding characteristics of protein and antibody, and available antibodies are all appropriate. A sandwich ELISA for trait purity testing starts with an antibody that is immobilized to a well surface, the target analyte (protein) extracted from seed samples is then added to the well, resulting in a formation of an antibody–protein complex. With the addition of a labeled second antibody, a sandwich of antibody–protein–antibody complex is formed. After appropriate incubation, washing, and reagent addition, a detectable signal is generated, which is directly related to the amount of target trait protein in the seed extract samples.

As the sandwich ELISA is a two-site assay in which the analyte is sandwiched between the capture antibody and the detector antibody, appropriate antibody pair should be available, which can bind trait protein through a repeating epitope or two discrete antigenic determinants with minimum steric hindrance. Both polyclonal antibody and monoclonal antibody can be used in sandwich ELISA assay, poly–poly, mono–mono, and mono–poly pairs are very commonly seen in sandwich ELISAs for trait purity testing (Chapter 6). The binding characteristics of antibody and target trait protein can be evaluated in many ways such as epitope binning, binding kinetics evaluation, and checkerboard titration. The ultimate antibody pair to be used for trait purity testing should assure that the assay is reliable, sensitive, specific, and easy to perform.

Microplate ELISAs, either competitive or sandwich format, can be performed in well-equipped laboratories or in a field or other nonlaboratory environment in a qualitative or quantitative format. Qualitative assays provide a simple positive or negative result for a sample. The cut-off between positive and negative is determined experimentally and may be statistical. In quantitative ELISA, the amount of colored product is proportional to the amount of labeled enzyme-linked antibody that binds, which is in turn related to the amount of target trait protein present in the seed samples. If known amounts of target trait protein are added, a standard curve, which is typically a serial dilution of the target, can be constructed by plotting each standard signal versus the standard concentrations. The concentration of each unknown target trait protein can then be determined from the standard curve. Whether qualitative or quantitative assay, the result should reflect the relevant absence/presence or the amount of target trait protein in the seed sample.

13.2.2.2 Lateral flow strips

LFS immunoassay (Posthuma-Trumpie et al., 2009) is a porous membrane-based assay. Although a lot of membrane sticks are available commercially either based on hydrophobic nitrocellulose or cellulose acetate membrane, they are in the same/similar setup. LFS provides more of a qualitative or semiquantitative detection of analyte for trait purity testing. A typical run time of an LFS assay is around 15–30 min, which makes LFS a perfect format for trait protein testing especially in the field with the consideration of time, effort, and expense.

Similar to ELISA, LFS can be run in a sandwich or competitive format. In the sandwich format for trait purity testing, the sample application pad is dipped into a tube or well containing an extract of the crop sample. The extract containing the target trait protein is placed in the LFS device for analysis according to the manufacturer's instructions. At the end of the sandwich LFS run, the presence of only one (control) line on the membrane indicates a negative sample, and the presence of two (both test and control) lines suggests a positive sample. If no lines showed up which means the assay is invalid, then some troubleshooting is needed to figure out the specific issues before the repeat is executed.

For the competitive format of LFS, it is also primarily for detecting small proteins (less than 10 kDa) and proteins with only one available epitope. If the trait protein-specific antibody is labeled and preloaded onto the conjugate pad, then the Test zone is comprised of immobilized trait protein. The trait protein in the extract of the seed sample will compete with the immobilized trait protein in the test zone to bind to the labeled antibody. In another configuration, if the trait protein is labeled and preloaded onto the conjugate pad, then the trait protein-specific antibody is immobilized at the test zone. The free trait protein in the extract and labeled trait protein in the conjugate pad will compete for the same binding sites on the antibody immobilized at the test zone. As the competitive format LFS is dealing with relatively small protein, it represents more challenges in establishing such assays. Special care should be taken to interpret the results because the test signal is negatively correlated to the trait protein in the extract. Two visible lines in both test and control zones mean a negative result, while a single control line visible is a positive result.

LFS immunoassay is a protein-based method. It is easy to perform and cost-effective and able to produce reliable results within a few minutes. Once the assay is established, a single step is enough to perform the assay. Commercial strips have been developed to detect either insect-resistant Bacillus Thuringiensis protein or herbicide-tolerant proteins (Lipton et al., 2000; Fagan et al., 2001). Even though the current commercially available LFS are limited to some GM products, multiplexed LFS is being developed to simultaneously detect different trait proteins from the same extract. The multiplexed LFS is the method of choice for purity testing of combined traits in the same sample of GM crop.

13.2.2.3 Other immunoassay formats

With the advancement of biotechnology and plant breeding, a variety of traits conferring herbicide tolerance, insect control, drought tolerance, and disease resistance have been introduced into different commodity crops. Different trait purity testing methodologies are needed to detect single traits, double stacks, triple stacks, and even eight or more trait stacks. In the past few years, many technologies have emerged to tackle challenges associated with combined traits. Out of so many immunoassay-based multiplexing assay platforms, two of them offer great advantages in simultaneously detecting multiple trait proteins from the same sample: MSD and bead-based Luminex assay.

Meso scale discovery

As a variation on ELISA with flexible multiplexing capability, MSD is a multiarray-based technology with a proprietary combination of electrochemiluminescence (ECL) detection and pattered arrays. The ECL detection technology uses SULFO-TAG labels containing an electrochemiluminescent Ru^{2+} compound that emits light upon electrochemical stimulation initiated at the surface of the electrode that is built into the bottom of the plate. The emitted light signal is with a high signal-to-noise ratio and minimal no-specific noise and minimal matrix interference. Multiarray plate formats include 96- and 384-well plates, with 96-well as the most commonly used multiplexing format with up to 10 spots per well for multiplexing detection of trait proteins.

MSD assay can be set up in either competitive or sandwich formats, depending upon the target protein size, unique feature of different crop/tissue samples, and the characteristics of available antibodies. For a sandwich MSD multiplexing assay, the capture antibodies, each against individual trait proteins, are coated on the bottom of the wells of a specialized ECL compatible 96-well plate. The seed sample extract containing all the target trait proteins is added to the well. The trait proteins of interest bind to each individual antibody immobilized on the working electrode surface to form antibody−protein complexes. After the addition of labeled detection antibodies, antitrait protein antibody labeled with an MSD SULFO-TAG, the formed antibody−protein complex will recruit the labeled detection antibody to complete the antibody−protein−antibody sandwich. An MSD read buffer is added, and the plate is then loaded into an MSD SECTOR instrument for data capture and analysis. A voltage is applied to the plate electrodes causing the labels bound to the electrode surface to emit light. The instrument measured the intensity of emitted light to provide a measure of trait proteins present in the sample. For quantitative measurement, both quality control sample and unknown sample concentrations are extrapolated from a standard curve prepared in the appropriate biomatrix using purified trait proteins as the reference standard. For qualitative evaluation, the signal of unknown samples will compare to the limit of detection of the assay with a positive quality control sample as reference.

Customized MSD plates coated with antibodies against different trait proteins ensure consistent performance of the assay in use. It can be easily adapted for automation and high throughput analysis. Other than the featured multiplexing capabilities, the most recent advancement in MSD technology with streptavidin-coated plate provides quick trait protein quantification with a duration of less than 1 h.

Microbead-based array-Luminex

In 1977, a multiplexing bead array-based assay was first reported for single-cell analysis (Horan and Wheeless, 1977). Efforts since then have resulted in a lot of progress in bead-based multiplexing. The separation of beads can be conducted through color coding and/or the size of the beads, or via magnetic force. The detection system is based on the specific usage of the assay; fluorescence is very common because it typically detects samples with a high signal-to-noise ratio. In the late 1990s, scientists at Luminex invented xMAP® Technology, a major advance in multiplexed biological assays. xMAP Technology draws from the strengths of solid-phase separation technology without the typical limitations of solid-phase reaction kinetics. Currently, the most seen commercial available bead-based arrays for protein biochemistry are based on Luminex or Luminex 100/200 with xMAP platform.

Luminex xMAP is a flow cytometry-based platform that combines microfluidics, optics, and digital processing with magnetic microbead technology. The microscopic size and low density of these beads used in the Luminex assay allow reactions to exhibit virtually solution-phase kinetics while the incorporated magnetic properties further simplified the washing step. Also, the solid-phase characteristics allow each bead to be analyzed discretely after the assay is complete. A highly multiplexed Luminex immunoassay can measure potentially hundreds of analytes simultaneously.

Many multiplex biological Luminex assay kits have been developed in the clinical diagnostics, pharmaceutical, and life science research markets. The usage of Luminex in the agriculture field starts to pick up the pace. A multiplexed sandwich assay for quantification of target proteins in genetically modified crops using the Luminex FLEXMAP 3D system can be set up as follows: First, the antibodies against the target trait proteins were screened and the best performers were coupled to microbeads. Then, the crop samples were extracted, diluted, and added to the coupled microbeads to facilitate the interaction between antibody and target trait. After the addition of biotinylated detection antibody and reporter, the reactions were analyzed on the Luminex FLEXMAP 3D system at the high reporter gain setting. Typically, the assay takes <2 h with excellent reproducibility (CVs of $\leq 15\%$) and a broad dynamic range (up to five logs).

In theory, Luminex can detect up to 500 analytes simultaneously from a single sample with FlexMAP 3D system. Reading a Luminex plate can be very flexible compared to ELISA. For example, after the assay is complete, the plate has to be read within a certain time period (<0.5 h for the HRP-based detection system), while the Luminex fluorescence system is rather stable, the plate still generates a very good signal after a few days storage at 4°C. Reading time for each Luminex plate is relatively long, and it is about 30−45 min for Luminex 200 or 15−20 min for FlexMAP 3D system. Another limitation of Luminex is the proper storage and handling of the microbeads. Freeze−thaw will damage the beads, which may incur some practical inconvenience for transferring conjugated microbeads.

For trait purity testing, ELISA and LFS are the two major assay platforms for protein expression detection. As future alternatives, multiplexing assay platforms, for example, MSD and Luminex, offer advantages in detecting multiple trait proteins simultaneously from a single sample. Still, a lot of challenges have to be circumvented before a multiplexing assay is implemented. One of the major challenges is the cross-reactivity between antibodies, from nontarget trait proteins or other proteins in the assay mixture. In addition, some protein−protein interactions (between target trait proteins, target to nontarget, antibody to antibody, etc.) may poise an issue for multiplexing. Matrix effects, the effect from all the components except the target trait protein, and the dilutional agreement of sample extract have to be evaluated to ensure assay accuracy. Satisfactory extraction efficiency for all target trait proteins utilizing a universal extraction buffer and a common extraction method is key to multiplexing, which sometimes are time consuming and challenging steps. The wide expression spectrum of target trait proteins represents another unique bottleneck for multiplexing, which need a multiplexing platform with a very broad dynamic range.

13.3 Confirming the absence of unintended traits

Long before the introduction of biotech crops, the existence of "off types," plants or seeds that are different than the intended variety, has been unavoidable due to the nature of agriculture and the biology of crop plants. Hence, it is recognized that seed for production agriculture rarely achieves 100% varietal purity (Roberts et al., 2015).

The term adventitious presence (AP) refers to the unintentional presence and accidental commingling of trace amounts of "off types" of seed or grain in a parcel of seed or grain, which is an inevitable consequence of the production (e.g., cross-pollination, volunteer plants, etc.) and distribution systems for seed and grain such as commingling during transport, storage or processing (Demeke et al., 2005). As such, and regardless of its regulatory status, AP refers to the unintended and incidental trace presence of an impurity occurring in a material as a result of the natural variability in the controlled process used to generate the particular material type. Impurities may include an undesired seed variety, trait, event, or other off-type.

Low-level presence (LLP) as used in relation to biotech crops, refers to a situation of AP where there is the incidental presence of genetically modified material in food, feed, or grain at levels that are consistent with generally accepted agricultural and manufacturing practices. This unintended low-level presence of genetically modified seeds has been approved for unrestricted cultivation in at least one country but not in the country of import (OECD, 2013).

AP and LLP can be minimized but cannot be avoided due to the biological characteristics of the crops, open nature of the field, and the technical management along the supply chain from sowing in the field to crop management, harvest, storage, transportation, and processing. AP is not unique to biotech crops and does not comprise product quality. Although allowances of AP have been recognized in laws and regulations, different stakeholders still need a solid detection method to monitor AP of biotech products. For example, seed companies, farmers, food and feed industry, enforcement authorities, and testing laboratories are still in need of relevant methodologies for AP testing to monitor products after commercial release, to assure segregation of products, and to check compliance with legislation. The variability of sampling and testing protocols for the purpose desired makes the analytical strategies require thoughtful considerations of when a method is better adapted than another for a specific objective.

Analytical methods for GMO detection, identification, and quantification are at three levels: qualitative, semiquantitative, and quantitative. Qualitative analysis reports the presence or absence of GMO traits, basically making positive or negative calls; semiquantitative analysis compares GMO content to the predefined thresholds (above or below); whereas, the quantitative provides quantification of GMO content in the tested samples. AP testing can be performed on individual seeds, bulked seeds, or grouped seeds. Individual seed samples make it possible to distinguish breeding stacks from mixtures of two or more single events. If two or more traits are detected in a single seed, the seed contains the stack. Single seed testing is labor-intensive, time-consuming, and can be very expensive considering the whole process from sampling through data interpretation.

As AP testing is to detect low levels of the traits or trait combinations, pooled seed samples are preferred. A proper pooling strategy is critical to avoid sample nonhomogeneity, hence both sample size and sampling procedures are important factors. The sample size must be sufficient to allow sensitive detection of specific trait proteins, while the sampling plan will help ensure the sample is the real representative of the larger lot of the material and minimize error originating from sampling and handling. Many government bodies have established guidelines for AP testing of GM materials, including the United States Department of Agriculture, Grain Inspection, Packers and Stockyards Administration, the International Seed Testing Agency (ISTA), and the Codex Alimentarius Commission of the UN. A statistical sampling scheme and a proper testing plan (Laffont et al., 2005) need to be defined on how to handle samples of GM products for testing for AP of transgenic materials. Proper sampling and handling to prepare test samples are keys to generating reliable and informative analyses for traits detection. It is critical to adopt appropriate detection methods (Anklam et al., 2002) to minimize the main sources of error in the detection of GMO contamination.

13.3.1 Immunoassay

An ideal analytical method for specific GMO detection should be science-based and validated. Immunoassays can detect trace amounts of trait proteins from biotech crops qualitatively or quantitatively over a wide range of concentrations. Some guidelines for validating assay performance (Lipton et al., 2000) and certain method validation criteria (Grothaus et al., 2006) are available for both qualitative and quantitative formats. For qualitative assessment, assay sensitivity, specificity, and false-negative/positive rates are the major performance indicators; while for quantitative determination, assay repeatability, reproducibility, and assay variation including protein standard deviations are key performance indicators. ELISA (Mason, 1992) and LFS (Hermanson et al., 1992) are the most commonly used immunoassay platforms for AP testing of genetic trait proteins in crop samples.

13.3.1.1 Enzyme-linked immunosorbent assay

A typical ELISA procedure for AP testing for crop seed semiquantitatively or quantitatively starts with preparation of seed samples. Calibrators can be commercially available or can be made in-house as follows: prepare a certain number (e.g., 4 or 5) of pools of nontransgenic seeds, the pool size can vary from a few hundreds to thousands of seeds depending upon the assay plan (see Chapter 12). These pools are spiked with transgenic seeds in various proportions, expressed as ratios, for example, 1:5000, 1:10, or expressed as a percentage, for example, 0.02%, 10%. These spiked pools will be ground and homogenized in the extraction buffer. After clarification by centrifugation or filtration, the aliquots of the extract supernatant can be used as calibrators in the assay. The test seed samples to be evaluated for the presence of transgenic traits will be collected and pooled according to the sampling procedure. After the addition of extraction buffer, the test samples will be ground and homogenized, and the clarified supernatant for all the test samples will be aliquoted and used in the assay.

The aliquots of test samples together will the aliquot of calibrators will then be used in an ELISA assay. Determination of trait protein is accomplished by comparing the absorbance values of test samples to that of calibrators. If a semiquantitative determination is needed, then a range call (for example, between 0.05% and 0.2%) of transgenic seeds present in the test samples is enough. If quantification is required, then the AP of transgenic seeds in the test samples will be interpreted from the standard curves generated from the calibrators with the known proportion of transgenic to nontransgenic seeds. ELISA for AP testing provides results in hours with great sensitivity. For example, Lipp et al. (2000) confirmed that the validated ELISA method for CP4 5-enolpyruvylshikimate-3-phosphate synthase (EPSPS) could reliably detect transgenic soybean in homogenized samples at a portion of 0.3%. Depending upon the specific trait protein, the effectiveness of detecting the presence of transgenic seeds in nontransgenic seeds samples with detection limits of ~0.1% (Christianson et al., 2008).

ELISA for AP testing can be established in-house after assay parameter optimization and assay validation. For propriety proteins of the seed company, the assays have to be developed and validated within that company to protect intellectual property. For some publicly available trait proteins, commercial ELISA kits are available, and the assays are well established. A typical commercial kit contains key biological reagents (antibody-coated plate, enzyme-labeled secondary antibody), calibrators (standard trait protein with known concentrations), controls (positive and negative), and buffers (sample extraction buffer, washing buffer). Simply following the manufacturer's instructions will help generate satisfactory data for AP testing.

13.3.1.2 Lateral flow strips

LFS immunoassay commonly adopts a sandwich format and can be used to determine AP of unintended trait proteins in different crop tissue samples, including leaves, seeds, and grains. Antibody selection and pairing are key to developing an LFS assay. The antibodies that worked in a microplate-based ELISA under a more controlled environment do not necessarily guarantee their performance in LFS. More evaluation, comparison, and optimization are needed to identify the proper antibody pair for LFS immunoassay under different sometimes rough conditions.

Once the antibody pair is determined, an LFS kit can be developed and validated. The kits for detection of GMO trait proteins are available from different manufacturers, for example, Romer Labs (Union, Mo, USA) and EnviroLogix (Portland, ME, USA). Although differences may exist for the kits from different suppliers, the principles are the same. Following the manufacturer's instruction, an LFS assay can be run with tissue extract. At the end of the run, if two lines appear, the result is positive; if only the control line appears, the result is negative; a missing control line indicates the malfunction of the strip and the result is invalid.

Because of its simplicity, accuracy, and reliability, LFS immunoassay has been extensively used in the field, at the elevator, and in the lab on a large scale in a rapid and cost-effective manner. LFS is designed for qualitative or semiquantitative testing. LFS immunoassay for detecting AP of GMO trait proteins also starts with sample preparation. Commonly, a few sizes of seed samples (e.g., 200 seed, 400 seeds) are tested, and calibrators are prepared by adding different GM seeds to non-GM seeds at different levels (e.g., 0.0%, 0.5%, 1.0%, and 5.0%), with many replicates to ensure randomized experimental design and facilitate semiquantitative detection. Seed samples and calibrators are ground for a few seconds to minutes depending upon the grinding methods and equipment used. Immediately after grinding, a certain volume of water or other extraction buffer is added and a quick shake or mix (e.g., 10 s) follows to thoroughly wet and extract the samples. The extracts are allowed to settle for a few minutes to clarify the particulate materials, and the supernatant can be saved in separate clean containers and ready for assessment of AP of trait proteins. The specific LFS kit supplier has detailed information on how to perform the assay and how to interpret the test results. The effectiveness of LFS immunoassay for detecting the presence of GM seed in non-GM seeds samples can go down to 0.1% (EnviroLogix-AQ/AS 076 TC) depending upon the trait. And further optimization of the available kit can improve the capability of detecting GM contamination down to 0.033% level (Mazzara et al., 2013).

The sensitivity of both ELISA and LFS is limited by the unique properties of the protein to be detected, the quality of the protein extracted from the test samples, the protein expression level in a particular crop tissue, the specificity of antibodies used for the testing, the way in which sample extract is prepared, and the type of signal amplification chemistry employed. The practical limit of detection (LOD) is sometimes different than the method LOD that was established under optimal conditions. The practical LOD can only be determined from the results of analyses of real samples. Many factors may lower the sensitivity of the methods, including the biochemical composition of the sample, different varieties, and different environmental conditions. Some of these factors are difficult or impossible to control.

Both ELISA and LFS are protein-based methodologies for AP testing of GM traits in non-GM samples, and proper statistical analyses are needed to ensure relevance of the data from start to finish, including sampling, data generation and the final data interpretation. Specific business needs will help define which assay platform to use, including sample size, assay sensitivity, skill requirements, and cost and so.

13.3.1.3 Other formats

Besides ELISA and LFS, other protein-based methods have gained a lot of development in handling large numbers of samples in AP testing of GMO traits. In addition to multiarray-based MSD and microbead-based Luminex, several different homogeneous proximity assays have been exploited to detect analytes in complex sample matrices. These antibody-based homogeneous assays eliminate washing steps and reduce assay time and unnecessary instrumentation and are preferred assay types for screening large numbers of samples. Some antibody-free methodologies also provide new tools for GMO traits detection at the protein level.

Antibody-based homogeneous proximity assays

In a homogeneous assay, the analytes commonly serve as the bridges, while antibodies specific to the analyte, with unique labels, will be brought together by the bridge in close enough proximity to facilitate energy transfer, thus generating detectable signals. A number of homogeneous assays have been developed for detecting analytes in complex sample matrices. Scintillation proximity assay (SPA) detects proteins when the radiolabeled molecule is brought to the proximity of the beads and stimulates the scintillant inside the beads (Park et al., 1999). Time-resolved fluorescence resonance energy transfer relies on the really long decay half-life of the donor fluorophores to minimize the background fluorescence and maximize the signal/background ratio (Comley, 2006). Amplified Luminescent Proximity Homogeneous Assay (ALPHA)-based immunoassay-AlphaLISA requires a donor bead to generate singlet oxygen that, in turn, interacts with acceptor beads to generate a chemiluminescent signal (Beaudet et al., 2008). Proximity ligation assay and proximity extension assay are nucleic acid-based proximity assays (Fredriksson et al., 2002; Lundberg et al., 2011). These assays utilize antibody-labeled oligonucleotides or nucleic acids to either enhance specific recognition of the target analyte, to boost the detection signal, or to minimize the interferences from nonspecific sample components. Among the proximity homogeneous assays, spatial proximity analyte reagent capture luminescence (SPARCL) technology enables assays to be miniaturized for high throughput screening while maintaining sensitive results with good dynamic range.

Spatial proximity analyte reagent capture luminescence

SPARCL technology is a proximity-dependent, nonseparation, chemiluminescent detection method. In a SPARCL assay, a chemiluminescent substrate (acridan) is brought into the proximity of an oxidative enzyme (horseradish peroxidase, HRP) through the specific antigen/antibody interaction. A flash of light proportional to the quantity of analyte present in the sample is generated upon the addition of a trigger solution containing H_2O_2 and para-hydroxycinnamic acid.

The SPARCL technology is applicable to both sandwich and competitive assays with and without a solid phase. In the format with a solid phase, both the acridan compound and a specific capture antibody are coupled to solid phases such as microparticles or microtiter plates. The capture antibody will bring analyte-labeled (in a competitive assay) or antibody-labeled (in a sandwich assay) enzyme close to the immobilized acridan compound. Besides the benefits of detecting the analyte homogeneously, this format does offer an option of separating the complex matrix and excess reagents and detecting the analyte heterogeneously. SPARCL solid phase formats exhibited both good sensitivity and good dynamic range. While for solution-phase SPARCL, the solid phase is omitted, the capture antibody is directly labeled with the acridan compound, the interaction between antibody and analyte occurs in solution. And a background reducing agent can be added to minimize the background signal from unbound reactants and to enhance the signal-to-noise ratio.

Solution-phase SPARCL is fast, and it usually takes 0.5−1 h to complete an assay. In a typical solution-phase SPARCL assay, two set of antibodies were labeled: one set with the acridan and the other with HRP. No subsequent purification of unbound acridan was performed. The assay mixture, in a plate, contained 20 μL acridan-labeled antibody, 20 μL complementary HRP-labeled antibody, and 30 μL calibrator (analyte samples, QC or buffer) solutions. The mixture was incubated for 30−60 min at room temperature. A solution of ascorbic acid (10 μL) was added followed by an injection of trigger solution (100 μL). Signal was read and integrated for 2 s port triggering on a plate luminometer.

SPARCL technology enables rapid immunoassay development and desirable performance characteristics and allows for simple reagent preparation, considerable savings in labor, disposables, and capital equipment. Existing software and luminometers can be used for a SPARCL data acquisition, the data can be easily managed by the existing LIMS. The SPARCL offers significant time savings and simplification of assay mechanics, which makes it an attractive technology for high throughput screening and general life science research, including Agricultural biotechnologies.

Antibody-free detection methods

The current in-use methods are limited by the availability of antibodies of high quality. To circumvent the hurdles associated with antibodies, some new technologies have been evaluated, which appear to be promising areas for GMO detection. Chromatography methods and near-infrared (NIR) spectroscopy are technologies among many others. Chromatography methods are more amendable to detect the composition of GMO ingredients instead of monitoring the trait proteins directly. High-performance liquid chromatography or gas chromatography coupled with mass spectrometry is suitable for investigating the differences in the chemical profile between GM and conventional crops. This methodology is especially useful for evaluating some quality traits for example high stearic acid canola oil by monitoring the changes in triacylglycerols patterns. Chromatography is applicable only when significant changes occur in the composition of GM crops compared to conventional counterparts. NIR transmittance spectroscopy on the other hand is a nondestructive technology that has the potential application for the analysis of moisture, protein, oil, fiber, and starch. Roussel et al. (2001) employed this technique to distinguish Roundup Ready soybean from conventional soybean and were able to make 93% accurate calls. As NIR does not identify and analyze a specific compound, a large number of samples are needed to generate calibration spectra for GMO event prediction. Although NIR is sensitive to major organic compounds, its accuracy in detecting whole seed/grain is low. The capacity of NIR to detect a small amount of GMO in conventional events is also assumed to be low. The disadvantages of this technology may limit its wide application for GMO detection; however, there are still a lot of potential uses for the NIR technique as it is fast, cheap, and nondestructive.

13.4 Interpreting results

13.4.1 Introduction

When testing for genetic traits, a sample of seeds is typically taken from a much larger lot of seeds. No matter how accurate the assay, the results of the testing will only be as good as the sample collected. If the sampling approach is inappropriate then the result of the testing will most likely be inaccurate. For details on appropriate seed sampling, see the AOSA Rules for Testing Seeds (2014) edition.

Still one must remember that even the results from an appropriately collected sample will almost assuredly differ from the actual presence of the genetic trait in the lot. The difference between the results of the sample and the true presence in the lot is largely a result of random sampling variability. Thus, the only way to know the true presence of the genetic trait would be to test the entire lot. Of course, testing the entire lot is almost never possible. This is where statistics prove to be a useful tool for interpreting testing results.

Multiple statistical considerations must be considered when interpreting results. Before conducting the analysis, it is important to ensure the sample size was large enough to accomplish the goal of the testing. Once the sample size is set, the sampling strategy will affect the sources of variability that need to be accounted for in the analysis and ultimately where the inference of the results can be applied. Here, the focus will be on using a sample to make inference to a much larger lot of seeds. Finally, the method used will also affect the statistical analysis. This is because different statistical methods are used for qualitative and quantitative methods. The statistical methods for developing testing plans and summarizing testing results, summarized here, are described fully for qualitative methods in Remund et al., (2001) and quantitative methods in Laffont et al., (2005). These statistical methods can be implemented in Seedcalc8, a Microsoft Excel spreadsheet application freely available on the International Seed Testing Association (ISTA) Website (http://seedtest.org/en/stats-tool-box-_content—1–1143.html). The use of SeedCalc and similar approaches for GMO testing is covered in International Standards Organisation (2021).

13.4.2 Qualitative test

Qualitative testing can be performed on either individual seeds or on pools of seed. Testing individual seeds can be thought of as a pool size of one seed. Qualitative testing plans are created based on three key decisions. (1) How many pools of seed to test? (2) How many seeds are in each pool? (3) How many deviant pools can be observed in the sample before the lot is rejected? The definition of a deviant pool depends on the goal of the testing. If the goal is to test for LLP, then a deviant pool would be a positive result for an unintended trait. However, if the goal is to test for trait purity, then a deviant pool would be a negative result for the intended trait. SeedCalc8 can be used to implement the qualitative seed testing plan design methodologies summarized here by utilizing the "Qual Plan Design" worksheet. Sampling Statistics is further discussed in Chapter 9.

Remund et al. (2001) utilize acceptance sampling logic and terminology as the framework for developing and evaluating qualitative testing plans. See Montgomery (2013) for an in-depth discussion of acceptance sampling. One of the most important terms to remember in qualitative seed testing is the lower quality limit (LQL). The LQL can be simply thought of as the quality threshold. If a seed lot is accepted, then there should be a high level of confidence that the LQL has been met. The probability of accepting a lot that does not meet the LQL is called the consumer's risk. The exact definition of the LQL will differ based on the purpose of the testing. For purity testing, the LQL is the lowest level of purity in the seed lot that would be considered acceptable to the consumer. For LLP testing, the LQL is the highest level of LLP in the seed lot that would be considered acceptable to the consumer.

Another important term is the acceptable quality level (AQL). While the LQL is a level that should be rejected, the AQL is a level that should be accepted. The AQL is a level of quality that meets the consumer's requirements but falls short of perfection because in production the standard of perfection is often unrealistic. The probability of rejecting a lot at the AQL is called the producer's risk. The exact definition of the AQL will differ based on the purpose of the testing. When purity testing, the AQL is a level of purity in the seed lot that is less than 100% but would be considered acceptable to the consumer. For LLP testing, the AQL is a level of LLP in the seed lot that is greater than 0% but would be considered acceptable to the consumer. When designing a testing plan, the sample size required will increase as the AQL gets closer to the LQL.

Good testing plans will maintain low levels of both consumer's and producer's risk. A graphical tool used to evaluate testing plans is an operating characteristic (OC) curve. An OC curve for a seed testing plan will have the true purity/impurity in the lot on the horizontal axis and the probability of accepting the lot on the vertical axis. For qualitative testing, the calculation of the probability of accepting the lot is based on the binomial distribution. As an example, let the LQL be 1% for an LLP study. If the true impurity was 1% and 300 seeds were tested, then there would be less than a 5% probability of observing no deviant seeds. Thus, a reasonable testing plan would be to test 300 seeds with no deviants allowed to accept the lot. The OC curve for this testing plan, displayed in Fig. 13.1, shows the consumer risk is low because there is less than a 5% probability of accepting a lot with 1% impurity.

Plans that result in rejection when even a single deviant is observed are called zero-tolerance plans. Zero-tolerance plans are used frequently because of reluctance to accept lots that have any sign of impurity. However, again assume that the consumer requires an LQL of 1% but also assume that the producer's average lot contains 0.25% LLP. If the plan proposed in the previous example was used, the average lot would have less than a 50% probability of being accepted. In this case, more than half of the producer's lots would be rejected, even though the lots were well below the acceptable threshold of 1% LLP.

In this scenario, a better plan might be to test 1050 seeds and allow up to five deviants seeds to accept the lot. The resulting OC curve, displayed in Fig. 13.2, shows that the plan would still control the consumer's risk by having less than a 5% probability of accepting a lot with 1% impurity. However, by allowing for deviants, the plan would also control the producer's risk by having about a 95% probability of accepting a lot containing 0.25% LLP.

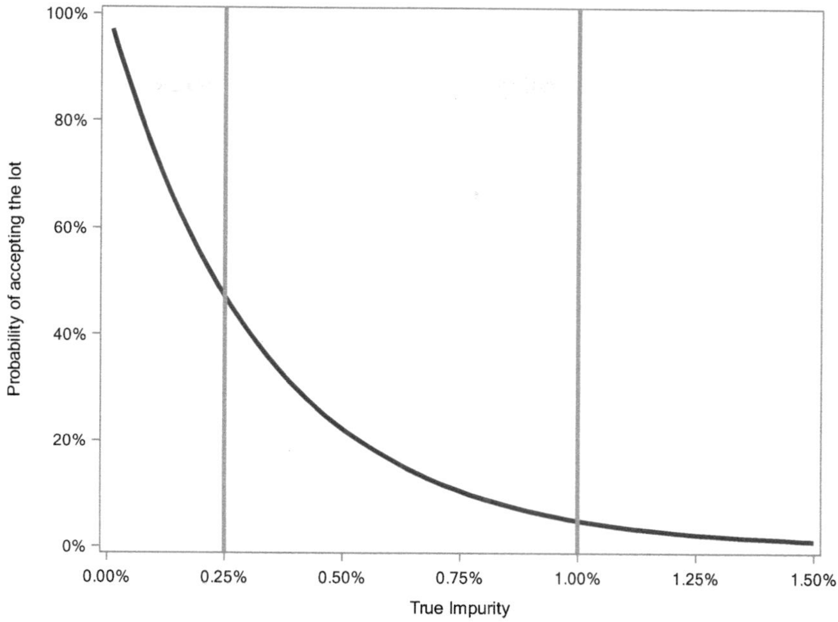

FIGURE 13.1 OC curve for plan in which 300 seeds are tested and zero deviants are allowed.

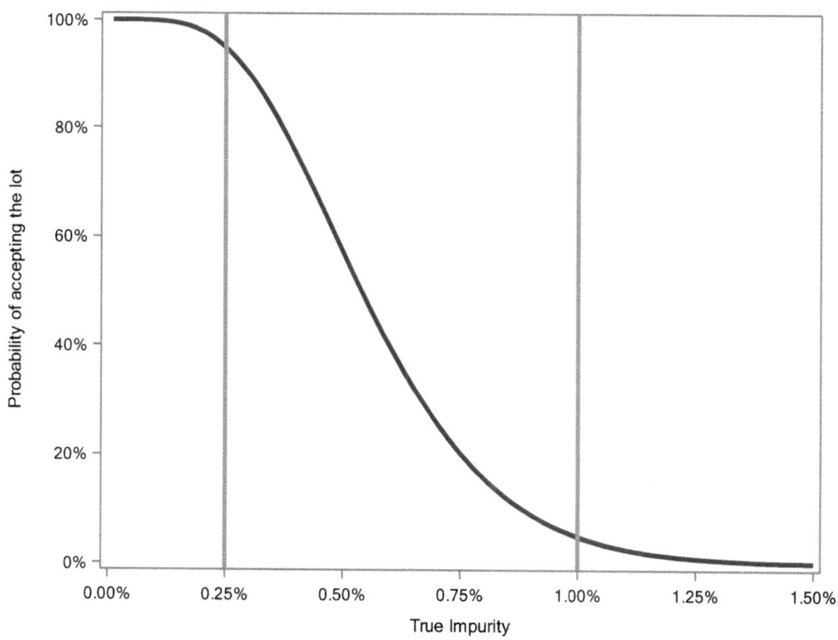

FIGURE 13.2 OC curve for plan in which 1050 seeds are tested and five deviants are allowed.

The previous example illustrates that a zero-tolerance plan does not necessarily control the consumer risk any better than a plan that allows for deviants. However, zero-tolerance plans typically have a very high producer's risk. Zero-tolerance plans are also commonly misinterpreted. When no deviants are observed, that does not mean there is no LLP in the seed lot. Again, the only way to be sure the lot contains absolutely no impurity is to test the entire lot.

In the examples presented, the error rates for the assay were assumed to be zero. However, in practice, an assay will have some nonzero false-positive and false-negative rates. These error rates will affect the producer's and consumer's risks. Error rates can be estimated by testing a randomized series of known positive and known negative samples. The false-positive and false-negative rates can be accounted for in the test plan design, but all efforts should be made to minimize the error rates of an assay.

SeedCalc8 can be used to summarize the results from a qualitative seed testing plan by utilizing the "Qual Impurity Estimation" worksheet. When summarizing the results of a qualitative test, it is often of interest to estimate the true presence of the genetic trait in the lot. As an example, again assume 300 seeds are tested individually and no seeds containing the trait of interest are observed. The observed LLP in the sample is thus 0% or 0/300. Again this does not mean the lot contains 0% LLP. However, the result of the sample can be used to estimate an interval that will contain the actual LLP in the lot with a specified level of confidence. This is typically done by calculating a one-sided confidence limit for the true LLP. For example, in this case, the LLP would be less than 1% with 95% confidence. Note that the confidence interval is not interpreted as the lot containing 1% LLP. Instead, the interpretation should be that if the true LLP was 1%, then there would be a 95% chance of observing at least one deviant. Because no deviants were observed, the evidence suggests that the LLP in the lot is less than 1% with 95% confidence. Note that the confidence limit is directly related to the LQL of the testing plan. If an LLP testing plan has an LQL of 1% and the confidence limit exceeds 1%, then the lot would be rejected.

As another example, again assume 300 seeds are tested. However, now the testing is done in two pools of 150. If one pool is positive and one pool is negative, then the LLP in the lot is less than 2.42% with 95% confidence. This result may seem counterintuitive at first. When one pool of 150 seeds is positive, then all seeds in that pool could potentially be positive. Thus there could be 50% positive seeds in the sample. However, as long as appropriate sampling procedures are followed, the two pools of 150 seeds both represent random samples from the lot. Thus the probability that one sample of 150 seeds is completely negative, but the other pool is completely positive would be incredibly small. In fact, the estimated percentage in the sample would be 0.46% LLP. When testing pools of seed, it is the negative pools that provide the majority of the information. For further discussion see Chapter 9.

13.4.3 Quantitative test

Quantitative testing is usually performed on pools of seed. Quantitative testing plans are created based on five key decisions. (1) How many pools of seed to test? (2) How many seeds are in each pool? (3) How many flour subsamples to test per pool? (4) How many measurements per flour subsample? (5) What is the Acceptance Limit (AL)? Laffont et al. (2005) define the term AL to be the highest level of estimated impurity of the lot that will not result in rejection. In qualitative testing, AL is used along with the terminology utilized for qualitative testing (LQL, AQL, consumer risk, and producer risk) to develop testing plans. SeedCalc8 can be used to implement the quantitative seed testing plan design methodologies summarized here by utilizing the "Quant Plan Design" worksheet.

An OC curve is also a useful graphical tool for evaluating quantitative testing plans. Again an OC curve for a seed testing plan will have the true purity/impurity in the lot on the horizontal axis and the probability of accepting the lot on the vertical axis. The OC curve for a quantitative test is created from the AL and the variance of the estimated impurity of the lot. SeedCalc8 accounts for three major sources of variability in estimating the variance of the estimated impurity of the lot: seed sampling; flour subsampling; and measurement variability. Seed sampling variability is derived from the binomial distribution. Although other methods may be preferable, SeedCalc utilizes the method of moments estimation to estimate the flour and measurement variability.

As an example, let the LQL be 1% and AQL be 0.25% for an LLP study. First, a pilot study is run to estimate the flour subsampling and measurement variability. In the pilot study, a pool of 1000 seeds was spiked with 0.3% impurity. The pool of seeds is ground and two flour subsamples are taken. Each subsample is measured in triplicate. The "Quant Impurity Estimation" worksheet in SeedCalc8 can be used to estimate the variance parameters of the estimated impurity of the lot. The resulting variability estimates are then used to evaluate proposed testing plans. In the testing plan, the AL is set to 0.618%. The test design is to grind two pools of 1000 seeds, take two flour subsamples from each pool, and measure each subsample in triplicate. From the OC curve, displayed in Fig. 13.3, it is clear that this testing plan would have a low probability of accepting a lot with true LLP at 1%. The plan would also control the producer's risk, given that the probability of accepting a lot with true LLP at 0.25% is also high.

When summarizing the results of a quantitative test, it is often of interest to estimate the true presence of the genetic trait in the lot. As an example, again assume that the test design is to grind two pools of 1000 seeds, take two flour subsamples from each pool, and measure each subsample in triplicate. If the estimated LLP in the sample was 0.2%, this does not mean the lot contains 0.2% LLP. However, the result of the sample can be used to estimate an

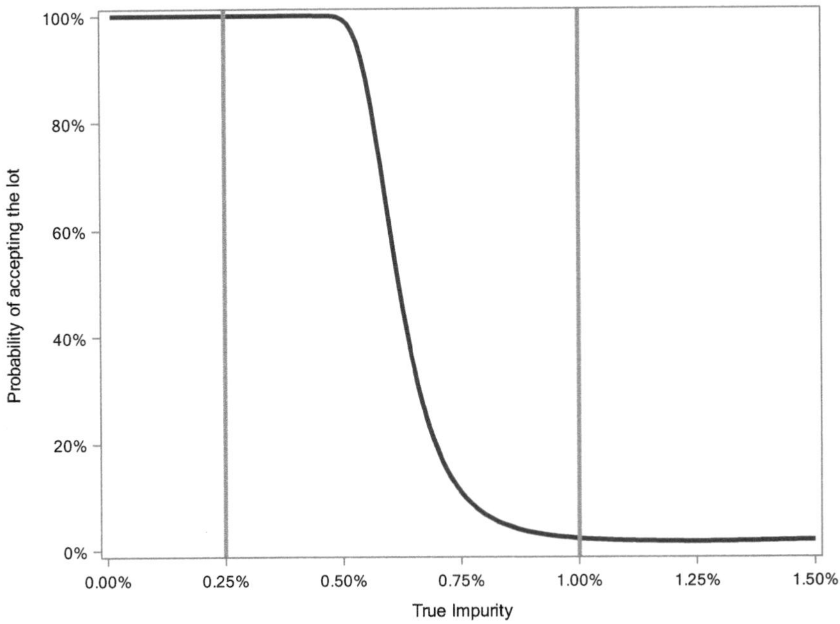

FIGURE 13.3 OC curve for plan to grind two pools of 1000 seeds, take two flour subsamples from each pool, and measure each subsample in triplicate. The AL is set to 0.618.

interval that will contain the actual level of trait present in the lot with a specified level of confidence. This is typically done by calculating a one-sided confidence limit for the true trait presence. For example, in this case, the trait purity would be less than 0.4% with 95% confidence. This interval is estimated based on seed sampling variability and the observed flour subsampling and measurement variability. The "Quant Impurity Estimation" worksheet in SeedCalc8 can be used to estimate the LLP in the sample and the associated 95% one-sided confidence interval for the LLP in the lot.

It should be noted that the units for qualitative and quantitative tests are not always going to be directly comparable. In qualitative testing, the units are percent seed. In quantitative testing, the units are percent DNA. Lipp et al. (2005) provide an in-depth description of the biological sources that can prevent this direct comparison. SeedCalc8 can incorporate a *b*-factor into the calculations in an attempt to make the units more comparable. For more information on the *b*-factor, see Laffont et al. (2005).

References

Anklam, E., Gadani, F., Heinze, P., Pijnenberg, H., Van Den Eede, G., 2002. Analytical methods for detection and determination of genetically modified organisms in agricultural crops and plant-derived food products. European Food Research and Technology 214, 3–26.

AOSA Rules for Testing Seeds, 2014a. AOSA Rules for Testing Seeds Volume 1. Principles and Procedures. http://www.aosaseed.com. edition).

AOSA Rules for Testing Seeds, 2014b. AOSA Rules for Testing Seeds Volume 4. Seedling Evaluation. http://www.aosaseed.com. edition).

Beaudet, L., Rodriguez-Suarez, R., Venne, M.-H., Caron, M., Bédard, J., Brechler, V., Parent, S., Bielefeld-Sévigny, M., 2008. AlphaLISA immunoassays: the no-wash alternative to ELISAs for research and drug discovery. Nature Methods 5.

Brett, G.M., Chambers, S.T., Huang, L., Morgan, M.R.A., 1999. Design and development of immunoassays for detection of proteins. Food Control 10, 401–406.

Christianson, J., McPherson, M., Topinka, D., Hall, L., Good, A.G., 2008. Detecting and quantifying the adventitious presence of transgenic seeds in safflower, Carthamus tinctorius L. Journal of Agricultural and Food Chemistry 56, 5506–5513.

Comley, J., 2006. TR-FRET Based Assays: getting better with age. Drug Discovery World Spring 22–37.

Demeke, T., Perry, D.J., Scowcroft, W.R., 2005. Adventitious presence of GMOs: scientific overview for Canadian grains. Canadian Journal of Plant Science XX, 1–23.

EnviroLogix-AQ/AS 076 TC. http://www.envirologix.com/artman/publish/article_320.shtml.

Fagan, J., Schoel, B., Haegert, A., Moore, J., Beeby, J., 2001. Performance assessment under field conditions of a rapid immunological test for transgenic soybeans. International Journal of Food Science and Technology, 36, 1–11.

Fredriksson, S., Gullberg, M., Jarvius, J., Olsson, C., Pietras, K., Gústafsdóttir, S.M., Ostman, A., Landegren, U., 2002. Protein detection using proximity-dependent DNA ligation assays. Nature Biotechnology 20, 473–477.

Goggi, A.S., Stahr, M.G., 1997. Roundup™ pre-emergence treatment to determine the presence of the Roundup Ready™ gene in soybean seed: a laboratory test. Seed Technology 19, 99–102.

Grothaus, G.D., Bandla, M., Currier, T., Giroux, R., Jenkins, G.R., Lipp, M., Shan, G., Stave, J.W., Pantella, V., 2006. Immunoassay as an analytical tool in agricultural biotechnology. Journal of AOAC International 89, 913–928.

Gutormson, T.J., 1998. Bioassay procedure for determining the presence of the Roundup Ready™ gene in corn. In: Zaworski, F. (Ed.), Testing Methodologies for Hybrid Parentage and Trait Determination. Corn, Sorghum and Soybean Technology. 1999 a Special Publication of Seed. Trade News.

Hermanson, G.T., Mallia, A.K., Smith, P.K., 1992. In: Immobilized Affinity Ligand Techniques. Academic Press, Inc., San Diego, CA. Harcourt Brace Jovanovich, USA.

Horan, P.K., Wheeless Jr., L.L., 1977. Quantitative single cell analysis and sorting. Science 198, 149–157.

International Standards Organisation, 2021. ISO 22753:2021Molecular biomarker analysis—Method for the statistical evaluation of analytical results obtained in testing sub-sampled groups of genetically modified seeds and grains—General requirements. International Standards Organisation. In press.

ISAAA, 2019. Global Status of Commercialized Biotech/GM Crops: 2019. ISAAA Brief No. 55. ISAAA, Ithaca, NY. https://www.isaaa.org/resources/publications/briefs/55/executivesummary/default.asp. accessed June 6, 2022.

Laffont, J.L., Remund, K.M., Wright, D., Simpson, R.D., Gregoire, S., 2005. Testing for adventitious presence of transgenic material in conventional seed or grain lots using quantitative laboratory methods: statistical procedures and their implementation. Seed Science Research 15, 197–204.

Lipp, M., Anklam, E., Stave, J.W., 2000. Validation of an immunoassay for detection and quantitation of a genetically modified soybean in food and food fractions using reference materials: interlaboratory study. Journal of AOAC International 83, 919–927.

Lipp, M., Shillito, R., Giroux, R., Spiegelhalter, F., Charlton, S., Pinero, D., Song, P., 2005. Polymerase chain reaction technology as analytical tool in agricultural biotechnology. Journal of AOAC International 88, 136–155.

Lipton, C.R., Dautlick, J.X., Grothaus, G.D., Hunst, P.L., Magin, K.M., Mihaliak, C.A., Rubio, F.M., Stave, J.W., 2000. Guidelines for the validation and use of immunoassays for determination of introduced proteins in biotechnology enhanced crops and derived food ingredients. Food and Agricultural Immunology 12, 153–164.

Lundberg, M., Eriksson, A., Tran, B., Assarsson, E., Fredriksson, S., 2011. Homogeneous antibody-based proximity extension assays provide sensitive and specific detection of low-abundant proteins in human blood. Nucleic Acids Research 39, e102. https://doi.org/10.1093/nar/gkr424.

Mason, M.M. (Ed.), 1992. Immunochemical Protocols (Methods in Molecular Biology. Humana Press, Totowa, NJ vol. 10.

Mazzara, M., Paoletti, C., Corbisier, P., Grazioli, E., Larcher, S., Berben, G., De Loose, M., Folch, I., Henry, C., Hess, N., Hougs, L., Janssen, E., Moran, G., Onori, R., Van den Eede, G., 2013. Kernel lot distribution assessment (KeLDA): a comparative study of protein and DNA based detection methods for GMO testing. Food Analytical Methods 210–220.

Montgomery, D.C., 2013. Introduction to Statistical Quality Control, seventh ed. Wiley, Hoboken, NJ.

Non-Tolerant Seedling Study Guide. https://analyzeseeds.com.

Organisation for Economic Co-operation and Development, 2013. Low Level Presence of Transgenic Plants in Seed and Grain Commodities: Environmental Risk/safety Assessment, and Availability and Use of Information. Series on Harmonisation of Regulatory Oversight in Biotechnology No. 55. ENV/JM/MONO(2013)19.

Park, Y.W., Cummings, R.T., Wu, L., Zheng, S., Cameron, P.M., Woods, A., Zaller, D.M., Marcy, A.I., Hermes, J.D., 1999. Homogeneous proximity tyrosine kinase assays: scintillation proximity assay versus homogeneous time-resolved fluorescence. Analytical Biochemistry 69, 94–104.

Pomeranke, G.J., Nickell, C.D., 1988. Inheritance of Chlorimuron Ethyl Sensitivity in the Soybean Strains BSR 101 and M74-462 Crop Science, pp. 59–60 vol. 28.

Posthuma-Trumpie, G.A., Korf, J., van Amerongen, A., 2009. Lateral flow (immuno)assay: its strengths, weaknesses, opportunities and threats. A literature survey. Analytical and Bioanalytical Chemistry 393, 569–582.

Remund, K.M., Dixon, D.A., Wright, D.L., Holden, L.R., 2001. Statistical considerations in seed purity testing for transgenic traits. Seed Science Research 11, 101–119.

Roberts, A., Finardi-Filho, F., Hedge, S., Kiekebusch, J., Klimpel, G., Krieger, M., Lema, M., Macdonald, P., Nari, C., Rubinstein, C., Slutsky, B., Vicien, C., 2015. Proposed criteria for identifying GR crop plants that pose a low or negligible risk to the environment under conditions of low-level presence in seed. Transgenic Research 24 (5), 783–790. https://doi.org/10.1007/s11248-015-9899-z.

Roussel, S.A., Hardy, C.L., Hurburgh, C.R., Rippke, G.R., 2001. Detection of Roundup Ready™ soybeans by near-infrared spectroscopy. Applied Spectroscopy 55, 1425–1430.

Savoy, B.R., Berkey, D.A., Johnson, P.G., 2001. Rolled towel bioassay to identify Roundup Ready trait in cotton planting seed. Seed Technology 23, 165–168.

GlobalSCST Seed Technologist Training Manual Chapter 14 Update, 2016. https://analyzeseeds.com/product/seed-technologist-training-manual-chapter-14-update. (Accessed 6 June 2022).

Sebastion, S.A., Chaleff, R.S., 1987. Soybean mutants with increased tolerance for sulfonylurea herbicides. Crop Science 27, 948–952.

Sinning, L.A., 1995. Genetically altered corn tests. Society of Commercial Seed Technologist News 69, 34–38.

Stave, J.W., 2002. Protein immunoassays methods for detection of biotech crops: application, limitations and practical considerations. Journal of AOAC International 85, 780–786.

Wilson, W., Dahl, B., 2016. Potential economic impacts of low level presence (LLP) in the global wheat market. In: Kalaitzandonakes, N., Phillips, P.W.B., Wesseler, J., Smyth, S.J. (Eds.), The Coexistence of Genetically Modified, Organic and Conventional Foods, pp. 241–256.

Chapter 14

Detection methods for genome-edited crops

Sonia Herrero[1], Sherry Whitt[2], Farhad Ghavami[3], Margit Ross[4], Donna Houchins[5] and Ray Shillito[2]
[1]*Syngenta Crop Protection, LLC, Durham, NC, United States;* [2]*BASF Corporation, Morrisville, NC, United States;* [3]*Eurofins BioDiagnostics Inc., River Falls, WI, United States;* [4]*Corteva Agriscience, Johnston, IA, United States;* [5]*Romer Labs, Inc., Union, MO, United States*

14.1 Introduction

Genome editing (also called gene editing) has been used in an increasing number of applications in agriculture (Zhang et al., 2018; EuropaBio, 2020) and is an additional tool in the plant breeder's toolbox. Genome editing can be used to produce targeted changes in the plant's DNA, from small mutations to larger genetic rearrangements (e.g., inversions) and insertions. The techniques are more efficient in terms of time and precision compared to the traditional approaches of random mutagenesis and selection of beneficial variants (Holme et al., 2019) or in the case of insertion of transgenic DNA, the original Genetically Modifed (GM) technology relying on random integration.

The application of genome editing relies on knowledge of gene function followed by scientific expertise in bioinformatics, genetic transformation, tissue culture, molecular analysis, and breeding. Once this information and competencies are available, the technology can be applied to a diverse range of plants wherever whole plants can be regenerated from single edited cells. The use of genome editing in plants is rapidly increasing, both as a research and crop development tool. While genome editing can be used for both academic research and commercial product development, the regulatory environment, as well as social acceptance, could be limiting factors for the broad application of this technology in agriculture (CAST, 2018; Whelan et al., 2020). Some forms of gene-edited plants are scientifically indistinguishable from those produced through conventional breeding.

Genome editing tools represent technologies used to create targeted changes in the DNA of an organism by leveraging the inherent DNA repair and recombination processes within the cell. Tools such as clustered regulatory

Application of Sampling and Detection Methods in Agricultural Plant Biotechnology
https://doi.org/10.1016/B978-0-323-99293-0.00002-7

interspaced short palindromic repeat (CRISPR)-associated protein (Cas) (i.e., CRISPR/Cas), transcription activator-like effector endonucleases (TALENs), and zinc finger nucleases (ZFNs) are also being used in plant breeding. Although CRISPR/Cas (Jaganathan et al., 2018; Chen et al., 2019) is the most recent and best-known system used for this purpose, it is not the first nor the only genome-editing tool available.

The ease of detecting an edit in the genome depends on the type of edit performed. While detection of an introduced novel or significantly changed protein can be possible via immunoassays, many traits introduced via genome editing do not involve changes to the trait protein but rather are the result of sequence deletion of a target gene resulting in no protein expression, impact the protein expression level but not the structure, or make edits in DNA small enough to create a significant third-order structural change in the corresponding protein, and thus an antibody approach will not be able to distinguish the edited and unedited proteins. Moreover, not all matrices have protein remaining in processed products. This makes small edits of a gene not amenable to detection by immunoassay methods. Detection of a genome-edited trait will therefore in almost every case need to rely on the examination of the DNA change.

In this chapter, we review current methodologies that can be used to develop detection methods for these types of products and discuss the technical and regulatory challenges that will have to be overcome for such methods.

14.2 What is genome editing—the key technologies

Naturally occurring site-directed nucleases (SDNs) able to specifically cut long DNA sequences (i.e., meganucleases) were originally used for genome editing (Xu et al., 2015), but have since been replaced by relatively easy to engineer chimeric SDNs that include ZFNs, TALENs, and CRISPR/Cas (Malzahn et al., 2017; Mahfouz and Li, 2011; Podevin et al., 2013; Kamburova et al., 2017). It should be noted that a nonnuclease-based method known as oligonucleotide-directed mutagenesis (ODM) has also been used to edit plant genomes (Beetham et al., 1999; Zhu et al., 1999).

The basic applications of genome editing by ZFNs, TALENs, or CRISPR/Cas have been classified as SDN1, SDN2, and SDN3 based on the type of DNA repair process and the resulting DNA change (Podevin et al., 2013). SDN1 changes usually lead to variants with a single or a few nucleotide changes including insertions and deletions (INDELs) and occasionally translocations and inversions; using two DNA double-strand break sites would lead to deletion of the DNA sequence in between the sites. SDN2 relies on the presence of donor DNA harboring specific edits that can be used to introduce desired mutations in a targeted fashion. The third type, SDN3, also relies on the presence of donor DNA and results in the targeted insertion of a specific sequence at a specific place in the genome. Some DNA changes resulting from genome editing have been compared to the DNA changes that result spontaneously or from induced random mutagenesis, for example, gamma irradiation, fast neutron, or Ethyl methanesulfonate (EMS) mutagenesis (Oladosu et al., 2016; Podevin et al., 2013; Kumawat et al., 2019; Graham et al., 2020); however, as indicated by Holme et al. (2019) "*SDN technology improves precision and reduces the extent of mutations in a crop where a specific trait is pursued.*"

While ZFNs, TALENs, and CRISPR/Cas nucleases can generate either SDN1, SDN2, or SDN3 edits, their modes of DNA recognition differ significantly. In the case of ZFNs, each of the three DNA-binding domains in the nuclease can recognize three specific nucleotides; however, the DNA-binding domains in TALENs contain variable amino acid residues that allow these TALEN variants to bind to different and specific single nucleotide targets. The DNA-binding domains of ZFNs and TALENs have been engineered to introduce targeted edits in plant genomes and are discussed in several reviews (Kamburova et al., 2017; Zhang et al., 2018; Holme et al., 2019). Conversely, the target specificity of the Cas nuclease in the CRISPR/Cas system depends on the presence of a guide ribonucleic acid (gRNA) that binds to complementary genomic sequences that are processed by the Cas nuclease. The ease of modifying the specificity of gRNAs and the increasing number of Cas enzymes that have been identified and engineered to improve the genome-editing machinery and the overall genome-editing efficiency have made the CRISPR/Cas system the most attractive for generating genome-edited plants (Jaganathan et al., 2018; Chen et al., 2019; Van Eck, 2020).

Genome editing relies on the introduction of double-strand breaks (DSBs) by these SDNs, and subsequent DNA repair through either nonhomologous end-joining (NHEJ) or homologous recombination (HR). Depending on the desired edit, DSBs can be introduced at a single site or at multiple sites. The repair mechanisms for SDN1 are based on NHEJ, which due to the lack of fidelity can result in randomly introduced or deleted nucleotide(s). Conversely, repair mechanisms for SDN2 and SDN3 are based on HR between the genome target sequence and an exogenously delivered donor DNA engineered with intended edits. Even though the edits are more specific, the HR-based repair mechanism is rather inefficient in plants (Chen et al., 2019; Van Eck, 2020). For certain applications, CRISPR base editors that do not rely on the generation of DSBs in the DNA are being added to the toolkit. Although the targeting components of the nuclease are intact, the nuclease is modified to function as a nickase and single base pair changes can be readily introduced (i.e., cytosine and adenine base editors convert C to T, and A to G, respectively) (Eid et al., 2018).

The genome editing toolkit is rapidly evolving toward robust methods that allow engineering crops that are efficacious on target editing and contain few or even no off-target effects (Graham et al., 2020), which can be removed via plant breeding. The genome-editing machinery can be delivered into a host plant by the stable transformation of individual components (i.e., *Agrobacterium* harboring a plasmid with Cas and gRNA) that can be removed through breeding once the desired edits are identified such that the resulting plant is not considered transgenic. Alternatively, the genome-editing machinery can be deployed using transient expression (Graham et al., 2020; Shillito et al., 2021). In addition, DNA-free genome editing systems based on protein-RNA complexes have become more attractive because recombinant DNA is not used (Woo et al., 2015; Svitashev et al., 2016; Kanchiswamy, 2016; Metje-Sprink et al., 2019). These RNA-guided nucleases and ribonucleoproteins (RNPs) complexes can be delivered by transient expression of mRNAs encoding for TALENs or Cas nucleases and gRNA, or by direct delivery of the RNPs.

14.3 The drivers for detection methods

In general, methods to detect, verify, or quantify agricultural biotechnology products (i.e., genome-edited crops) serve different purposes depending on where they are used in the value chain and on the drivers. Whether in the lab or in field trials, developers use detection methods during research and development during production and commercialization and at the postharvest stage (i.e., processing of crops into food/feed and their commercialization), to analyze, track, and trace materials, for stewardship and compliance purposes, for variety and identity preservation, and intellectual property activities. Other players in the value chain (e.g., grain traders, processors, food manufacturers, distributors) also depend on the availability of detection methods to comply with legal labeling and traceability obligations or to meet the requirements of private standards. Food manufacturers may be interested in detection methods to meet the demand of specific consumer groups, to facilitate informed choice as regards methods of manufacture or production. For such cases, provision of specific information on food and feed components may be required based on traceability processes, including the use of detection methods for verification purposes.

The regulatory status in different jurisdictions for genome-edited products (Whelan and Lema, 2015) may drive the interest of the supply chains in the detection of genome-edited crops and derived products (Chapter 3) to reduce the risk of trade disruptions. However, even if detection methods for genome-edited products are required by law in certain jurisdictions, method-specificity remains a challenge because a priori knowledge is necessary to establish a relationship between a method and a product (Ribarits et al., 2021).

14.4 Detection method approaches

14.4.1 PCR

Multiple approaches are available to detect a DNA modification, such as conventional or real-time polymerase chain reaction (rtPCR), digital PCR (dPCR), or isothermal PCR methods. PCR amplification relies on DNA recognition and requires prior knowledge of the relevant nucleotide sequence (Chapter 4). The differentiation of a single nucleotide change using restriction endonucleases (e.g., cleavage-amplified polymorphic sequence or CAPS) or conventional PCR coupled with visualization of the DNA and fragment size-based discrimination has been utilized successfully for decades. The standard approach to identify a single nucleotide change involves methods that rely on competitive allele amplification, such as allele specific primer extension (ASPE) or amplification refractory mutation system (Newton et al., 1989). In this case, the relevant allelic mismatches can be positioned at the 3′ end of the primers. A primer with a perfect match to the DNA template will hybridize and extend; whereas, the alternate or nonspecific allele primer will not hybridize or may hybridize but extends less efficiently. The efficiency of the hybridization and prevention of off-target binding is dependent on the specific base changes targeted. CAPS and ASPE are limited in throughput and sensitivity relative to today's technologies. However, the ASPE technique has recently been modified by LGC Biosearch[1] using fluorescence resonance energy transfer chemistry and branded kompetitive allele-specific PCR (KASP)[1] to detect single nucleotide polymorphisms (SNPs) in a high throughput system. Quantitative real-time PCR (qPCR) has been adapted for the detection of SNPs or small INDELs using specialized TaqMan[2]chemistries. Also, specificity has been improved with an RNase-H2-dependent PCR (rhPCR) method that uses blocked primers to minimize dimer formation coupled with cleavage by RNAse H2 (Broccanello et al., 2018).

In a typical rtPCR assay for the detection of an SNP or small INDEL, fluorogenic probes targeting each allele (e.g., susceptible/wildtype and tolerant/mutant) are enhanced with chemical moieties that bind the minor groove of the DNA template. The modifications afford much greater specificity than a standard probe used, for example, for the detection of low concentrations of the target material. One such moiety is the locked nucleic acid (LNA). LNAs are nucleic acid analogs that display unprecedented hybridization affinity toward complementary DNA and RNA. Incorporation of a single LNA provides a substantial increase in duplex stability, which has led to the design of LNA-incorporated nucleic acid probes and primers (Karkare and Bhatnagar, 2006). LNA oligonucleotides have been successfully used in the design of probes and primers in qPCR to enhance specificity and sensitivity needed to discriminate between closely related sequences such as SNPs (Johnson et al., 2004; Maertens et al., 2006; Solbach and Gieffers, 2006; Miyazawa et al., 2008).

---

1. Biosearch and KASP are registered trademarks of LGC, Teddington, United Kingdom.
2. Taqman is a registered trademark of Roche Diagnostics, Inc. Basel, Switzerland.

Such methods have been used very successfully in the routine detection of INDELs and SNPs to confirm zygosity in single plants or small, genetically uniform pools of seed. However, the ability to quantitate a low amount of product with a small sequence difference in a bulk seed or grain sample remains very challenging. The amplification products will be dominated by the most prevalent material. Hybridization of one or both probes (e.g., wildtype and mutant) to the most abundant allele will occur despite existing mismatches.

The challenge has been partly circumvented with the utilization of a peptide nucleic acid (PNA) discovered in the 1990s by Peter E. Nielsen and his team at the University of Copenhagen (Nielsen, 1999). The function of PNA is to prevent amplification by forming a stable duplex with the DNA template; thereby suppressing amplification of the highly abundant alleles present as the major proportion of the trait positive (e.g., gene edited) bulk matrix. PNAs are synthetic, nucleic acid analogs that consist of a pseudopeptide (polyamide) backbone rather than a sugar (phosphate ribose) backbone. PNAs bind strongly with excellent sequence specificity to complementary DNA or RNA (Karkare and Bhatnagar, 2006) and can be used as clamps in PCR, where they prevent the unspecific binding of primers or probes, or to inhibit primer extension and thus the amplification of the unwanted target (Ørum 2000; Peano et al., 2005). PNA clamps have been successfully used in SNP PCR assays (Urata et al., 2004; Miyazawa et al., 2008). In gene editing applications, the use of a PNA can potentially lead to a substantial decrease not only in the amplification of the gene-edited allele, but also adversely decrease the amplification of the non-edited allele. Careful design and validation of several PNAs in combination with several LNAs is advised.

Marker-assisted selection routinely employs SNP analysis for use in breeding. Strategically chosen polymorphisms in close linkage with desired traits can be used advantageously during selection. The selected polymorphisms are typically the easiest to assay in single plants or seeds. In the case of certain agronomic traits such as target site-based herbicide tolerance, the SNP of interest is the causative one. The SNP of interest conferring the desired phenotype may be introduced by nature (spontaneous mutagenesis), random induced mutagenesis (e.g., EMS), conventional breeding, or the targeted genome edit. Thus, there is no freedom to randomly select SNP markers as with, for example, varietal identification. While that limits assay flexibility and adds to the challenge, numerous design tactics including incorporation of deliberate mismatches or PNA clamps, chemistries such as LGC Biosearch's BHQ*plus* or Thermo Fisher-Applied Biosystems' Taqman™ MGB probe and enzymes in conjunction with altered cycling or experiment run conditions can help offset the difficulties.

14.4.1.1 Challenges

PCR-based detection of transgenic products is based on the detection of transgenic sequence(s) as those are absent in unmodified plants. For unique product-specific detection, event-specific PCR is used, which amplifies a unique sequence combination created between the inserted sequence and the flanking genomic DNA in the transformation event. Likewise, in genome-edited organisms, the SDN3 insertions can be relatively easy to detect if they similarly create a novel combination of genetic material absent in non-edited plants. Changes introduced in SDN1 and SDN2 products can be difficult to detect, and even if detected, their potential occurrence through spontaneous or induced random mutagenesis would not allow to distinguish a genome-edited mutation from those other sources of mutations.

In addition, crop genomes present unique challenges not present in the human genetics space (Maagd et al., 2020). The genome copy number or ploidy varies in plants from diploid (e.g., maize, rice or cultivated sunflower) to tetraploid (Brassica sp., durum wheat) and to hexaploid (bread wheat) or greater. Often this means the gene of interest will be present either as a gene family or duplicated on other chromosomes, and these gene sequences may be highly homologous. For example, the three acetohydroxyacid synthase genes in bread wheat are greater than 98% homologous at the nucleotide level (Pozniak and Hucl, 2004). That translates into a need for additional considerations related to specificity when developing a diagnostic method.

As noted in the previous section, a targeted site SNP limits the assay flexibility. That is due primarily to having to design primers within a fixed 150–400 base pair nucleotide window or amplicon size depending upon the assay type. The SNP may lie in a region of unfavorable G-C content, repetitive sequence, or near polymorphisms not related to the trait that is present in certain haplotypes.

The difference in PCR efficiency between single base pair changes varies greatly depending on the physical nucleotide change (Rejali et al., 2018), which can be from approximately 40 fold (for G-T change) to 20,000 fold (for G-A) difference in PCR amplification rate for a 3′ mismatch. That is only for the first PCR cycle, thereafter all products amplify at the same rate as they originate from the amplified amplicons. Consequently, optimization of chemistry, reaction conditions, and cycling parameters are essential to achieve success.

Thus, detection of a single nucleotide change in a bulk sample may be very difficult or unpredictable. In these cases, detection methods will have to be pushed to their limits that are likely to lead to limitations in assay specificity and sensitivity. For these reasons, it is expected to be a challenge to meet regulatory requirements in bulk grain samples, where expectations are that PCR can detect large changes (e.g., GM events) at concentrations less than 1/3000 to 1/10,000. A group testing (semiquantitative) or pooling strategy

(Laffont et al., 2005; Mano et al., 2011; Freese et al., 2015) as described in Chapter 13 can also be used to provide a reliable alternative to quantitative PCR. This enables estimation of the content of a genome edit where whole seeds/grains are the sample material. This would be particularly useful where the detection method has poor sensitivity.

Certain regulatory systems (e.g., EU) require that a validated method be provided to quantify the presence of at least 0.09% (~ 1/1100) GM content in seeds or grain based on copy number (ENGL 2015), which may be difficult to achieve for small edits using PCR. Single base mutations can be detected at some level but not to the specifications required by EU for low-level presence needed for submissions.

14.4.2 Digital PCR

Digital PCR (dPCR) was introduced as a third-generation PCR method about a decade ago after the existing methodology was significantly improved (Kinzler, 1999; Diehl et al., 2006; Demeke et al., 2018). Since then, various commercial platforms have become available, but the basic principle remains the same and as with PCR, dPCR technology utilizes Taq polymerase, primers, and probe within a standard PCR reaction to amplify a DNA target. Detailed information on dPCR is presented in Chapters 4 and 14, but in summary, before amplification, the PCR reaction volume is separated into many individual partitions using droplets or chambers that each theoretically contains a single copy of the target DNA. Following the PCR reaction, end-point measurements are obtained from each partition based on fluorescence. Therefore, the quantitative nature of dPCR is achieved by qualitative (i.e., presence or absence) PCR in each partition, and using Poisson distribution statistics, the entire droplet/chamber population can be used to quantify the amount of the target in a sample.

There are two different platforms: the Droplet Digital PCR (ddPcR) and the chip-based dPCR (cdPCR). These platforms have been reviewed in Chapter 14, but specific information on how they compare with each other to quantify genome-edited products is included here. In the droplet-based RainDrop platform (RainDance Technologies), each reaction is divided into millions of droplets making this platform quite sensitive for the detection of rare mutations and genome-edited targets in a large pool of, for example, wildtype (WT) targets. Conversely, the BioRad QX100 or QX200 dPCR platforms are able to generate approximately 20,000 droplets, which are adequate for many applications, but may have limitations in sensitivity for detection of rare targets in large WT pools. However, these platforms can handle 96 samples at once, which enables higher throughput compared to the RainDrop platform that only handles eight samples at each run (Demeke et al., 2018). The Qiagen QIAcuity® Digital PCR platform is a nanoplate-based microfluidic (i.e., chip-based) dPCR system that offers five color multiplexing and can generate

8–30K partitions. There are other options also available that have been described and compared in detail elsewhere (Demeke et al., 2018). Based on the user's needs, which may include the limit of detection and the volume of the sample pool, the user can decide on the appropriate platform that best fits the purpose. Digital PCR platforms continue to evolve.

The advantages that dPCR offers over qPCR have also been discussed in Chapter 14 and briefly, they are (1) the absolute quantification of a target without a need of a standard curve; (2) designing the assays for quantification would be relatively easy for point mutations and small changes (i.e., genome edits by SDN1 and SDN2) as it uses end-point assays; (3) the results are highly precise, sensitive, and accurate, especially for targets present in low copy numbers; (4) inhibitors in the partitioned droplet or chamber are reduced, and therefore the detection and quantification of low-frequency mutations in a high background of WT DNA is increased (Košir et al., 2019); and (5) qualitative or quantitative PCR assays optimized for detection of a given target have the potential to be used in dPCR.

While dPCR has been proposed as a method for the detection and quantification of gene-edited crops, most published applications have focused on the medical field (Findlay et al., 2016; Mock et al., 2016; Falabella et al., 2017); however, described methods could be adapted to plant matrices (Miyaoka et al., 2018). Additionally, dPCR has been used to screen genome-edited cells (Findlay et al., 2016) to detect diverse gene editing events in vivo, including SNPs (Falabella et al., 2017), and to assess gene-editing frequencies (Mock et al., 2016). Recently, Jouanin et al. (2020) used ddPCR assays to detect INDELs (1–50 bp) and large deletions (>300 bp) in polyploid wheat and concluded that ddPCR "*is suitable for high-throughput screening of copy number variation and gene-editing-induced mutations in large gene families*." It should be noted that any qPCR assay that is optimized for the detection of a given target will have the potential to be used in dPCR for quantification. A competitive allele-specific PCR assay or a zygosity assay that can detect the edited as well as the WT target can be adapted to a ddPCR platform. This makes dPCR a more attractive platform to qPCR (Findlay et al., 2016) as it avoids the presence of competing alleles in the same reaction.

14.4.2.1 Challenges

The challenges of using a dPCR method for the detection of genome edits are similar to the ones discussed above for PCR. In addition, it should be mentioned that some of the challenges pertain to the practicality and implementation of dPCR platforms in testing laboratories and include (1) more expensive instrumentation and need of highly skilled analysts; (2) limited multiplexing options, although as previously mentioned, the Qiagen platform offers five color multiplexing; (3) limited high throughput, and (4) limited dynamic range because this is a function of a finite number of partitions.

14.4.3 Sequencing

Sequencing costs have been steadily decreasing over the years due mainly to a shift from automated high throughput Sanger sequencing to the invention of new revolutionary paralleled sequencing methods known as next-generation sequencing (NGS) that has been described in Chapter 14. The advent of NGS platforms has led not only to a decrease in costs, but also to an increase in DNA sequencing power (Dijk et al., 2014).

NGS applications have rapidly increased in different areas ranging from de novo whole-genome sequencing (WGS) to whole-genome resequencing, transcriptome analysis using RNAseq, genotyping by sequencing (GBS), and targeted genotyping by sequencing (tGBS). Recently, GBS has been widely used for plant and animal genotyping since a large number of samples can be barcoded and pooled in one sequencing reaction, thus significantly reducing the cost of genotyping per each sample (Dijk et al., 2014). The flexibility and low cost of GBS have made it an excellent tool for recent plant and animal breeding activities like marker-assisted selection and genomics selection.

Different methods have been used in GBS. In all these methods the goal is to reduce the complexity of the genome and reduce the amount of genome that needs sequencing. One of the early methods in GBS was to use two different restriction enzymes to reduce the complexity (i.e., restriction fragment length polymorphisms or RFLPs) and only sequence the fragments that were cut by both enzymes (Poland et al., 2012). Restriction site-associated DNA sequencing (RADSeq) is used to find and score thousands of genetic markers across the genome even when limited genome information is available (Davey et al., 2010). However, when a reference genome is available or when there is enough information about specific regions in the genome, tGBS can be a cost-effective approach for genotyping. In this case, genome reduction and target enrichment can be achieved using multiplex PCR (i.e., amplicon based), or hybrid capture (i.e., probe based) (Mamanova et al., 2010).

Although the amplicon-based tGBS approach can be very affordable and practical for high throughput zygosity assays, for different edited targets, its application for quantification of low concentrations of a target will face multiple challenges. One challenge is that the PCR amplifications, especially in the multiplexed format, may have different PCR efficiencies that bias the quantitative output of the sequencing and the results. The other challenge is the semiquantitative nature of the sequencing methods. Even the Illumina sequencing platforms with minimal error rate (10^{-3}) are rarely used for quantitative tests unless the depth of the sequencing and the quality of the target selected for sequencing are high (Aloisio et al., 2016).

Sequencing errors are one of the key components that affect the detection of rare mutations or genome edits in bulk samples, as the error rate may be close to the limit of detection for those rare mutations in the sample. The error rate depends on the nucleotide substitution(s) as some are more prone to error than others, thus making the selection of the target important for its downstream detection. However, by eliminating the low-quality reads and also making some experimental and computationally adjustments, it can be reduced to 10^{-4} to 10^{-5} (Ma et al., 2019). Another factor that significantly affects the quantitative nature of detection for rare mutations or gene edits in a massive number of WT targets is the sequencing depth and coverage (Aloisio et al., 2016). In these situations, high sequence coverage will be needed for quantitative tests as compared to the regular genotyping or zygosity tests. This will significantly increase the cost of quantitative NGS tests compared to qPCR or dPCR tests.

There is a lack of data at this time on the use of NGS for the detection of small sequence differences in bulk samples of seeds, grain, or food. Technologies that can provide some insight into the detection of rare alleles are differential expression analysis or microbiome analysis. Although 10× to100× sequencing depth (approximately 3–30K reads per target) is adequate for genotyping by sequencing, differential expression analysis requires 10–25 million reads. High sequencing depth has been shown to increase the statistical power of detecting differentially expressed genes, but also increases the cost of sequencing (Liu et al., 2014). Microbiome analysis (using 16S target) or metagenome shotgun sequencing (i.e., WGS) uses 25–100 million reads to detect the relative abundance of the bacteria, but this is not sufficient to quantify them (Zaheer et al., 2018). Combining the precision of dPCR with 16S targeted amplicon-based tGBS has achieved some advances in the absolute quantification of microbes (Barlow et al., 2020). Theoretically, a similar approach may also be useful in quantitative tests for genome-edited crop products.

14.4.3.1 Challenges

Sequencing can easily distinguish single base pair changes (Délye et al., 2015) as well as small INDELs. Although increasing sequence coverage may result in the necessary statistical power to detect low-frequency variants in a heterogenous sample and make NGS a semiquantitative detection method, statistical challenges for making sequencing a truly quantitative method still remain. The same challenges were encountered when using NGS in GM quantitation. NGS has not been used frequently for this purpose due to numerous challenges such as uneven coverage of the genome, size of the host genome, and the complexity of the food and feed samples that likely contain several species. Despite all these challenges, some experiments show that target capture sequencing eliminates the complexity and the challenge of uniform amplification of the targets as it has been reported for amplicon-based GBS. Consequently, some success in semiquantitative sequencing methods for GM detection has been achieved (Debode et al., 2019).

Recent technology developments have focused on reducing the error rate in sequencing, optimizing target-capture sequencing protocols, and combining different technologies like droplet single amplification or fluidic compartmentation of the master mixes, which facilitate sequencing multiple targets. Also improving the bioinformatics tools and algorithms will reduce the amount of false positive or negatives, and therefore lower the limit of detection. In summary, NGS will be an additional tool available in the toolkit for developing methods to detect genome edits. However, without a priori sequencing knowledge of products with the same mutation(s) and without a high-quality reference genome or pan-genome, the use of NGS for the identification and detection of, for example, unauthorized genome-edited products is limited and remains a challenge.

14.5 Differentiation of genome edits and spontaneous or induced untargeted mutations

While detection methods may identify mutations, it is still not possible to determine what method was used to generate the mutation by only examining the DNA change. Detection methods that identify a nucleotide change cannot distinguish between the product of directed, spontaneous, or chemically/radiation-induced mutagenesis if that change is below a certain size that has been estimated as 14–17 nucleotides (Grohmann et al., 2019), and "*most modifications produced by genome editing are very small, down to the substitution, deletion, or insertion of one single nucleotide, which might also occur naturally in non-genome-edited plants*." The variation in plant genomes found in nature is often much larger than this. A typical field will contain millions of plants and each will contain one or more mutations in the germline (Wang et al., 2019), and, for example, a study of cultivars in soybean (such as may be expected to be present in a large shipment) exhibit sequence variation in hundreds of genes, including significant deletions (Anderson et al., 2016).

If it is not possible to determine how the sequence change arose, then the edit will be indistinguishable from mutations that arise spontaneously, or that result from other breeding methods (including induced mutagenesis, and ongoing "natural" mutagenesis) (Graham et al., 2020). This presents a difficult enforcement issue for regulators, especially for the detection of bulk consignments. Where there is public information that a developer has commercialized a specific edit, the same edit could have been made by another developer without public knowledge or arisen by other induced or spontaneous conventional mutagenesis routes. Thus, it may not be possible to uniquely identify a product. Indeed, the EU's Joint Research Center has acknowledged that detection methods cannot discriminate intentional from background spontaneous mutations (Emons et al., 2018) and thus may provide a regulatory and legal challenge especially in bulk shipments and stated "*Although it is technically possible to detect specific DNA alterations, without prior knowledge, none of the techniques described are able to distinguish whether the Single Nucleotide Variation or Insertion/Deletion is caused by genome editing, by classical breeding technologies or by natural mutation*" (ENGL, 2019).

Certain interest groups have suggested that induced and native mutations can be distinguished. Yves Bertheau (2019) hypothesizes that one can unequivocally identify whether mutations in the plant genome originate from natural mutagenesis processes or whether they were generated by genome editing. Bertheau further claims that it would just be a matter of establishing appropriate methods for this issue. Several criticisms of this suggestion have been made by the Central Committee on Biological Safety (ZKBS, 2019), which is a voluntary expert panel responsible for evaluating GMOs, sponsored by the German Federal Office of Consumer Protection and Food Safety. In particular, the ZKBS stated that the option suggested by Bertheau to retroactively identify an edited base within a plant genome and the technique used to generate it is nonexistent. The proposed methods are not based on current scientific knowledge and furthermore involve highly variable biological parameters (like epigenetic changes) that are not a reliable basis for identification (ZKBS, 2019).

Chhalliyil et al. (2020) claimed that they could detect a genome edit in plants using qPCR and that they were able to develop a detection method to discern between a single nucleotide change generated via chemical mutagenesis from that induced by gene editing. Through the use of Sanger sequencing specifically of the *AHAS* gene family across 20 canola varieties, combined with published and/or disclosed information from the developer and through a process of comparison and elimination based on variety development, a detection method specific to the Cibus SU canola developed through ODM. However, they demonstrated that they could not identify the targeted allele as unequivocally being the product of gene editing because the canola product and specifically the mutation that was tested was apparently not the product of a directed gene edit (BVL, 2020).

For those jurisdictions where validated methods for the detection of GMOs or genome-edited products are required by law, proving method specificity unequivocally for the detection of a given product will be challenging, especially if the sequence changes are small. Specificity is one of the parameters that are evaluated during method validation, and a priori knowledge of sequences surrounding the region of interest is necessary for method development. This will include not only knowing sequences of all plant varieties but also having sequence information of all commercial products. Obtaining such information is practically impossible for multiple reasons, because most commercial varieties, including those that may have resulted from mutagenesis, are not fully sequenced, intellectual property barriers and the availability of robust reference genome information, which ideally should be based on a pan-genome reference sequence.

Despite these challenges, various approaches for the detection of genome-edited products have been proposed. However, and as previously stated, the major roadblock is not the technology, but the inability to discern without a priori knowledge how a change in the nucleotide sequence was generated.

14.6 Conclusions

Methods for detection of genome editing crop products may be needed across the supply chain and to comply with biosafety protocols (e.g., detection in bulk grain/seed or food and feed samples). Large inserts, deletions, and substitutions of DNA sequences can be detected using similar approaches to those developed for traditional biotechnology products. As the nucleotide change(s) gets smaller, the ability to detect small quantities of these variants in bulk seed and grain or in food and feed with high confidence decreases. Depending on the base pair change made and the sensitivity and specificity of the methods available, single base pair changes may be technically difficult to detect, especially in bulk samples or food and feed products. Moreover, it will not be possible to determine the origin of the sequence change (i.e., whether it is the result of genome editing, mutagenesis, or of spontaneous mutation). These factors make it difficult to apply detection methods for genome-edited products as a category for their unique detectability and traceability, though we can predict that there will be significant efforts to apply either adapted PCR methods or NGS sequencing to meet this need.

References

Aloisio, M., Licastro, D., Caenazzo, L., Torboli, V., D'Eustacchio, A., Severini, G.,M., Athanasakis, E., 2016. A technical application of quantitative next generation sequencing for chimerism evaluation. Molecular Medicine Reports 14 (4), 2967–2974. https://doi.org/10.3892/mmr.2016.5593.

Anderson, J.E., Michno, J.M., Kono, T.J.Y., Stec, A.O., Campbell, B.W., Curtin, S.J., Stupar, R.M., 2016. Genomic variation and DNA repair associated with soybean transgenesis: a comparison to cultivars and mutagenized plants. BMC Biotechnology 16, 41. https://doi.org/10.1186/s12896-016-0271-z.

Barlow, J.T., Bogatyrev, S.R., Ismagilov, R.F., 2020. A quantitative sequencing framework for absolute abundance measurements of mucosal and lumenal microbial communities. Nature Communications 11 (1), 1–13. https://doi.org/10.1038/s41467-020-16224-6.

Beetham, P.R., Kipp, P.B., Sawycky, X.L., Arntzen, C.J., May, G.D., 1999. A tool for functional plant genomics: chimeric RNA/DNA oligonucleotides cause in vivo gene-specific mutations. Proceedings of the National Academy of Sciences USA 96 (15), 8774–8778. https://doi.org/10.1073/pnas.96.15.8774.

Bertheau, Y., 2019. New breeding techniques: detection and identification of the techniques and derived products. In: Laurence Melton, Fereidoon Shahidi, Peter Varelis, Encyclopedia of Food Chemistry. ELSEVIER SCIENCE BV, AMSTERDAM, NLD, pp. 320–336. https://doi.org/10.1016/B978-0-08-100596-5.21834-9.

Broccanello, C., Chiodi, C., Funk, A., McGrath, J.M., Panella, L., Stevananto, P., 2018. Comparison of three PCR-based assays for SNP genotyping in plants. Plant Methods 14 (28). https://doi.org/10.1186/s13007-018-0295-6.

BVL, 2020. Bundesamt für Verbraucherschutz und Lebensmittelsicherheit: 'Neue Nachweismethode verspricht spezifische Detektion genom-editierter Rapslinien – was kann das Verfahren tatsächlich leisten? accessed September 16th 2020. www.bvl.bund.de/SharedDocs/Fachmeldungen/06_gentechnik/2020/2020_09_09_Fa_Nachweismethode-genomeditierte-Pflanzen.html.

CAST, 2018. Genome Editing in Agriculture: Methods, Applications, and Governance — A Paper in the Series on the Need for Agricultural Innovation to Sustainably Feed the World by 2050. Issue Paper 60. Council for Agricultural Science and Technology, Ames, Iowa.

Chen, K., Wang, Y., Zhang, R., Zhang, H., Gao, C., 2019. CRISPR/Cas genome editing and precision plant breeding in agriculture. Annual Review of Plant Biology 70, 667–697. https://doi.org/10.1146/annurev-arplant-050718-100049.

Chhalliyil, P., Ilves, H., Kazakov, S.A., Howard, S.J., Johnston, B.H., Fagan, J., 2020. A real-time quantitative PCR method specific for detection and quantification of the first commercialized genome-edited plant. Foods 9, 1245. https://doi.org/10.3390/foods9091245 (Correction published on 18 February 2022, see Foods 2022, 11(4), 585).

Délye, C., Causse, R., Gautier, V., Poncet, C., Michel, S., 2015. Using next-generation sequencing to detect mutations endowing resistance to pesticides: application to acetolactate-synthase (ALS)-based resistance in barnyard grass, a polyploid grass weed. Pest Management Science 71 (5), 675–685. https://doi.org/10.1002/ps.3818.

Davey W., J., Blaxter L., M., 2010. RADSeq: next-generation population genetics. Briefings in Functional Genomics 9 (5–6), 416–423. https://doi.org/10.1093/bfgp/elq031.

Debode, F., Hulin, J., Charloteaux, B., et al., 2019. Detection and identification of transgenic events by next generation sequencing combined with enrichment technologies. Scientific Reports 9 (1), 1–9. https://doi.org/10.1038/s41598-019-51668-x (2019).

Demeke, T., Dobnik, D., 2018. Critical assessment of digital PCR for the detection and quantification of genetically modified organisms. Analytical and Bioanalytical Chemistry 410 (17), 4039–4050. https://doi.org/10.1007/s00216-018-1010-1.

Diehl, F., Li, M., Kinzler, K.W., Vogelstein, B., Dressman, D., 2006. BEAMing: single-molecule PCR on microparticles in water-in-oil emulsions. Nature Methods 3 (7), 551–559. https://doi.org/10.1038/nmeth898.

Dijk, E.,L., Auger, H., Jaszczyszyn, Y., Thermes, C., 2014. Sequencing technology. Trends in genomics. Ten years of next-generation sequencing technology. Trends in genetics 30 (9), 418–426. https://doi.org/10.1016/j.tig.2014.07.001, 2014.

Eid, A., Alshareef, S., Mahfouz, M.M., 2018. CRISPR base editors: genome editing without double-strande d breaks. Biochemical Journal 475 (11), 1955–1964. https://doi.org/10.1042/BCJ20170793.

Emons, H., Broothaerts, W., Bonfini, L., Corbisier, P., Gatto, F., Jacchia, S., Mazzara, M., Savini, C., 2018. Challenges for the Detection of Genetically Modified Food or Feed Originating from Genome Editing, EUR 29391 EN. Publications Office of the European Union, Luxembourg, ISBN 978-92-79-96398-8. https://doi.org/10.2760/732526, 2018.

ENGL, 2015. European Network of GMO Laboratories, Definition of Minimum Performance Requirements for Analytical Methods of GMO Testing European Network of GMO Laboratories, 20 October 2015. https://gmo-crl.jrc.ec.europa.eu/doc/MPR%20Report%20Application%2020_10_2015.pdf. (Accessed 29 September 2020).

ENGL, 2019. European Network of GMO Laboratories, Detection of Food and Feed Plant Products Obtained by New Mutagenesis Techniques, 26 March 2019 (JRC116289). https://gmo-crl.jrc.ec.europa.eu/doc/JRC116289-GE-report-ENGL.pdf. accessed September 30th 2020.

EuropaBIO Report on Genome-Edited Products and Projects: Resources and Examples, August 2020. https://www.europabio.org/cross-sector/publications/genome-edited-products-and-projects-resources-and-examples. (Accessed 25 August 2020).

Falabella, M., Sun, L., Barr, J., Pena, A.Z., Kershaw, E.E., Gingras, S., Goncharova, E.A., Kaufman, B.A., 2017. Single-step qPCR and dPCR detection of diverse CRISPR-Cas9 gene editing events in vivo. G3 (Bethesda) 7, 3533–3542. https://doi.org/10.1534/g3.117.300123.

Findlay D., S., Vincent M., K., Berman, J.R., Postovit M., L., 2016. A digital PCR-based method for efficient and highly specific screening of genome edited cells. PLOS ONE 11 (4), e0153901. https://doi.org/10.1371/journal.pone.0153901.

Freese, L., Chen, J.W., Shillito, R.D., 2015. Sampling of grain and seed to estimate the adventitious presence of biotechnology-derived seeds in a lot. Cereal Foods World 60 (1), 9–15. https://doi.org/10.1094/CFW-60-1-0009.

Graham, N., Patil, G.B., Bubeck, D.M., Dobert, R.C., Glenn, K.C., Gutsche, A.T., Kumar, S., Lindbo, J.A., Maas, L., May, G.D., Vega-Sanchez, M.E., Stupar, R.M., Morrell, P.L., 2020. Plant genome editing and the relevance of off-target changes. Plant Physiology 183 (4), 01194. https://doi.org/10.1104/pp.19.01194, 2019.

Grohmann, L., Keilwagen, J., Duensing, N., Dagand, E., Hartung, F., Wilhelm, R., Bendiek, J., Sprink, T., 2019. Detection and identification of genome editing in plants: challenges and opportunities. Frontiers of Plant Science 10, 236. https://doi.org/10.3389/fpls.2019.00236.

Holme, I.B., Gregersen, P.L., Brinch-Pedersen, H., 2019. Induced Genetic Variation in Crop Plants by Random or Targeted Mutagenesis: Convergence and Differences. In: Frontiers in Plant Science, 10, p. 1468. https://doi.org/10.3389/fpls.2019.01468.

Jaganathan, D., Ramasamy, K., Sellamuthu, G., Jayabalan, S., Venkataraman, G., 2018. CRISPR for crop improvement: an update review. Frontiers of Plant Science 9, 985. https://doi.org/10.3389/fpls.2018.00985.

Johnson, M.P., Haupt, L.M., Griffiths, L.R., 2004. Locked nucleic acid (LNA) single nucleotide polymorphism (SNP) genotype analysis and validation using real-time PCR. Nucleic Acids Research 32 (6), e55. https://doi.org/10.1093/nar/gnh046.

Jouanin, A., Tenorio-Berrio, R., Schaart, J.G., Leigh, F., Visser, R.G., Smulders, M.J., 2020. Optimisation of droplet digital PCR for determining copy number variation of α-gliadin genes in mutant and gene-edited polyploid bread wheat. Journal of Cereal Science 92, 102903.

Kanchiswamy, C.N., 2016. DNA-free genome editing methods for targeted crop improvement. Plant Cell Reports 2016 (35), 1469–1474.

Kamburova, V.S., Nikitina, E.V., Shermatov, S.E., Buriev, Z.T., Kumpatla, S.P., Emani, C., Abdurakhmonov, I.Y., 2017. Genome editing in plants: an overview of tools and applications. International Journal of Agronomy 2017. https://doi.org/10.1155/2017/7315351.

Karkare, S., Bhatnagar, D., 2006. Promising nucleic acid analogs and mimics: characteristic features and applications of PNA, LNA, and morpholino. Applied Microbiology and Biotechnology 71 (5), 575–586. https://doi.org/10.1007/s00253-006-0434-2.

Kinzler, W.K., 1999. Digital PCR. Proceedings of the National Academy of Sciences of the United States of America 96 (16), 9236–9241. https://doi.org/10.1073/pnas.96.16.9236.

Košir, A.B., Demšar, T., Štebih, D., Žel, J., Milavec, M., 2019. Digital PCR as an effective tool for GMO quantification in complex matrices. Food Chemistry 294, 73–78. https://doi.org/10.1016/j.foodchem.2019.05.029.

Kumawat, S., Rana, N., Bansal, R., Vishwakarma, G., Mehetre, S.T., Das, B.K., Kumar, M., Yadav, S.K., Sonah, H., Sharma, T.R., Deshmukh, R., 2019. Expanding avenue of fast neutron mediated mutagenesis for crop improvement. Plants 8 (6), 164. https://doi.org/10.3390/plants8060164.

Laffont, J.L., Remund, K.M., Wright, D., Simpson, R.D., Gregoire, S., 2005. Testing for adventitious presence of transgenic material in conventional seed or grain lots using quatitative laboratory methods: statistical procedures and their implementation. Seed Science Research 15 (3), 197–204. https://doi.org/10.1079/SSR2005210.

Liu, Y., Zhou, J., White, K.,P., 2014. RNA-seq differential expression studies: more sequence or more replication? Bioinformatics 30 (3), 301–304. https://doi.org/10.1093/bioinformatics/btt688.

Ma, X., Shao, Y., Tian, L., et al., 2019. Analysis of error profiles in deep next-generation sequencing data. Genome Biology 20 (1), 1–15. https://doi.org/10.1186/s13059-019-1659-6.

Maagd, R.A., van de Wiel, C., Schouten, H.J., 2020. The Plasticity of Plant Genomes. Causes and Consequences: A Survey of Data on Structural Genome Variation in Plants. Wageningen Plant Research, Wageningen, The Netherlands. COGEM Report.

Maertens, O., Legius, E., Speleman, F., Messiaen, L., Vandesompele, J., 2006. Real-time quantitative allele discrimination assay using 3' locked nucleic acid primers for detection of low-percentage mosaic mutations. Analytical Biochemistry 359 (1), 144–146. https://doi.org/10.1016/j.ab.2006.07.039.

Mahfouz, M.M., Li, L., 2011. TALE nucleases and next generation of GM crops. GM Crops 2 (2), 99–103. https://doi.org/10.4161/gmcr.2.2.17254.

Malzahn, A., Lowder, L., Qi, Y., 2017. Plant genome editing with TALEN and CRISPR. Cell & Bioscience 7, 21. https://doi.org/10.1186/s13578-017-0148-4.

Mamanova, L., Coffey J., A., Scott E., C., Kozarewa, I., Turner H., E., Kumar, A., Howard, E., Shendure, J., Turner J., D., 2010. Target-enrichment strategies for next generation sequencing. Nature Methods 7 (2), 111–118. https://doi.org/10.1038/NMETH.1419.

Mano, J., Yanaka, Y., Ikezu, Y., Onishi, M., Futo, S., Minegishi, Y., Ninomiya, K., Yotsuyanagi, Y., Frank, S., Akiyama, H., Teshima, R., Hino, A., Naito, S., Koiwa, T., Takabatake, R., Furui, S., Kitta, K., 2011. Practicable group testing method to evaluate weight/weight GMO content in maize grains. Journal of Agricultural and Food Chemistry 59 (13), 6856–6863.

Metje-Sprink, J., Menz, J., Modrzejewski, D., Sprink, T., 2019. DNA-free genome editing: past, present and future. Frontiers of Plant Science 9. https://doi.org/10.3389/fpls.2018.01957.

Miyaoka, Y., Mayerl, S.J., Chan, A.H., Conklin, B.R., 2018. Detection and quantification of HDR and NHEJ induced by genome editing at endogenous gene loci using droplet digital PCR. In: Digital PCR. Humana Press, New York, NY, pp. 349–362. https://doi.org/10.1021/jf200212v.

Miyazawa, H., Tanaka, T., Nagai, Y., Matsuoka, M., Sutani, A., Udagawa, K., Zhang, J., Hirama, T., Murayama, Y., Koyama, N., et al., 2008. Peptide nucleic acid–locked nucleic acid polymerase chain reaction clamp-based detection test for gefitinib-refractory T790M epidermal growth factor receptor mutation. Cancer Science 99 (3), 598–600. https://doi.org/10.1111/j.1349-7006.2007.00706.x.

Mock, U., Hauber, I., Fehse, B., 2016. Digital PCR to assess gene-editing frequencies (GEF-dPCR) mediated by designer nucleases. Nature Protocols 11 (3), 598–615. https://doi.org/10.1038/nprot.2016.027.

Newton, C.R., Graham, A., Heptinstall, L.E., Powell, S.J., Summers, C., Kalsheker, N., Smith, J.C., Markham, A.F., 1989. Analysis of any point mutation in DNA. The amplification refractory mutation system (ARMS). Nucleic Acids Research 17 (7), 2503–2516. https://doi.org/10.1093/nar/17.7.2503.

Nielsen, P.E., 1999. Applications of peptide nucleic acids. Current Opinion in Biotechnology 10 (1), 71–75. https://doi.org/10.1016/S0958-1669(99)80013-5.

Oladosu, Y., Rafii, M.Y., Abdullah, N., Hussin, G., Ramli, A., Rahim, H.A., Miah, G., Usman, M., 2016. Principle and application of plant mutagenesis in crop improvement: a review. Biotechnology & Biotechnological Equipment 30 (1), 1–16. https://doi.org/10.1080/13102818.2015.1087333.

Ørum, H., 2000. PCR clamping. Current Issues in Molecular Biology 2 (1), 27–30. https://doi.org/10.21775/cimb.002.027.

Peano, C., Lesignoli, F., Gulli, M., Corradini, R., Samson, M.C., Marchelli, R., Marmiroli, N., 2005. Development of a peptide nucleic acid polymerase chain reaction clamping assay for semiquantitative evaluation of genetically modified organism content in food. Analytical Biochemistry 344 (2), 174–182. https://doi.org/10.1016/j.ab.2005.04.009.

Podevin, N., Davies, H.V., Hartung, F., Nogue, F., Casacuberta, J.M., 2013. Site-directed nucleases: a paradigm shift in predictable, knowledge-based plant breeding. Trends in Biotechnology 31 (6), 375–383. https://doi.org/10.1016/j.tibtech.2013.03.004.

Poland A., J., Rife W., T., 2012. Genotyping-by-Sequencing for plant breeding and genetics. The Plant Genome 5 (3), 92–102. https://doi.org/10.3835/plantgenome2012.05.0005.

Pozniak, C., Hucl, P., 2004. Genetic analysis of imidazolinone resistance in mutation-derived lines of Common wheat. Crop Science 44 (1), 23–30. https://doi.org/10.2135/cropsci2004.2300.

Rejali, N.A., Moric, E., Wittwer, C.T., 2018. The effect of single mismatches on primer extension. Clinical Chemistry 64 (5), 801–809. https://doi.org/10.1373/clinchem.2017.282285.

Ribarits, A., Eckerstorfer, M., Simon, S., Stepanek, W., 2021. Genome-edited plants: opportunities and challenges for an anticipatory detection and identification framework. Foods 10 (2), 430. https://doi.org/10.3390/foods10020430, 16.

Shillito, R.D., Whitt, S., Ross, M., et al., 2021. Detection of genome edits in plants—from editing to seed. In Vitro Cellular & Development Biology-Plant 57, 595–608. https://doi.org/10.1007/s11627-021-10214-z.

Solbach, W., Gieffers, J., 2006. Single-nucleotide-polymorphism-specific PCR for quantification and discrimination of Chlamydia pneumoniae genotypes by use of a "locked" nucleic acid. Applications in Environmental Microbiology 72 (5), 3785–3787. https://doi.org/10.1128/AEM.72.5.3785-3787.2006.

Svitashev, S., Schwartz, C., Lenderts, B., Young, J., Cigan, A.M., 2016. Genome editing in maize directed by CRISPR-Cas9 ribonucleoprotein complexes. Nature Communications 7 (13274). https://doi.org/10.1038/ncomms13274.

Urata, M., Wada, Y., Kim, S.H., Chumpia, W., Kayamori, Y., Hamasaki, N., Kang, D., 2004. High-sensitivity detection of the A3243G mutation of mitochondrial DNA by a combination of allele-specific PCR and peptide nucleic acid-directed PCR clamping. Clinical Chemistry 50 (11), 2045–2051. https://doi.org/10.1373/clinchem.2004.033761.

Van Eck, J., 2020. Applying gene editing to tailor precise genetic modifications in plants. Journal of Biological Chemistry 295 (38), 13267–13276. https://doi.org/10.1074/jbc.REV120.010850.

Wang, L., Ji, Y., Hu, Y., Hu, H., Jia, X., Jiang, M., et al., 2019. The architecture of intra-organism mutation rate variation in plants. PLoS Biology 17 (4), e3000191. https://doi.org/10.1371/journal.pbio.3000191.

Whelan, A.I., Lema, M.A., 2015. Regulatory framework for gene editing and other new breeding techniques in Argentina. GM Crops Food 6 (4), 253–265. https://doi.org/10.1080/21645698.2015.1114698.

Whelan, A.I., Gutti, P., Lema, M.A., 2020. Gene editing regulation and innovation economics. Frontiers in Bioengineering and Biotechnology 8, 303. https://doi.org/10.3389/fbioe.2020.00303.

Woo, J.W., Kim, J., Kwon, S., Corvalán, C., Seung Woo Cho, H.K., Kim, S.W., Kim, H., Kim, S.G., Kim, S.T., Choe, S., Kim, J.S., 2015. DNA-free genome editing in plants with preassemble CRISP-Cas9 ribonucleoproteins. Nature Biotechnology 33 (11), 1162–1164. https://doi: 10.1038/nbt.3389.

Xu, H., Liu, B., Pardinas, J., 2015. Biological and technological implications of meganucleases. Gene and Gene Editing 1 (1), 41–46. https://doi.org/10.1166/gge.2015.1005.

Zaheer, R., Noyes, N., Ortega Polo, R., et al., 2018. Impact of sequencing depth on the characterization of the microbiome and resistome. Scientific Reports 8 (1), 1–11. https://doi.org/10.1038/s41598-018-24280-8.

Zhang, Y., Massel, K., Godwin, I.D., Gao, C., 2018. Applications and potential of genome editing in crop improvement. Genome Biology 19 (1), 1–11. https://doi.org/10.1186/s13059-018-1586-y, 210.

Zhu, T., Peterson, D.J., Tagliani, L., St. Claire, G., Baszczynski, C.L., Bowen, B., 1999. Targeted manipulation of maize genes *in vivo* using chimeric RNA/DNA oligonucleotides. Proceedings of the National Academy of Science USA 96 (15), 8769–8773. https://doi.org/10.1073/pnas.96.15.8768.

ZKBS, 2019. On the Identifiability of Genome Editing in Plants - Commentary of the ZKBS on Y. Bertheau (accessed on August 27th 2020). http://www.zkbs-online.de/ZKBS/EN/01_Aktuelles/ZKBS-commentary on Y. Bertheau (2019)/ZKBS-commentary on Y. Bertheau (2019)_basepage.html.

Chapter 15

Future perspectives and challenges

Ning Zhou[1], Kazumi Kitta[2], Junichi Mano[3], Ryan Hill[4] and Kerrm Y. Yau[1]
[1]*Corteva Agriscience, Johnston, IA, United States;* [2]*National Agriculture and Food Research Organization, Tsukuba, Ibaraki, Japan;* [3]*Institute of Food Research, National Agriculture and Food Research Organization, Tsukuba, Ibaraki, Japan;* [4]*Eli Lilly and Company, Lilly Corporate Center, Indianapolis IN, United States*

15.1 Introduction

15.1.1 Issues and challenges

Since its commercialization in 1996, genetically engineered (GE) or genetically modified (GM) crops have entered a new era with more complex and sophisticated stacked-event products, asynchronous regulatory approvals worldwide, higher standards of quality control, and new emerging gene-editing technologies (See chapter 14 (Herrero et al., 2022)). This has raised new challenges in methods of detection and sample collection that are critical for data collection supporting new trait product development and trade.

Gene stacking of multiple traits in a single crop product has become common practice requiring detection methods capable of detecting multiple genes or proteins in a single assay (termed multiplexing). In addition, zero-tolerance policies for any unintended genetically modified organisms (GMOs) traits in almost every country require a detection sensitivity for GMO presence at 0.1% or lower. An ideal new detection technology that is suited for this purpose should be simple, rapid, sensitive, and able to multiplex. In the past few years, new technologies have emerged to overcome these challenges. In this chapter, a brief review of the development of new technologies for both DNA and protein-based detection is provided. The leading DNA-based detection technologies include qPCR array, digital PCR, isothermal amplification, and next-generation sequencing. For protein detection, this includes nonantibody-based LC-MS technology, antibody-based Meso Scale Discovery (MSD) multiplexing arrays, Biacore biosensor systems, high throughput AlphaLISA, and bead-based flow cytometry.

Application of Sampling and Detection Methods in Agricultural Plant Biotechnology
https://doi.org/10.1016/B978-0-323-99293-0.00010-6

15.2 New GMO detection technologies

15.2.1 DNA-based technologies

15.2.1.1 qPCR-based multiplex strategies

Multiplex qPCR: To improve screening throughput and reduce the number of reactions, multiplex qPCR strategies have been applied (Gaudron et al., 2009; Dorries et al., 2010) to detect the presence of several common elements in different GMOs. The design and optimization of multiplex PCR assays could be a challenge due to the potential interferences between primers and amplification products and competition for reagents within the reaction. Moreover, multiplexing in qPCR is limited by the number of filters in the instrument that can reliably detect the fluorescence of different fluorophores. To date, the simultaneous measurement of six genes to screen for the presence or absence of GMOs has been the highest level of multiplexing reported in this format (Bahrdt et al., 2010).

qPCR array: With the aim of improving screening throughput, real-time PCR arrays have been developed for multiple-target purposes. The real-time PCR array is a plastic PCR plate prespotted with a different primer-probe combination in each well. Multiple simultaneous PCRs on such a PCR plate provide more information on an analytical sample (Querci et al., 2009; Mano et al., 2009). Given the current environment for GMO testing, a real-time PCR array system that allows the simultaneous implementation of numerous validated methods is necessary. The real-time PCR array is an ideal system because it can be easily updated and customized depending on the situation. Particularly with respect to new approved GMO events, component PCR assays comprising the real-time PCR array can be developed and added to the existing assays with limited or no requirement of reevaluation of the whole system. The analyst can simply select the assays corresponding to the targeted GMO events and/or plant species. The multitarget qPCR array was successfully evaluated by multiple laboratories and demonstrated to be robust (Kluga et al., 2012; Rosa et al., 2016). Although the approach may not be efficient for screening high volume samples, the detection specificity, dynamic range, ease of manipulation, updatability, customizability, and time efficiency for assay validation are advantageous.

Alternative multiplex strategies: Various hybridization-based high-throughput nucleic acid methodologies were investigated for GMO detection, such as DNA microarray and Luminex platforms. Microarray is a solid-phase array incorporating a broad range of specific DNA molecules corresponding to specific DNA elements of GMOs that are immobilized separately on microarray chip. DualChip GMO microarray (Leimanis et al., 2008) contains 117 spots, targeting 14 specific DNA elements. The oligonucleotide microarray in MACRO system was reported to measure 91 targets covering a broad spectrum of GMOs (Shao et al., 2014). Although these technologies present a higher throughput, the sensitivity may not meet current GMO detection requirements. To ensure sufficient sensitivity, the samples may be amplified by PCR using target-specific and/or universal primers before hybridization steps. Various PCR strategies were developed to couple with microarray for GMO detection, such as multiplex quantitative DNA array-based PCR (Rudi et al., 2003), nucleic-acid-sequence-based amplification-implemented microarray analysis (Dobnik et al., 2010), and padlock probe ligation in combination with microarray detection (Prins et al., 2008). Luminex is a microfluidic bead-based array where GMO target-specific oligonucleotide probes are covalently attached to the fluorophore coded beads. The samples are amplified by single or multiplex specific primer pairs labeled with biotin. The hybridization signal of biotinylated PCR products to the probe-coupled bead is detected through flow cytometry. The liquid bead array is considered more sensitive and faster than the solid-phase array. Luminex technology was reported to simultaneously detect up to 500 different targets. Since the first report for GMO detection using Luminex in 2008, it has been applied to several studies on different GMO crops. The development of an integrated multiplex PCR-liquid bead array allowing identification of 13 GMO maize has also been reported (Fu et al., 2015).

Hybridization-based approaches have complex and laborious experimental steps, including PCR, hybridization, multiple washes, and fluorescence detection. The large-scale application of those technologies to GMO detection can be limited due to required specialized instruments and reagents.

15.2.1.2 Digital PCR technology

Digital PCR represents a class of PCR technologies performed by diluting a PCR reaction into large numbers of smaller partitions, hundreds to millions of minute reaction, so that a proportion contains either one or zero molecules of the analyte before amplification. After PCR reactions, each partition of the fractionated sample is determined as positive (amplified target observed) or negative (no amplified target observed) and these values are used to calculate the target concentration using binomial Poisson statistics. The concept of digital PCR (dPCR) can be traced back 1990s (Morley, 2014). The rapid development of emulsion and microfluidics chemistries recently enable digital PCR to become one of the most promising technical improvements. Current dPCR technologies are categorized into two main groups, microfluidic/chip-based dPCR (cdPCR) or droplet digital PCR (ddPCR) (Zhu et al., 2020), based on the partition technology used. cdPCR uses microfluid chambers that can partition up to a few thousand individual reactions. The droplet digital PCR (ddPCR) uses water–oil emulsion-based partitioning into several thousand, or even millions, of individual droplets that are counted using flow cytometry. The ddPCR is limited to end-point detection, whereas cdPCR provides both end-point and real-time measurement (Guiducci et al., 2013; Duewer et al., 2015). Separating DNA sampling into large numbers of compartments for single-molecule reactions avoids the issue of amplification bias. Both approaches demonstrated capability of absolute copy number estimations, especially at rare and low copy number range (Whale et al., 2013). Compared to conventional PCR and real-time PCR, dPCR has several advantages for the quantification of GMOs. It enables the determination of absolute copy number, without the use of standard curves (Demeke and Dobnik, 2018). It is more accurate and sensitive in low copy number ranges; Demeke et al. (2014) reported consistent repeatability of detection down to 0.01% OXY235 canola and DP305423 soybean events and potential ultralow detection at 0.001% spiked samples of two events using the RainDance system. dPCR is more tolerant to various types of inhibitors and components that may hinder the PCR assay efficiencies. Finally, the transfer of assays from the qPCR platform to the dPCR platform is relatively easy simplifying the implementation of dPCR platforms in laboratories compared with other approaches (Cao et al., 2017a; Demeke and Dobnik, 2018).

Digital PCR has many potential applications in GMO detection. The majority of the methods were developed to target one (single-plex) or two (duplex) DNA sequences within a single reaction. Different strategies have been described for multiplex dPCR development to precisely detect more than two targets in the same reaction by Whale et al. (2016). They predicted that "higher-order multiplexing" can be achieved with the advancement of the dPCR workflow in the number of partitions and detection channels. Additionally, Dobnik et al. (2016) reported development and evaluation of two 4-plex ddPCR assays for the quantification of GM maize events demonstrating the potential of multiplexing in ddPCR in terms of simultaneous repeatable and reliable absolute quantification of four targets.

Digital PCR was utilized to develop the strategy for simultaneous detection of a wide range of GMO targets. High throughput multiplex GMO detection utilizing Fluidigm (cdPCR system) microfluidic architecture to combine samples and primer-probe sets into hundreds to thousands of PCR reactions has been reported (Li et al., 2015b). It was demonstrated that the method allowed the simultaneous detection of 48 targets in 48 samples. The high-throughput methodology is suitable for efficiently and accurately identifying GMOs in samples with greatly increased target and sample throughputs.

The importance of reference materials diminishes with the use of dPCR because the quantification does not rely on the availability of a reference material. Given high resilience to inhibitors and good reproducibility of dPCR technologies, evaluation across different laboratories for their potential suitability as the method for value assignment of reference materials may be beneficial.

A number of commercial dPCR systems based on different partition technologies are available. The various systems offer different levels of flexibility and capability. Although dPCR is more precise and robust for certain applications, it can be more costly for reagent and consumables and may present high throughput limitations. With new development in microfluidic technologies and microfabrication, dPCR could be continuously improved and simplified. In addition, the application of dPCR could be simplified if other thermocycling-free amplification chemistries, such as isothermal amplification and lens-free imaging could be integrated.

15.2.1.3 Isothermal amplification

Real-time PCR is the most widely known nucleic acid amplification detection approach and has been broadly applied to GMO detection and quantification due to its dominant performance over other quantification methods in terms of accuracy, specificity, repeatability, and dynamic range. Nevertheless, real-time PCR suffers from drawbacks such as the requirement for large and expensive thermal cyclers and long reaction times limiting the application of PCR in resource-limited settings and on-site analysis. Isothermal DNA amplification is an alternative to PCR-based amplification which does not require thermal cycling. These features greatly simplify the isothermal amplification implementation process and make on-site analysis possible. An international standard that describes many of these isoPCR approaches, and guidance in their validation is available (ISO22942). In this section, several isothermal amplification techniques used for GMO detection will be discussed.

Loop-mediated isothermal amplification (LAMP). LAMP is a sensitive and specific isothermal amplification method. LAMP requires the use of four or more different primers specific to six distinct regions on the target DNA template and a DNA polymerase with strand displacement activity. Under isothermal conditions, an inner primer containing sequences of the sense and antisense strands of the target DNA initiates LAMP. The following strand displacement DNA synthesis primed by an outer primer releases a single-stranded DNA. This serves as a template for DNA synthesis primed by the second inner and outer primers that hybridize to the other end of the target, leading to the production of a stem-loop DNA structure. The subsequent cycles, comprising elongation and recycling steps, lead to final products of LAMP constituted by a mixture of stem-loop DNAs that have various stem lengths and cauliflower-like structures with multiple loops formed by annealing between alternately inverted repeats of the target in the same strand. Because LAMP recognizes the target by six distinct sequences initially and by four distinct sequences afterward, it is expected to amplify the target sequence with high selectivity (Notomi et al., 2000).

The LAMP reaction can be performed at a constant temperature between 60 and 65°C. A simple isothermal instrument, such as water bath and heat block, is adequate for LAMP amplification. LAMP provides a high amplification efficiency, with replication of the original template copy 10^9-10^{10} times during a 15−60 min reaction. The amplification can be visualized as a change in color using nucleic acid staining or fluorescent dye or real time monitored by measuring fluorescence (Randhawa et al., 2013). Due to its competitive advantage, the LAMP method has been increasingly utilized for GMO detection (Li et al., 2015a). With a simple and fast DNA extraction method, LAMP assay can be applied in the lab and on-site analysis. The most commonly cited disadvantage of LAMP is the design of multiple complex primers to cover six regions of the target DNA, however web-based software for designing candidate LAMP primers and loop primers has increased the popularity of the LAMP assay. LAMP is a difficult amplification method to multiplex, that is, to amplify more than one target sequence at a time, although it is reported to be extremely specific due to the multiple primers that must anneal to the target to further the amplification process.

Helicase-dependent amplification (HDA). HDA is one of the simplest approaches for DNA amplification (Zhao et al., 2015). The process uses a mixture of enzymes for DNA strand separation and polymerization: a helicase enzyme to unwind double-stranded DNA (dsDNA) to which forward and reverse primers can bind, the DNA polymerase then extends the 3′ ends of each primer using free deoxynucleotides (dNTPs) to produce two DNA replicates. Following elongation, helicase can again act on the freshly synthesized dsDNA and the cycle asynchronously repeats, resulting in exponential amplification of the target sequence. HDA can amplify both DNA and RNA. The primary appeal is the relative simplicity of the reaction and amplification at 60−65°C without further temperature steps. Various microfluidic devices using HDA are discussed in the literature (Ramalingam, 2009). Moura-Melo et al. (2015) reported the development of an electrochemical platform for detection of HDA amplicons with improved detectability, allowing Yes/No detection of GMOs with a limit of detection of approximately 30 copies of the CaMV35S genomic DNA implying the possibility of on-site GMO detection kits using HDA in the future.

Recombinase Polymerase Amplification (RPA). RPA is a single tube, low temperature (37−42°C) isothermal amplification method in which DNA amplification is achieved by three enzymes: a recombinase, a single-stranded DNA-binding protein, and a strand-displacing polymerase with the presence of target-specific primers (Zhao et al., 2015). The recombinase pairs oligonucleotide primers with a homologous sequence in the target DNA. The single-stranded DNA-binding protein then binds to the displaced strand of DNA and prevents the dissociation of primers. DNA synthesis is carried out by the strand displacing polymerase where the primer has bound to the target DNA. The RPA reaction is tolerant to the fluctuation of the incubation temperatures ranging from 37 to 42°C). This makes RPA an excellent candidate for developing rapid low-cost on-site GMO tests where precise temperature control is often challenging.

The RPA product can be measured in real-time using different probes with a fluorescence detection device. TwistDX (Cambirdge, UK) developed a portable fluorescence detector on which the test time can be as short as 15 min. The RPA assays that specifically detect the CaMV-35S promoter (P−35S) (CaMV) and the 3′ nontranslated region of the nopaline synthase gene (T-nos) were evaluated in rice, cotton, maize, and soybean (Xu et al., 2014). It can reliably detect 100 copies or more of the targets, equivalent to detecting GMO content at the level of 0.1%. Chandu et al. (2016) reported a duplex RPA assay detecting RR2Y soybean at a 0.5% level. The RPA assay in combination with the portable detection device offers a significant breakthrough for the development of DNA detection methods for on-site use.

Nicking and extension amplification reaction (NEAR). NEAR is an emerging isothermal amplification method that exponentially synthesizes short oligonucleotides (18−28 nt) using two enzymes—a DNA polymerase and a DNA-nicking enzyme at a temperature of approximately 60°C (Van Ness et al., 2003). NEAR employs a linker strand consisting of complementary sequences (8−15 nucleotides in length) to the target DNA downstream of a nicking enzyme binding site and a nicking site. Upon binding to a target, the linker is cleaved by the DNA-nicking enzyme at the nicking site from which polymerase elongation can initiate. This primed template continues to generate oligonucleotide products. To amplify a double-stranded nucleic acid target sequence, a pair of linkers are designed to target both the sense and antisense strands. This can produce bind-free templates and reenter the NEAR reaction resulting in geometric amplification. This NEAR process is extremely rapid and sensitive, enabling the detection of small target amounts in minutes. The amplification product can be detected by a variety of methods including gel electrophoresis, SYBR fluorescence, Förster Resonance Energy Transfer (FRET), and molecular beacon detection.

Envirologix (Portland ME, USA) has developed DNAble chemistry, incorporating the NEAR and a sequence-specific fluorescent dye-labeled molecular beacon for detection. Using a molecular beacon for detection not only improves detection specificity, but also implies the potential multiplex ability of a DNAble assay by labeling with several molecular beacons with different fluorophores. The DNAble CP4 EPSPS and DNAble PAT detection assays can be performed under 10 min using crude extracts without further purification. Moreover, Envirologix demonstrated detecting amplification products using a lateral flow version DNAble kit for plant pathogen test. With the continuous improvement of DNAble chemistry and potable equipment, it is a promising alternative isothermal technique for on-site nucleic acid detection.

15.2.1.4 Next-generation sequence

The emergence and rapid evolution of next-generation sequencing (NGS) technologies over the past few years enable sequencing at an unprecedented speed, scalability, and low cost compared to classical Sanger sequencing. NGS provides massive parallel DNA fragment sequencing resulting in millions of sequencing reads (Ansorge, 2009). This allows deep sequence and sufficient coverage of the plant genome, especially across difficult-to-sequence or repetitive regions in the plant genome. With different barcoding strategies, the high throughput of NGS allows the possibility to simultaneously sequence multiple different samples. The majority of agricultural plants for which transgenes are developed have been extensively studied with complete or draft genome assemblies to be publicly available (ncbi.nlm.nih.gov/genome).

Increasing numbers of literature are published on applications of NGS in GMO detection (Pauwels et al., 2015). Two main NGS approaches, whole-genome sequencing (WGS) and target capture sequencing, were investigated for their application in GMO detection. For WGS, the process uses random shearing of the transgenic plant DNA to produce fragments of predetermined size which together make a sequencing library. The data generated from the entire library are processed through computational analysis. The WGS approach allows the characterization of a sample with or without prior knowledge of the transgenic cassette. The target capture sequencing (TCS) approach involves the selection of sequences of interest from the whole genome library by using specific capture probes. TCS reduces sequence complexity by only selecting the sequence transgene insert and its flanking regions.

NGS has been studied for molecular characterization of transgenic events, compared to traditional methods like PCR and Southern blot analysis. Kovalic et al. (2012) utilized whole-genome sequences in combination with junction sequence analysis to successfully characterize transgenic soybean events. The approach was able to detect complex cases of multiple T-DNAs and rearrangements of inserted DNA in a GMO. Another study by Zastrow-Hayes et al. (2105) reported an approach of TCS, named as Southern by sequence, for rapid molecular characterization and selection of transgenic events during trait development. They demonstrated the TCS approach could determine the copy number and intactness of the inserted DNA and the presence or absence of plasmid backbone sequences and can identify small fragments independent of the primary insertion site, similar to the WGS approach. Guttikonda et al. (2016) recently compared WGS and TCS approaches with regard to generating data for regulatory submissions. Data generated by both NGS approaches can meet the requirement for risk assessment and possess sensitivity in revealing small size insertion and modifications that are not detected by Southern blot analysis. This feature becomes more attractive in characterizing new biotechnology products using new breeding approaches, such as precision genome modification, which result in small changes at the nucleic acid level. In addition, NGS can be a good tool for stacked event characterization in regard to reduced analytical cost as resequencing of the stacked DNA-inserts and flanking regions is required to assess their integrity.

NGS offers the ability for the detection and identification of the presence of GMOs, with limited or no information of transformation constructs since the assembly of readouts can be performed without a reference sequence. Several research teams have investigated strategies for GMO detection. Yang et al. (2013) proposed two NGS data analysis modules to characterize unknown GMO events under diverse a priori knowledge conditions. In those approaches, the data generated from WGS were mapped to the host genome and the results with both ends matching the host genome were filtered out for further analysis. When no a priori knowledge of transformation constructs is available, de novo assembly must be applied to the results that are partially matched or unmatched with the endogenous plant-species reference genome (Module 3 in the paper). Successful detection in Module 3 relies on efficient de novo assembly. The de novo assembly approach was reported to successfully apply to the characterization of transgenic rice TT51-1 and T1c-19 events and transgenic rice LLRICE62 events. It is noted that the short reading frames that most NGS platforms offer may pose some challenges to accurate assembly; the advancement in sequencing platforms that enable longer read lengths can mitigate the challenge. The other module (Module 2 in the paper) to identify an unknown GMO was designed for use when a transgene sequence database of genetic elements and transformation constructs for known GMOs is available for use as a reference library. Module 2 has the potential to serve as an alternative GMO-screening approach. Successful detection of the transgene, in this case, depends on matches between the transgene and the reference library. De novo assembly may be used to fully characterize the insert. The approach was demonstrated on transgenic rice TT51-1 and T1c-19 events. In addition to WGS, target capture sequencing can be employed to determine an unknown GMO. A set of capture probes can be designed using a transgene sequence database and used to capture the unknown insert sequence in a test sample. The DNA fragments containing matching sequences can be sequenced. As previously stated, the short reading frames of most NGS platforms may pose challenges to accurate assembly; however, conventional PCR and Sanger sequence approaches may be applied to verify and validate the identified inserts. The advancement in sequencing platforms that enable longer read lengths can mitigate the challenge in the future.

NGS is a promising alternative tool to detect and characterize GMOs from plant samples. Due to its relatively high cost and requirement of adequate computer infrastructures and qualified analysts in bioinformatics, implementation of NGS in routine GMO analysis may not be affordable for some laboratories. It is expected that the technology continues to advance, and the cost associated with NGS will continue to decrease over time. Further development of bioinformatics tools for data analysis and standardization of workflows may facilitate NGS in routine analysis. Importantly, criteria for NGS data generation and interpretation need to be established to ensure data are generated, analyzed, and presented in a way that allows scientifically sound assessments.

15.2.2 Protein-based technologies

15.2.2.1 LC-MS multiplexing technology

Accurate quantitative measurement of expressed transgenic proteins from a variety of complex matrices (plate tissues, soils, oils, insects, etc.) is necessary to support product development and regulatory risk and safety assessments of genetically engineered crops. Transgenic protein quantitation has traditionally been performed using uniplex enzyme-linked immunosorbent assays (ELISA) that incorporate highly specific antibodies to detect and measure target protein. However, ELISA method development is often time consuming and challenging to achieve exhaustive protein extraction with labile ELISA compatible buffers as well as resolving antibody cross-reactivity issues caused by sequence homology of endogenous proteins and target proteins. The advancement of agricultural biotechnology toward genetically engineered crops expressing multiple transgenes to combat pests, weeds, and diseases has presented a challenge to accommodate ELISA method development timelines as well as to generate the amount of data per sample in a uniplex format. With the increasing complexity of stacked transgenic products, alternative analytical methods for protein detection and quantitation have been investigated. When coupled with chromatographic separation, the quantitation of signature peptides as surrogates for intact protein measurements by mass spectrometry techniques is emerging as a possible solution.

The most typical application of surrogate peptide analysis utilizes liquid chromatography with detection by tandem mass spectrometry or LC-MS/MS. The advantages of LC-MS/MS include highly multiplexing, short method development timelines, compatibility with harsh protein extraction techniques, large detection ranges (3–5 orders of magnitude), no need to develop antibodies, as well as specificity and selectivity capable of resolving introduced traits as well as endogenous proteins. Surrogate peptide analysis termed "bottom-up proteomics" utilizes a protease of known cleavage specificity and knowledge of a protein's primary sequence to cleave high-molecular-weight proteins into smaller predicated peptide chains suitable for analysis by tandem mass spectrometry. Trypsin is typically chosen for quantitative methodologies due to its high specificity to cleave at the C terminus of lysine and arginine residues; however, a variety of alternative proteases are available depending on the application and target analyte. The foundation for surrogate peptide analysis relies on the assumption that protease digestion is 100% efficient. In a multiplexing assay, this can be challenging as diverse protein chemistries require a digestion protocol (denaturation, alkylation, proteolysis, etc.) that is optimal for all target proteins. Once extraction and digestion protocols have been developed and validated, the quantitation of stacked traits can be analyzed in a single analysis in a high-throughput format. The type of reference standard employed is another consideration unique for surrogate peptide analysis, unlike ELISA measurements where recombinant transgenic protein will be used for as the reference. Intact recombinant proteins, synthetic natural abundance peptides, or structurally identical heavy labeled peptides (AQUA) may be used as reference standards during quantitative analysis. The addition of structurally identical heavy-labeled peptide internal standards may alternatively be used to normalize all samples and synthetic natural abundance reference peptides to reduce the impact of matrix effects (suppression/enhancement) and other technical issues during analysis.

Current limitations of surrogate peptide analysis center around the requirement of proteolysis that impacts sensitivity, throughput, and must be validated. Mass spectrometry is a highly sensitive technique; however, peptide bonds are very stable affecting ionization leading to sensitivity challenges for certain peptides. Currently, the most robust proteolysis protocols require overnight digestion that is a rate-limiting step in terms of throughput. As instrumentation advances, future applications for the technology are moving toward intact measurement of transgenic and endogenous proteins with high-resolution accurate mass techniques (HRM) eliminating the need for proteolysis. In addition to simplifying method development and analysis, HRM may provide additional qualitative information that would not be observed in a targeted MS/MS analysis.

15.2.2.2 Immunoassays

The very first immunoassay was developed for insulin by Solomon Berson and Rosalyn S. Yalow in 1959. In the format of the radioimmunoassay for insulin, they pioneered the development of immunoassays for countless other analytes in the century to come. To meet the continuous and large quantity demand of polyclonal antibodies by manufacturers for in vitro diagnostic distribution and use, an alternative for the production of antibodies has been enduring. In 1975 Kohler and Milstein described the fusion of a mouse myeloma cell line— with B-cell lymphocytes isolated in the spleen of mice that have been immunized with an antigen. Once the cells have fused, a screening process is necessary to isolate the cells producing the specific monoclonal antibodies of interest. Once identified, the cell is cloned and allowed to grow in culture indefinitely (Koehler and Milstein, 1975). In 1984, Kohler and Milstein received the Nobel Prize in Medicine for the discovery of this process. Subsequently, variations of immunoassays, for example, noncompetitive (sandwich) immunoassays, were developed because of the unlimited supply of immune reagents. In contrast to the competitive assay format, these immunoassays make use of two antibodies in nonlimiting quantities. The first antibody is used to capture the analyte and the second is labeled and used to measure the concentration of the analyte.

One of the first applications for the sandwich immunoassay technique was described by Belanger et al. for the analysis of α-fetoprotein (1973). To date, sandwich immunoassays are the most amenable not only for automation, but also for multiplexing purposes.

Monitoring the expression of transgenic traits in key crops has rapidly expanded over recent years aided by the progressive development of research tools, in particular the different multiplex technologies allowing simultaneous measurement of multiple analytes. It is foreseeable that such technology will have an integral role in not only regulating the trials of these potential products in the field, and their eventual market launch (The World Bank, 2003), but also to help researchers to correlate expression data to field performance (Cao et al., 2017b). More and more multiplex and high throughput platforms are required to enable researchers to determine which of these are best for a particular application, as different platforms have their unique strength, as well as downfalls (Pavkovic et al., 2014). Picking the right platform is critical to match the needs and objectives of researchers. Below, we will discuss four unique multiplex, yet all are antibody-based, detection platforms. Our intention is not to exhaust everything available in the market, but to expose readers to different technologies that complement each other to answer different scientific questions (Table 15.1).

General working principle of these multiplex assay platform based on a transducer that makes use of a physical change accompanying the binding of the antibodies to the analytes. This may be changes in electrical or electronic output (Electrochemical), light output or light absorbance difference between the reactants and products (Optical), or based on the mass of the reactants or products (piezo-electric). Another source of signal, which will be discussed somewhere else, can be obtained through Red-Ox reactions (Lee et al., 2018), heat absorption or release (Ma et al., 2010), or gold particle precipitation (Zhou et al., 2015). To make the assay sensitive without false signals, identifying and reducing background "noise" is fundamental. Commonly, control samples of known concentration of the analyte of interest are used to establish the level of detection and level of quantification of the assay (Vial and Jardy, 1999). Subsequently, a standard curve with a linear detection range would be established to ensure the assay is accurate and precise. By reducing unwanted noise, amplifying the signal from the assay output, which was in turn converted into a digital signal and passed to a microprocessor (the assay instrument). The data are then processed, converted into concentration units, and output to a display device or data store (Chaplin, 2004).

There are a few challenges commonly faced by the assay developers of multiplex immunoassays. Cross-reactivity between analytes with detecting antibodies within the same multiplex assay would produce false signals. Preliminary single-plex assays are required to select the right antibodies to test for and eliminate any potential cross-binding between antibodies and analytes. In addition, the matrix effect is a common problem with immunoassays. Using

TABLE 15.1 Comparison of different multiplex immunoassays.

Characters	Surface plasmon resonance	Chemi-luminescence	Bead based	Time-resolved fluorescence
Key vendor	BiaCore	MSD	Luminex	AlphaLISA (PE)
Real time	Yes	No	No	No
Sensitivity	High	Med to high	Med to high	High
Quantitative	Yes	Semi	Semi	Semi
Qualitative	Yes	Yes	Yes	Yes
Multiplexing	No	Yes (6–10)	Yes (100)	Yes
Development time	Short (1–2 days)	Med (>a week)	Med (>a week)	Med (> a week)
Throughput	Low (one sample at a time, a minute per sample)	High (microtiter plates)	Med (one sample at a time, few seconds per sample)	High (microtiter plates)
Features	Measurement of real time binding kinetics	Multiplex in microtiter wells. Amenable to automation.	Bead-based technology that allows high degree of multiplexing. Amenable to automation.	Homogenous assay that increase the throughput of analysis.

the appropriate extraction buffer to recover the desired analytes from the sample effectively not only improves the accuracy, but also the sensitivity of the assay. When there are multiple analytes to be extracted within the same sample, using the right buffer for extraction is paramount to success. Third, in many cases, transgenic proteins expressed in a plant do not present at the same level. In some cases, they can be a couple of orders of differences in concentrations (personal communications). In a single-plex assay, the routine method is to dilute the samples to the right level to fit the linear range of quantification of the standard curve of the assay. Only within the linear range of the quantification would the sample be quantified accurately. When there are multiple analytes present in one sample, the diluting factor needs to be predetermined with control samples to identify the optimal median, so all diluted samples will fall within the range of every single assay, before going into production mode. Multiplex assay development and validation hence required a significant investment of time and resources upfront (Tighe et al., 2013).

MSD multiplexing array (electrochemical luminescence)

A specifically rapid method to quantify the expression of multiple target analytes within the same sample is always desirable, as it will save both time and resources. Currently, quantification of multiple proteins uses separately administered ELISA, with each test requiring an average of 120 min to perform. The increased demand for protein quantification is beyond the capacity of the ELISA platform. A multiplexed analysis platform in which multiple proteins can be analyzed simultaneously in a convenient 96-well format is required for the multitude of samples needing analysis.

MSD electro-chemiluminescent technology passes an electrical impulse from the assay plate through a capture antibody, protein of interest, detection antibody, and SULFO-TAG molecule to produce a light output (Bard et al., 2000). The proprietary SULFO-TAG molecule reacts with the ruthenium ions in the read buffer, creating an emission of light at 620 nm. The light output from each well is captured via a high-resolution camera produced specifically for use with MSD assay plates. The light intensity from the image can be used to quantify and predict protein expression levels per sample. The ability to quantify multiple proteins per sample (multiplexing) is possible with electro-chemiluminescence because multiple capture antibodies can be coated in designated spots within each well, called a microarray. The signal from each microarray can be quantified to predict multiple specific protein levels per well of the assay plate. Other benefits of electro-chemiluminescence technology include a large dynamic range, low background signals, and the potential for homogeneous assays that do not require a wash step.

Biacore biosensor system (piezo-electric)

Affinity reaction was first observed in real time using the surface plasmon resonance (SPR) technique (BoLiedberg et al., 1983). Using a label-free receptor/ligand interaction, the SPR biosensor is a real-time analytical device, which works on the optical phenomenon in which plasmon waves are excited at the metal-dielectric interface. The energy produced is conducive to the refractive index changes sensitively at the sensor surface and is proportional to the sample mass. A small change at the surface due to binding or dissociation of biomolecules brings variation in SPR signal, leading to refractive index changes. Antibodies are the commonly used biomolecules for the capture of the analytes from the sample. For a consistent assay of biomolecular interaction in an SPR biosensor, one species should be immobilized, through chemistries, to allow covalent bonding of biomolecules at the sensor surface (Syed et al., 2011). SPR biosensors offer the unique opportunity of rapid, label-free, real-time, and cost-effective detection and identification of biomolecules.

Schmid et al. (2006) developed a simple method for the immobilization of antibodies on gold substrates for SPR applications. The gold surface on a glass slide was modified with Protein A with cross-linker dithiobissuccinimide propionate to get a uniform, stable, and sterically accessible antibody coating. Antibodies were then in turn bound to Protein A with respect to protein concentration, buffer pH, buffer type, and reaction time. The modified gold surface was stable for several weeks and the reproducibility was satisfactory. Thus this technique can be applied to construct immunosensors or biochips.

High throughput AlphaLISA (optical through FRET)

AlphaLISA is a combination of two different principles: time-resolved fluorometry and fluorescence resonance energy transfer (FRET) (Clapp et al., 2006). FRET involves two fluorophores, a donor and an acceptor. Excitation of the donor by an energy source (e.g., laser) produces an energy transfer to the acceptor if the two are within a given proximity to each other. The acceptor in turn emits light at its characteristic wavelength (Yan and Marriott, 2003). Several components of the system are stringently abide for the assay to work, namely spectral overlap and the proximity of the fluorophores involved, wherein energy transfer occurs only when the distance between the donor and the acceptor is small enough. The effective distance between the fluorophores and the acceptor molecule, defined as Förster's radius (R_0), will determine if the desired interactions have taken place. It is a fast and simple operation with high sensitivity and specificity. In other words, this unique signal production through FRET allows the detection of binding without the separation of bound and unbound analytes in the assay, leading to the development of a homogeneous assay, resulting in reduced assay time and cost (Wallrabe and Periasamy, 2005).

There are naturally limitations to the development of the AlphaLISA technology to be applicable to all life science applications. The many different components coextracted from different part of the plants commonly analyzed in these research applications contain many compounds and proteins which are naturally fluorescent, or conversely inhibits fluorescence (Berg and Beachy, 2008). Either way, the use of conventional, steady-state fluorescence measurement presents serious limitations in assay sensitivity. Unique fluorescence properties of Lanthanide ion complexes (Ln(III) chelates or cryptates) were some of the highly desired fluorophores due to their large Stokes shifts and extremely long emission lifetimes (from microseconds to milliseconds) compared to more traditional fluorophores (e.g., fluorescein, allophycocyanin, phycoerythrin, and rhodamine) (Vollmer et al., 1994). Long-lived fluorophores, such as lanthanides, combined with time-resolved detection (a delay between excitation and emission detection) minimizes prompt fluorescence interference.

Bead-based multiplex technology (optical—fluorescence with flow cytometry)

With some applications, multiple molecules of interest needed to be analyzed quantitatively all at once. With a limited sample volume, conducting multiple assays to obtain all the necessary data is unviable. Bead-based flow cytometry technology offers a solution. The advancement in producing fluorescence microbeads of different sizes and emitting lights at different wavelengths offers a large matrix of up to 100 combinations of output signals, hence theoretically a 100-plex assay in one run (Schmitt et al., 2006).

Briefly, the sample is added to a mixture of color-coded beads, precoated with analyte-specific capture antibodies. The antibodies bind to the analytes of interest. Biotinylated detection antibodies specific to the analyte of interest are added and form an antibody−analyte−antibody sandwich. Phycoerythrin (PE)-Streptavidin is added. It binds to the biotinylated detection antibodies. Polystyrene beads are read on a dual-laser flow-based detection instrument, such as the Luminex or Bio-Rad Bio-Plex analyzer. One laser classifies the bead and determines the analyte that is being detected. The second laser determines the magnitude of the PE-derived signal, which is in direct proportion to the amount of analyte bound. In addition to the previously listed instrument using fluorescence as readout, magnetic beads can be read using magnetic-based analyzer. A magnet in the analyzer captures and holds the magnetic beads in a mono-layer, while two spectrally distinct light-emitting diodes illuminate the beads. One light identifies the analyte that is being detected, and the second light determines the magnitude of the PE-derived signal. Each well is imaged with a CCD camera.

15.3 Closing remarks

The advancement of bioanalytical technologies in the past decade has provided flexibility and a broader selection of methods for specific GM detection needs. This chapter gives a glimpse of new DNA- or protein-detection technologies that have been applied in ag-biotechnology, which include qPCR array, digital PCR, isothermal amplification, and next-generation sequencing, LC-MS technology, MSD multiplexing arrays, Biacore biosensor systems, AlphaLISA, and bead-based flow cytometry. It is expected that new analytical technologies will continue to emerge and may offer novel solutions for future GM detection challenges. This book has provided practical guidance for basic requirement, method validation, and other considerations before the adoption of a new methodology.

References

Ansorge, W.J., 2009. Next-generation DNA sequencing techniques. New Biotechnology 25, 195–203.

Bahrdt, C., Krech, A.B., Wurz, A., Wulff, D., 2010. Validation of a newly developed hexaplex real-time PCR assay for screening for presence of GMOs in food, feed and seed. Analytical and Bioanalytical Chemistry 396 (6), 2103–2112.

Bard, A.J., Debad, J.D., Leland, J.K., Sigal, G.B., Wilbur, J.L., Wohlstadter, J.N., 2000. Chem-iluminescence, electrogenerated. Encyclopedia of Analytical Chemistry 9842–9849, 2000.

Belanger, L., Sylvestre, C., Dufour, D., 1973. Enzyme-linked immunoassay for alpha-fetoprotein by competitive and sandwich procedures. Clinica Chimica Acta 48 (1), 15–18.

Berg, R.H., Beachy, R.N., 2008. Fluorescent protein applications in plants. Methods in Cell Biology 85, 153–177.

BoLiedberg, B., Nylander, C., Lunström, I., 1983. Surface plasmon resonance for gas detection and biosensing. Sensors and Actuators 4, 299–304.

Cao, L., Cui, X., Hu, J., Li, Z., Choi, J.R., Yang, Q., Lin, M., Ying Hui, L., Xu, F., 2017a. Advances in digital polymerase chain reaction (dPCR) and its emerging biomedical applications. Biosensors and Bioelectronics 90, 459–474.

Cao, Z., Zhang, W., Ning, X., Wang, B., Liu, Y., Li, Q., 2017b. Development of monoclonal antibodies recognizing linear epitope: illustration by three Bacillus thuringiensis crystal proteins of genetically modified cotton, maize, and tobacco. Journal of Agricultural and Food Chemistry 65 (46), 10115–10122.

Chandu, D., Paul, S., Parker, M., Dudin, Y., King-Sitzes, J., Perez, T., Mittanck, D.W., Shah, M., Glenn, K.C., Piepenburg, O., 2016. Development of a rapid point-of-use DNA test for the screening of genuity(R) roundup ready 2 yield(R) soybean in seed samples. BioMed Research International 3145921, 2016.

Chaplin, C.F., 2004. Controlling Brightness of Light Sources Used for Data Transmission. US Granted Patent US8798479B2.

Clapp, A.R., Medintz, I.L., Mattoussi, H., 2006. Förster resonance energy transfer investigations using quantum-dot fluorophores. ChemPhysChem 7 (1), 47–57.

Demeke, T., Gräfenhan, T., Holigroski, M., Fernando, U., Bamforth, J., Lee, S.-J., 2014. Assessment of droplet digital PCR for absolute quantification of genetically engineered OXY235 canola and DP305423 soybean samples. Food Control 46, 470–474.

Demeke, T., Dobnik, D., 2018. Critical assessment of digital PCR for the detection and quantification of genetically modified organisms. Analytical and Bioanalytical Chemistry 410 (17), 4039–4050.

Dobnik, D., Morisset, D., Gruden, K., 2010. NAIMA as a solution for future GMO diagnostics challenges. Analytical and Bioanalytical Chemistry 396 (6), 2229–2233.

Dobnik, D., Stebih, D., Blejec, A., Morisset, D., Zel, J., 2016. Multiplex quantification of four DNA targets in one reaction with Bio-Rad droplet digital PCR system for GMO detection. Scientific Reports 6, 35451.

Dorries, H.H., Remus, I., Gronewald, A., Gronewald, C., Berghof–Jager, K., 2010. Development of a qualitative, multiplex real-time PCR kit for screening of genetically modified organisms (GMOs). Analytical and Bioanalytical Chemistry 396, 2043–2054.

Duewer, D.L., Kline, M.C., Romsos, E.L., 2015. Real-time cdPCR opens a window into events occurring in the first few PCR amplification cycles. Analytical and Bioanalytical Chemistry 407, 9061–9069.

Fu, K., Huang, W., Deng, T., Li, F., Liu, H., Chen, Y., 2015. Multiplex PCR assay and liquid bead array for detection of 13 lines genetically modified maize. Journal of Chinese Institute of Food Science and Technology 15 (1), 188e197.

Gaudron, T., Peters, C., Boland, E., Steinmetz, A., Moris, G., 2009. Development of a quadruplex-real-time-PCR for screening food for genetically modified organisms. European Food Research and Technology 229, 295–305.

Guiducci, C., Spiga, F.M., 2013. Another transistor-based revolution: on-chip qPCR. Nature Methods 10, 617.

Guttikonda, S.K., Marri, P., Mammadov, J., Ye, L., Soe, K., Richey, K., Cruse, J., Zhuang, M., Gao, Z., Evans, C., et al., 2016. Molecular characterization of transgenic events using next generation sequencing approach. PLoS One 11, e0149515.

Herrero, S., Whitt, S., Ghavami, F., Ross, M., Houchins, D., Shillito, R., 2022. Detection Methods for Genome-edited Crops. In: Shillito, R., Shan, G. (Eds.), Application of Sampling and Detection Methods in Agricultural Plant Biotechnology. Elsevier. Chapter 14 In this issue.

Kluga, L., Folloni, S., Van den Bulcke, M., Van den Eede, G., Querci, M., 2012. Applicability of the "real-time pcr-based ready-to-use multi-target analytical system for GMO detection" in processed maize matrices. European Food Research and Technology 234 (1), 109–118.

ISO 22942-1:2022Molecular biomarker analysis — Isothermal polymerase chain reaction (isoPCR) methods — Part 1: General requirements

Koehler, G., Milstein, C., 1975. Continuous culture of fused cells secreting antibodies of predefined specificity. Nature 256, 495–497.

Kovalic, D., Garnaat, C., Guo, L., Yan, Y., Groat, J., Silvanovich, A., Ralston, L., Huang, M., Tian, Q., Christian, A., et al., 2012. The use of next generation sequencing and junction sequence analysis bioinformatics to achieve molecular characterization of crops improved through modern biotechnology. The Plant Genome Journal 0, 0.

Lee, G.-Y., Park, J.-H., Chang, Y.-W., Cho, S.-B., Kang, M.-J., Pyun, J.-C., 2018. Chronoamperometry-based redox cycling for application to immunoassays. ACS Sens 3 (1), 106–112.

Leimanis, S., Hamels, S., Nazé, F., Mbella, G.M., Sneyers, M., Hochegger, R., et al., 2008. Validation of the performance of a GMO multiplex screening assay based on microarray detection. European Food Research and Technology 227 (6), 1621–1632.

Li, R., Wang, C., Ji, L., Zhao, X., Liu, M., Zhang, D., Shi, J., 2015a. Loop-mediated isothermal amplification (LAMP) assay for GMO on: recent progresses and future Perspectives. OALib 02, 1–8.

Li, X., Wu, Y., Li, J., Li, Y., Long, L., Li, F., Wu, G., 2015b. Development and validation of a 48-target analytical method for high-throughput monitoring of genetically modified organisms. Scientific Reports 5, 7616.

Ma, L., Wang, C., Hong, Y., Zhang, M., Su, M., 2010. Thermally addressed immunosorbent assay for multiplexed protein detections using phase change nanoparticles. Analytical Chemistry 82 (4), 1186–1190.

Mano, J., Shigemitsu, N., Futo, S., Akiyama, H., Teshima, R., Hino, A., Kitta, K., 2009. Real-time PCR array as a universal platform for the detection of genetically modified crops and its application in identifying unapproved genetically modified crops in Japan. Journal of Agricultural and Food Chemistry 57 (1), 26–37.

Morley, A.A., 2014. Digital PCR: a brief history. Biomol Detect Quantif 1, 1–2.

Moura-Melo, S., Miranda-Castro, R., de-Los-Santos-Álvarez, N., et al., 2015. Targeting helicase-dependent amplification products with an electrochemical genosensor for reliable and sensitive screening of genetically modified organisms. Analytical Chemistry 18;87 (16), 8547–8554.

Notomi, T., Okayama, H., Masubuchi, H., Yonekawa, T., Watanabe, K., Amino, N., Hase, T., 2000. Loop-mediated isothermal amplification of DNA. Nucleic Acids Research 28 (12), e63.

Pavkovic, M., Riefke, B., Gutberlet, K., Raschke, M., Ellinger-Ziegelbauer, H., 2014. Comparison of the Mesoscale Discovery and Luminex multiplex platforms for measurement of urinary biomarkers in a cisplatin rat kidney injury model. Journal of Pharmacological and Toxicological Methods 69 (2), 196–204.

Pauwels, K., De Keersmaecker, S.C.J., De Schrijver, A., du Jardin, P., Roosens, N.H.C., Herman, P., 2015. Next-generation sequencing as a tool for the molecular characterisation and risk assessment of genetically modified plants: added value or not? Trends in Food Science & Technology 45, 319–326.

Prins, T.W., van Dijk, J.P., Beenen, H.G., Van Hoef, A.A., Voorhuijzen, M.M., Schoen, C.D., et al., 2008. Optimised padlock probe ligation and microarray detection of multiple (non-authorised) GMOs in a single reaction. BMC Genomics 9, 584. https://doi.org/10.1186/1471-2164-9-584.

Querci, M., Foti, N., Bogni, A., Kluga, L., Broll, H., van den Eede, G., 2009. Real-time PCR-based ready-to-use multi-target analytical system for GMO detection. Food Analytical Methods 2, 325–336.

Ramalingam, N., San, T.C., Kai, T.J., Mak, M.Y.M., Gong, H.-Q., 2009. Micro-fluidic devices harboring unsealed reactors for real-time isothermal helicase-dependent amplification. Microfluid Nanofluid 7, 325–336.

Randhawa, G.J., Singh, M., Morisset, D., Sood, P., Zel, J., 2013. Loop-mediated isothermal amplification: rapid visual and real-time methods for detection of genetically modified crops. Journal of Agricultural and Food Chemistry 61, 11338–11346.

Rosa, S.F., Gatto, F., Angers-Loustau, A., Petrillo, M., Kreysa, J., Querci, M., 2016. Development and applicability of a ready-to-use PCR system for GMO screening. Food Chemistry 201, 110–119.

Rudi, K., Rud, I., Holck, A., 2003. A novel multiplex quantitative DNA array based PCR (MQDA-PCR) for quantification of transgenic maize in food and feed. Nucleic Acids Research 31, e62.

Schmid, A.H., Stanca, S.E., Thakur, M.S., Thampi, K.R., Suri, C.R., 2006. Site-directed antibody immobilization on gold substrate for surface plasmon resonance sensors. Sensors and Actuators B: Chemical 113 (1), 297–303.

Schmitt, M., Bravo, I.G., Snijders, P.J.F., Gissmann, L., Pawlita, M., Waterboer, T., 2006. Bead-Based Multiplex Genotyping of Human Papillomaviruses 44 (2), 504–512.

Shao, N., Jiang, S., Zhang, M., Wang, J., Guo, S.J., et al., 2014. MACRO: a combined microchip-PCR and microarray system for high-throughput monitoring of genetically modified organisms. Analytical Chemistry 86 (2), 1269–1276.

Syed, M.A., Bhatti, A.S., Li, C., Bokhari, H., 2011. Use of SPR biosensor for the study of proteolytic action of a serine protease enzyme. American Journal of Biomedical Science 3 (4), 253–257.

Tighe, P., Negm, O., Todd, I., Fairclough, L., 2013. Utilility, reliability and reproductibility of immunoassay multiplex kits. Methods 61 (1), 23–29.

Van Ness, J., Van Ness, L.K., Galas, D.J., 2003. Isothermal reactions for the amplification of oligonucleotides. Proceedings of the National Academy of Sciences of the United States of America 100 (8), 4504–4509.

Vial, J., Jardy, A., 1999. Experimental comparison of the different approaches to estimate LOD and LOQ of an HPLC method. Analytical Chemistry 71 (114), 2672–2677.

Vollmer, F., Rettig, W., Birckner, E., 1994. Photochemical mechanisms producing large fluorescence Stokes shifts. Journal of Fluorescence 4 (1), 65–69.

Wallrabe, H., Periasamy, A., 2005. Imaging protein molecules using FRET and FLIM microscopy. Current Opinion in Biotechnology 16 (1), 19–27.

Whale, A.S., Huggett, J.F., Tzonev, S., 2016. Fundamentals of multiplexing with digital PCR. Biomolecular Detection and Quantification 10, 15–23.

Whale, A.S., Cowen, S., Foy, C.A., Huggett, J.F., 2013. Methods for applying accurate digital PCR analysis on Low copy DNA samples. PLoS One 8 (3), e58177.

World Bank, 2003. Biosafety Regulation: A Review of International Approaches. April 2003. Agriculture and Rural development department. Report No. 26028.

Xu, C., Li, L., Jin, W., Wan, Y., 2014. Recombinase polymerase amplification (RPA) of CaMV-35S promoter and nos terminator for rapid detection of genetically modified crops. International Journal of Molecular Sciences 15, 18197–18205.

Yan, Y., Marriott, G., 2003. "Analysis of protein interactions using fluorescence technologies" (PDF). Current Opinion in Chemical Biology 7 (5), 635–640.

Yang, L., Wang, C., Holst-Jensen, A., et al., 2013. Characterization of GM events by insert knowledge adapted re-sequencing approaches. Scientific Reports 3, 2839.

Yalow, R.S., Berson, S.A., 1959. Nature 184, 1648–1649.

Zastrow-Hayes, G.M., Lin, H., Sigmund, A.L., Hoffman, J.L., Alarcon, C.M., Hayes, K.R., Richmond, T.A., Jeddeloh, J.A., May, G.D., Beatty, M.K., 2015. Southern-by-Sequencing: a robust screening approach for molecular characterization of genetically modified crops. The Plant Genome 8, 0.

Zhao, Y., Chen, F., Li, Q., Wang, L., Fan, C., 2015. Isothermal amplification of nucleic acids. Chemistry Review 115, 12491–12545.

Zhou, X., Hui, E., Yu, X.-L., Lin, Z., Pu, L.-K., Tu, Z., Zhang, J., Liu, Q., Zheng, J., Zhang, J., 2015. Development of a rapid immunochromatographic lateral flow device capable of differentiating phytase expressed from recombinant *Aspergillus Niger* phyA2 and genetically modified corn. Journal of Agricultural and Food Chemistry 63 (17), 4320–4326.

Zhu, H., Zhang, H., Xu, Y., Laššáková, S., Korabečná, M., Neužil, P., 2020. PCR past, present and future. Biotechniques 69 (4), 317–325, 69.

https://www.iso.org/standard/74245.html, 2022.

Further reading

Hill, R.C., Oman, T.J., Shan, G., Schafer, B., Eble, J., Chen, C., 2015. Development and validation of a multiplexed protein quantitation assay for the determination of three recombinant proteins in soybean tissues by liquid chromatography with tandem mass spectrometry. Journal of Agricultural and Food Chemistry 63 (33), 7450–7461.

Hill, R.C., Oman, T.J., Wang, X., Shan, G., Schafer, B., Herman, R.A., Tobias, R., Shippar, J., Malayappan, B., Sheng, L., Xu, A., Bradshaw, J., 2017a. Development, validation, and interlaboratory evaluation of a quantitative multiplexing method to assess levels of ten endogenous allergens in soybean seed and its application to field trials spanning three growing seasons. Journal of Agricultural and Food Chemistry 65 (27), 5531–5544.

Hill, R.C., Wang, X., Schafer, B.W., Gampala, S.S., Herman, R.A., 2017b. Measurement of lipid transfer proteins in genetically engineered maize using liquid chromatography with tandem mas.s spectrometry (LC-MS/MS). GM Crops & Food 8 (4), 239–252. https://doi.org/10.1080/21645698.2017.1349602.

Huggett, J.F., Cowen, S., Foy, C.A., 2015a. Considerations for digital PCR as an accurate molecular diagnostic tool. Clinical Chemistry 61, 79–88.

Huggett, J.F., O'Grady, J., Bustin, S., 2015b. qPCR, dPCR, NGS - a journey. Biomolecular Detection and Quantification 3, A1–A5.

Periasamy, A., 2005. Molecular Imaging: FRET Microscopy and Spectroscopy (Methods in Physiology Series). Academic Press, ISBN 0195177207, p. 336.

Settlage, S.B., Eble, J.E., Bhanushali, J.K., Cheever, M.L., Gao, A., Goldstrohm, D.A., Hill, R., Hu, T.X., Powley, C.R., Unger, A., Shan, G., 2017. Validation parameters for quantitating specific proteins using ELISA or LC-MS/MS: survey Results. Food Analytical Methods 10 (5), 1339–1348.

Willems, S., Fraiture, M.A., Deforce, D., De Keersmaecker, S.C., De Loose, M., Ruttink, T., Herman, P., Van Nieuwerburgh, F., Roosens, N., 2016. Statistical framework for detection of genetically modified organisms based on Next Generation Sequencing. Food Chemistry 192, 788–798.

Index

Note: 'Page numbers followed by "*f*" indicate figures those followed by "*t*" indicate tables and "*b*" indicate boxes.'

E

F

G

M

N

O

P

Q

R

S

FSC
www.fsc.org
MIX
Papier | Fördert gute Waldnutzung
FSC® C083411

Kolibri 360 GmbH
Ferdinand-Jühlke-Straße 7
99095 Erfurt, Deutschland
produktsicherheit@kolibri360.de